MARCELL MRSÁN

THE BIG BOOK OF FOOTWEAR UPPERS

FROM PATTERN MAKING TO THE COMPLETE PROCESS OF UPPER CONSTRUCTION

VOL 1.

2026

For information, please email to marcell.mrsan@gmail.com

Authors: Marcell Mrsan, Yaya Lee
Illustrations copyright © Marcell Mrsán
Photographer: Yaya Lee
Design and layout: Marcell Mrsán
Published by: AACG LLC

Editorial office
919 Square Street, Mount Joy
Pennsylvania, USA 17552

Picture Credits:

Page 99 (skivers), 100, Campbell Randall,

Page 87, 101, Comelz Spa,

Page 16, 32, 39, NewLast,

Page 98, 99 (sewing machines), Dürkopp Adler GmbH

All pattern illustrations were created using Caligola CAD.

ISBN 979-8-9863227-1-1

Library of Congress Control Number: 2026900851

FIRST EDITION

For my children

Content

VOLLEY
ZEGZUG.COM

Preface

I vividly remember my student years in Budapest, Hungary, in the mid-1980s. As shoemaking students, our primary focus was on making the shoe itself – learning various construction technologies, but always starting from a ready-made upper. Although we technically had a dedicated class for pattern design, upper making was considered a separate trade, taught in different classrooms and treated as an entirely different discipline.

Sounds boastful but it's true, that when it comes to footwear, I have a photographic memory. Hearing or seeing something once was often enough for me to recall years later. Even though many names and faces from that time have faded, the knowledge passed down to me during those years has remained. I remember every shoe I made – the lining details, the difficulties I had during lasting, and, interestingly, even techniques that were only briefly mentioned.

You don't realize the lapse in your memory or knowledge until the moment you try to conjure it when it becomes necessary.

As I was on my journey of the shoe making craft, I realized when it came to uppers, my memory was of little use, as if there was a high and almost impenetrable wall between shoemaking and upper making. The thorough education I received for making shoes, did not extend to barely anything that preceded the ready-made uppers. I started paying more attention to the uppers and I became curious about every aspect of this facet of the footwear making process.

Starting with books, I bought every single ones I could get my hands on. By then my craft, and with it the upper makers were a disappearing breed. Anything I could get was from decades earlier. Back then these textbooks served more as references than as tutorials. The knowledge was passed down personally, through demonstration and conversation. Reading alone was largely ineffective.

So I did what generations of makers did before me, I "stole" the knowledge. Meaning I pieced it together from watching professionals do it, asking questions when I could. Most upper makers I worked alongside were humble, generous, and who were willing to answer my likely newbie questions and seemed to enjoy talking about their craft. During those conversations I absorbed as much as possible.

Before I realized it, I had accumulated enough understanding to take the first steps in upper making myself. My early works, unsurprisingly, were not very great. Pieces I would never have used as a bespoke maker. I lacked experience but also the proper machinery, tools, and workshop setup. Even though I was already a professional shoemaker, my uppers were genuinely poor. But that only fueled my drive to not give up.

I returned to basics and kept practicing. Starting small, I made countless seam samples. Then, knowing better what I needed, I invested in better sewing machines, a skiving machine, and continued to practice relentlessly. It also helped immensely that I was able to employ an upper maker who was not only excellent in making but was willing to critique my early attempts. It took a while but eventually, upper making became a natural extension of my shoemaking vocabulary.

Although I never set out to become a pattern maker, I think it was only natural that it found me. As I was getting better at upper making, suddenly, pattern logic began to make sense: why certain lines exist, how designs could be improved, why some uppers are a nightmare to last, and how they can be constructed so lasting becomes a matter of minutes rather than a prolonged struggle.

I came to an important realization. Pattern makers who do not make uppers – or at least maintain close communication with upper makers – will have great difficulty improving and adopting better techniques. Pattern makers who understand and are able to execute the complete process have a significant edge over their peers, because they can learn directly from their own mistakes. For examples, a poorly shaped backline reveals itself immediately during lasting; insufficient lasting allowance becomes a physical struggle. Cause and effect are impossible to ignore when you're there to bear the grunt of it.

Working as a consultant accelerated the speed of my progression. Faced with new challenges, they forced me to extend the boundaries of my skills and click more puzzle pieces into their place in the grand scheme of the process and the industry itself. The more I learned about these three trades, the more holistic my approach become. All other auxiliary considerations, such as machinery, workshop organization, production logic, and countless other details had to work in sync both in theory and in actuality.

In recent years, I've started working towards passing down the knowledge, one book at a time. For the longest time I was intimidated by the prospect of broaching this subject. It all seemed too broad and vast for me to condense them down. But seeing that there haven't been any serious attempts at archiving the existing expertise in the past several decades, I felt it was my duty and legacy to see the know-how of my beloved industry preserved.

I deliberately put both pattern making and the upper construction in the same book, as they are, in my mind, inextricably intertwined.

This book will not simply provide a master pattern or a few pattern pieces; instead, it will guide the reader through the underlying logic of pattern making and then continue into the upper-making process itself.

My goal is to publish three books on this subject, bridging the gaps between the different trades involved in footwear making. In this first volume I put in what I deemed a reasonable foundation and a solid start for anyone working in the industry who has an interest in the complete upper making process. These are the basic models – the foundational constructions on which most uppers are built. The scope of this book has grown well beyond the originally planned page count, and although I had to edit out a considerable amount of material – due to time and cost constraints – , what remains is essential and valuable. I hope to have the opportunity to continue with volumes two and three, exploring more advanced patterns and upper constructions. In doing so, instead of "stealing" I was just safekeeping the information and now, I'm giving it back to the trade.

I hope you enjoy using this book as much as I enjoyed putting it together.

Marcell Mrsán

How to Use This Book

Pattern books are generally a heavy read, and much depends on how the reader approach using them. What makes this book unique is the detailed documentation of the hands-on production part. This way, the reader is not just presented with a final result, but also shown how it was reached. For most people, especially beginners, the best way to learn is simply to do the work. Make patterns, and mistakes, and learn from them. Compare your results with the examples, reread the chapter, and ask for feedback if you can. Cut the pattern in leather and assemble it. If something does not work, try to understand why and go back to the text – the answer is likely already there.

This volume covers eight models as thoroughly as possible, but the subject can always be taken further. Although the book spans around 400 pages, and highlights eight models, it could easily be spent homing in on just a couple of those. There are ever more advanced techniques to learn and more complicated ways to get a certain result. I understand the realities of learning anything footwear related skills, therefore I deliberately chose the most effective techniques that yield satisfying results compared to the invested efforts. For instance, I avoided heavily asymmetrical patterns. Even so, the material presented here is sufficient to produce professional-level patterns and footwear. Many constructions are based on these same principles and sequences. Learning one pattern often teaches far more than expected.

All pattern illustrations were created using Caligola software. This made it possible to generate many drawings and to explain each pattern gradually, including variations, common mistakes, and small but important details. Color coding and selective line visibility help keep the focus on what matters at each stage. If you are making these patterns by hand, the same logic applies. A line drawn on a screen or on paper is still a line, and a digital axis is no different from folding a sheet of paper. The main advantage of software is flexibility – it allows many variations to be shown quickly, something that would be too time-consuming to do by hand.

If there's anything worth learning from the trajectory of my career, it's this: stay curious and being open to improve. The moment we think our work cannot be further perfected is the beginning of a slow decline. Staying humble and accept criticism is the key to getting ahead. I'm constantly finding and welcoming new challenges in this beautiful trade and I urge you to do the same.

SECTION ONE

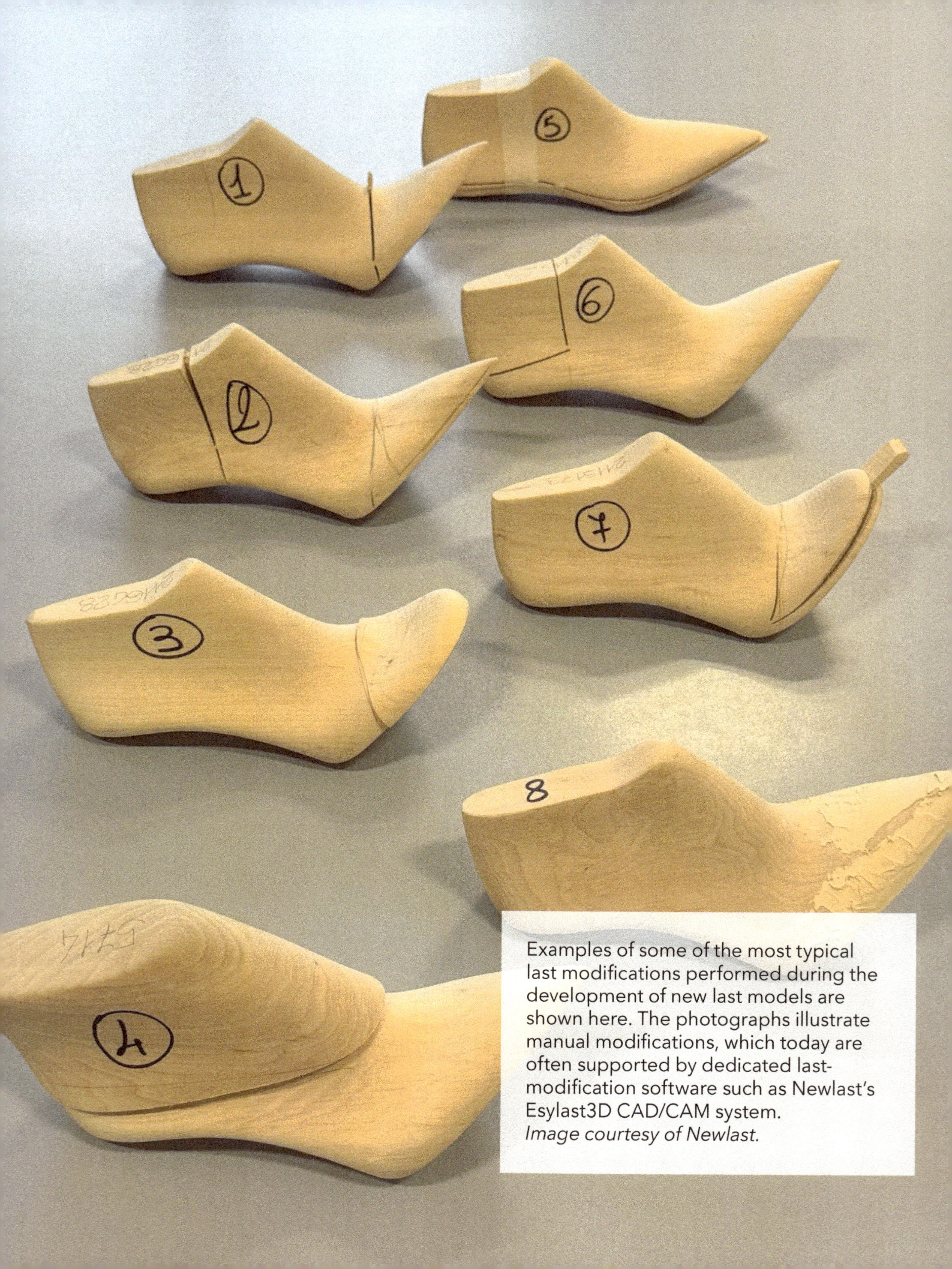

Examples of some of the most typical last modifications performed during the development of new last models are shown here. The photographs illustrate manual modifications, which today are often supported by dedicated last-modification software such as Newlast's Esylast3D CAD/CAM system.
Image courtesy of Newlast.

1 ABOUT SHOES AND LASTS

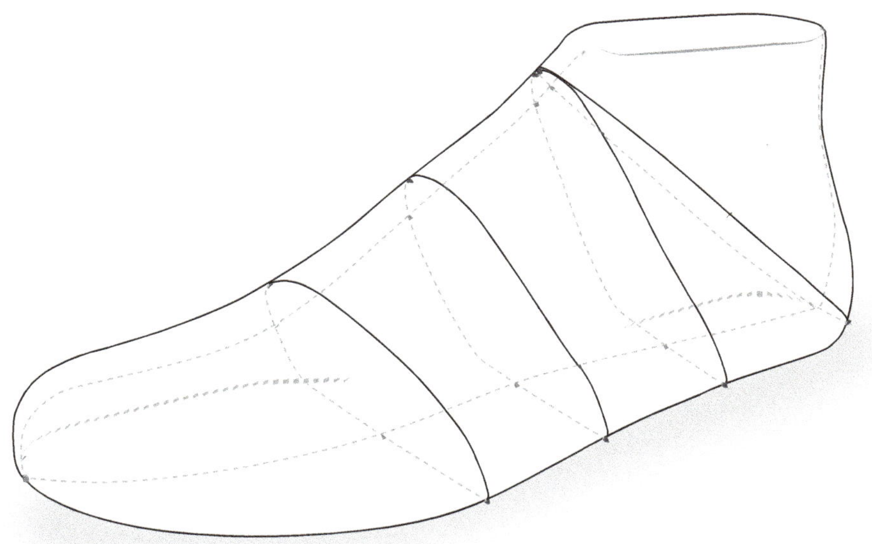

"The last comes first" is a cliché among the makers. However, it is indeed a universal truth: every stage of shoemaking – fit, design, pattern development, and construction – begins with the last. While an accurate last copy is essential, successful upper pattern making requires a deeper understanding of the last as a three-dimensional working medium rather than a simple form to trace.

The last is not intended to be an exact copy of the human foot, rather it represents the functional and anatomical translation of it, shaped by geometry, proportion, and intended use. This chapter introduces the essential vocabulary of shoes and lasts, outlines common last types, and explains basic geometric principles and the foot-last relationship. Together, these concepts provide a shared technical reference for pattern creation.

Although footwear styles change with fashion and economic trends, the core of any new design is always rooted in established construction principles and foot anatomy. The list of footwear discussed here offers a snapshot overview of the widely used shoe designs of today and industry-standard last geometry.

Footwear Vocabulary

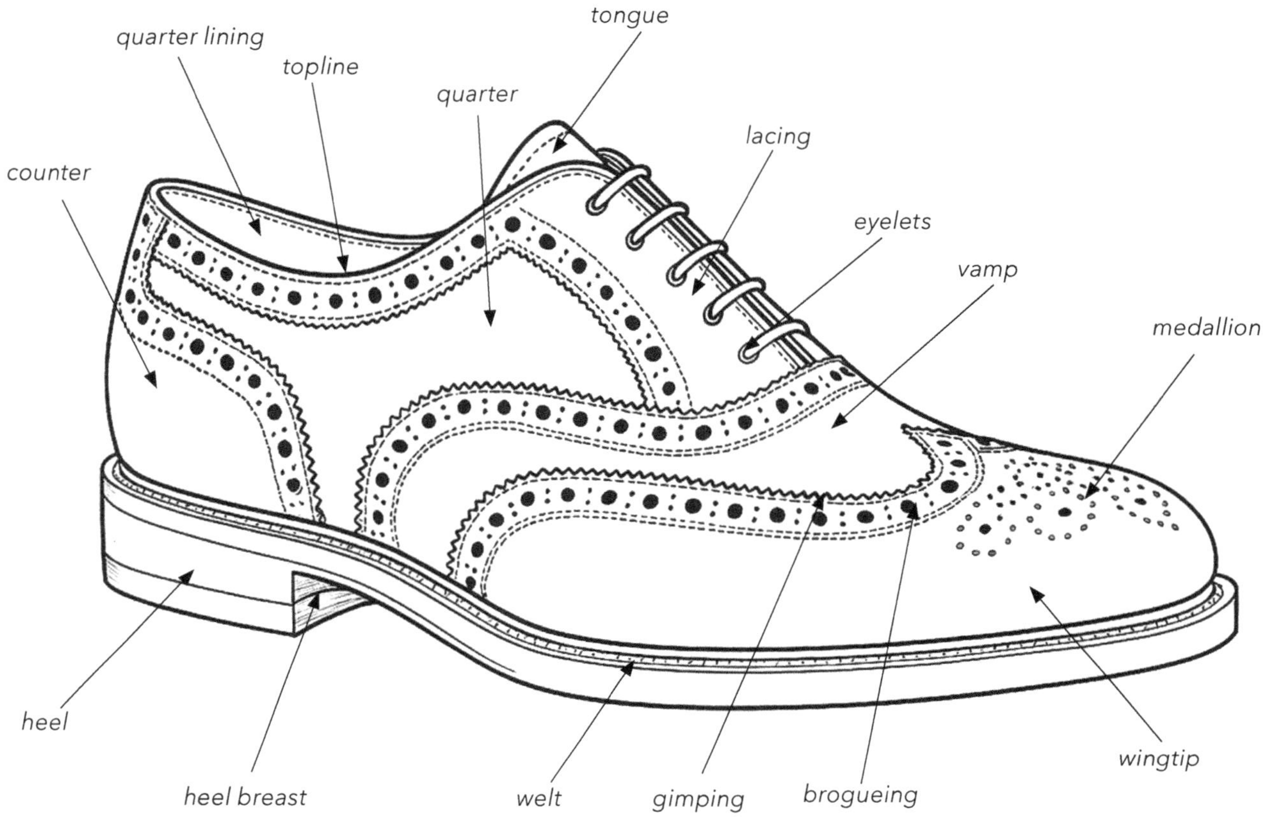

Aglet
The reinforced tip of a lace that prevents fraying and facilitates easy insertion through eyelets. It can be metal or plastic.

Ankle strap
A strap positioned around the ankle to provide additional support, retention, or decorative emphasis, commonly used in sandals and women's footwear.

Apron toe
A toe construction where the vamp is stitched to an overlay or apron piece, creating a raised seam line around the toe area, typical of moccasin-inspired shoes and loafers.

Backseam
The vertical seam at the rear of the upper that joins the quarters and defines the heel center line.

Bellows tongue
A tongue construction attached to the lacing along both sides to prevent debris or water from entering the shoe.

Brogueing
Ornamental perforations applied to specific pattern edges or panels.

Cap toe
A toe style featuring an additional leather panel stitched over the front of the vamp.

Counter
A stiffened component placed between the upper and lining at the heel to provide shape retention and rear-foot support. It can be made from leather or man-made material.

Eyelet
A reinforced hole or metal insert through which the lace passes, designed to resist wear and distribute tension.

Gimping
A decorative edge treatment consisting of small, regularly spaced notches cut along a seam or panel edge.

Heel
The rear bottom component of the shoe that elevates the foot and provides rear support and balance.

Heel breast
The front-facing surface of the heel that interfaces with the sole and influences heel pitch and stability.

Heel counter (stiffener)
The rigid internal structure that stabilizes the heel area and maintains the intended heel shape during wear.

Insole
The internal base layer of the shoe on which the foot rests, serving as the structural foundation for lasting.

Lace
A flexible cord or strip used to adjust fit and secure the upper around the foot through the lace holes (generally "lacing system").

Midsole
A layer positioned between the insole and outsole for structural support or technological reasons.

Moc toe
A toe construction inspired by moccasin footwear, characterized by a prominent hand- or similar looking machine-stitched seam around the toe.

Outsole
The bottommost layer of the shoe that contacts the ground and provides durability, traction, and wear resistance.

Quarter
The rear side panels of the upper that wrap around the heel and connect to the vamp.

Quarter lining
The lining component corresponding to the quarter, improving comfort, durability, and internal finish.

Sole
The complete bottom assembly of the shoe, typically consisting of insole, midsole, and outsole elements.

Tongue
The upper component positioned beneath the lacing that protects the instep and distributes lace pressure.

Topline
The upper edge of the shoe opening that defines entry, fit, and overall visual proportion.

Vamp
The front portion of the upper that covers the toes and instep and largely determines fit and style.

Welt
A strip of material stitched to the upper and insole that connects the upper to the sole, commonly used in welted constructions; it may be fully functional or "fake," in which case it serves a purely decorative purpose without structural involvement.

Classification of Footwear Uppers by Coverage

One way to classify types of footwear is by the proportion of the last – and therefore the foot – that the upper covers. This type of categorization isn't something the regular consumer would think twice about. It is, however, the consideration of a pattern maker. This scale begins at ¼ coverage and extends up to 7/1. These are just estimates and the practical relevance is to generalize a certain coverage. Each of these categories can lead to very different pattern making approaches, and the coverage does not necessarily correspond with the difficulty of the pattern.

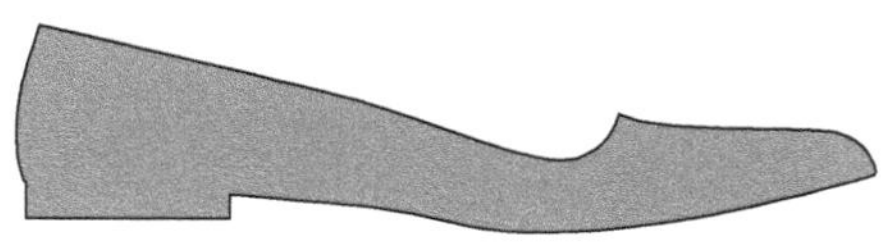

The **1/4-coverage** category represents the most minimal strap constructions – sandals, flip-flops, and similar open models. At first glance these footwear types may not seem like serious pattern-making subjects, and for one-of-a-kind handmade pieces, that assumption is often true. But in manufacturing, the story is entirely different.

Developing and engineering sandals pushes pattern accuracy to an extreme: tolerances often come down to fractions of a millimeter, and tiny adjustments across multiple points can make or break the fit and performance. With so little material controlling the foot, every strap angle, length, and attachment point becomes critical.

Sandals are not to be underestimated – from a product-development standpoint, they can be among the most complex and unforgiving footwear types a pattern maker will ever design.

The **1/2-coverage** category refers to footwear in which the upper rises at least to the vamp tack point. Even though complex pivoting may not always be required, these models are far from simple. They present their own technical challenges, especially when the chosen manufacturing method demands high precision. Slip lasting is a good example: it is widely used for this category because of its cost efficiency, yet it requires exceptionally accurate pattern work to achieve a clean, functional result.

A classic representative of this group is the ballerina shoe, which already incorporates numerous core pattern-making principles. These include unfolding techniques, compensation for the orange-peel effect, edge treatments, and the use of various seam constructions. Despite its minimal appearance, the ballerina offers a dense concentration of pattern fundamentals.

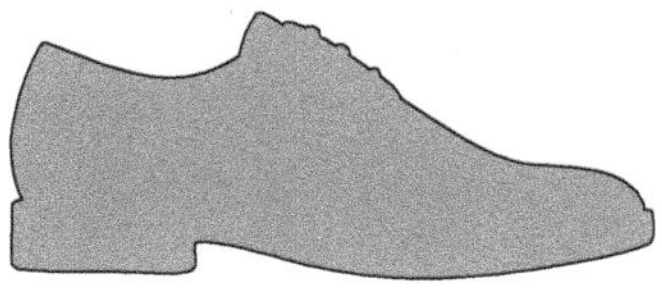

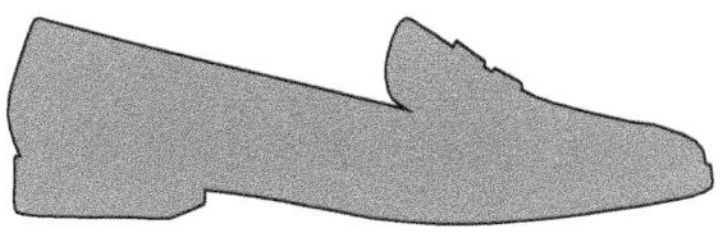

As the topline extends higher on the last, we reach the **3/4-coverage** and **1/1-coverage** categories. These are not dramatically different from each other in terms of overall complexity or technical challenges, but they do introduce additional considerations. Because the topline crosses over the vamp point, pivoting the last copy or selected pattern pieces becomes almost inevitable.

2/1 coverage - as the topline goes over the ankle bone, a whole new level of patternmaking comes in to the picture: crimping, heavy pivoting, the assembly order might get exponentially more complex and sometimes even necessitates special machinery.

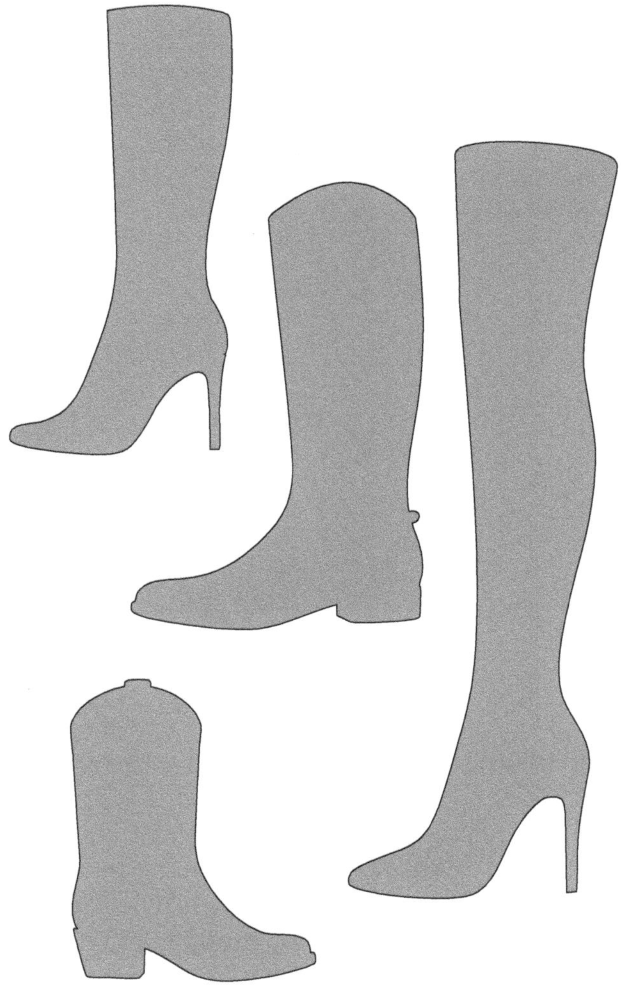

Boots, over 2/1 – **3/1, 4/1 all the way up to 7/1** (thigh high) means extra, special pattern making, technological and fitting challenges. Interestingly, the American heritage boot making tradition is occupied by a special niche, repeatedly using the patterns passed down, offering unchanging quality and design to their loyal customers. They are only altered to adjust to accommodate differing girths and other measurement. On the other end of the spectrum, the fashion boots, often change each season, which could mean intricate grading for manufacturing, with multiple (sometimes in the hundreds) variations for each model.

Classification of Footwear by Design

Let's review a few pattern designs without claiming to be exhaustive. These patterns cover the most of today's footwear fashion, in terms of technicality. A majority of footwear can be constructed following the general assembly order of these models. A good example for this is a minimalistic sneaker, which is technically very similar to a derby upper, with a few minor differences.

The mule owes its success and longevity in the history of footwear thanks to the simplicity of its pattern. It's fundamentally a slipper, originally designed to be worn in-house. It covers the toe and the instep, leaving the back half of the foot completely open. Since there isn't a counter to hold the foot in the footwear, this job remains to the grip of the sockliner, the instep and the correct last design.

The ballerina, or also known as the ballet flat – transitioned from the original dance footwear to a street shoe. They are lightweight, flexible shoes, with a low cut vamp, low heel, thin outsole and usually rounded toe. Most ballet shoes are minimalistic, but sometimes bows or other small embellishments are added.

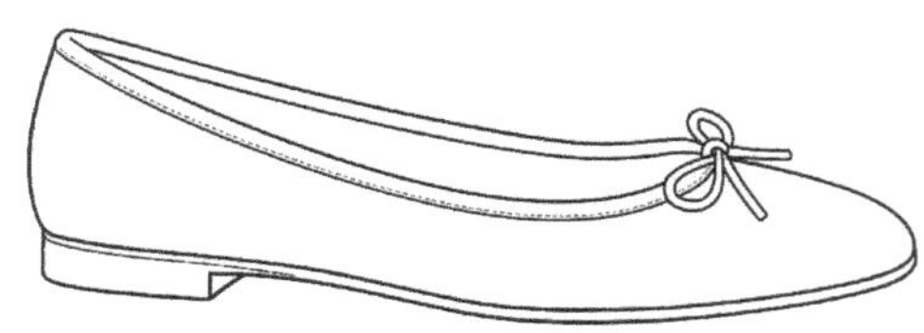

The high heel pump is a very recognizable footwear, defined by the elevated heel, the low cut topline and the slip on style. The high heel shifts most of the body weight from the heel to the forefoot. The upper can be made from one single piece of leather or alternatively two, adding an insert to the medial side. This shoe has an elegant, refined silhouette, which requires precise pattern construction.

Derby shoes are a classic category. The distinctive feature of the derby is that the quarters overlap the vamp. The modern era's derby started in the early 19th century but its roots extend back to several hundreds of years. Derby shoes are for everyday wear, semi-elegant or casual, made more for comfort than elegance. It has a bigger opening than an oxford shoe.

Oxford shoes are considered to be the most elegant footwear in men's fashion. Oxford is the right choice for ceremonies or other elegant occasions where dress shoes are required. The quarters are overlapped by the vamp, the lacing is "closed" – which refers to the fact that the bottom of the lacing cannot be opened.

The wholecut model is a very clean looking shoe, made from one single piece of leather, with no visible seam other than the backseam. The wholecut and the oxford are very closely related in its technical approach. The wholecut also features a closed lacing. Thanks to its clean look, the design can highlight the silhouette of the last, without the distraction of pattern pieces.

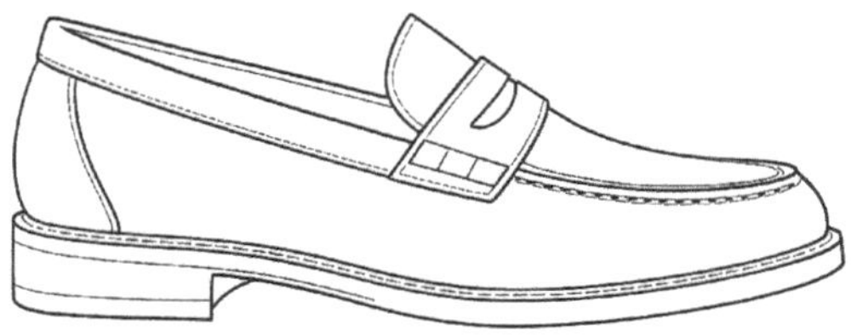

Loafers are slip-on shoes that are considered casual and semi-elegant footwear. They are usually built on a low-heel last with moderate toe spring. The vamp typically has a visible stitch line or an added pattern piece. Modern loafers originated from Norwegian farmer shoes and later evolved into styles such as the Belgian loafer and many other variations.

The blucher shoe has different versions, but the most prominent is probably this one here, with a one piece upper with an extra lacing piece and backstrap. The model often features French binding as a topline finish. Generally speaking, this is a casual footwear made with more robust components than its counterparts, the wholecut an the derby.

The Norwegian can refer to both a shoe style and a construction technology and these often come together, which can be confusing. The term referring to the upper style, is similar to a derby, but the lacing is made on the same piece of leather as the vamp. Another distinct feature is the "lake" or more often: "apron", which features an intricate hand stitching in the most visible part of the shoe that indeed, at times, poses a real challenge to the upper maker. The illustration shows a "split toe", which means a one piece upper, which unfold by the counter axis.

The monkstrap shoe is practically a derby-structured footwear where one or two buckle straps replace the traditional lacing. Monkstrap built on Oxford bases are also common occurrence. This style have become really popular in the recent decades. From a style perspective, the monkstrap occupies a unique position between formal and casual footwear.

The Balmoral model is a higher topline version of the oxford shoe, we can consider it as a boot version of it. The model has several variations. It falls into the elegant category of winter footwear.

The Chelsea boot has elastic side panels for better fit and easy slip on. This solution eliminates the need for other closures. It originated in the mid 19th century and remained mostly unchanged. The model can made to be elegant or a very practical outdoor footwear. In Australia the boot has been adopted and modified for heavy duty workboot, for farming, ranching, so very much the equivalent of the Western boot in the US.

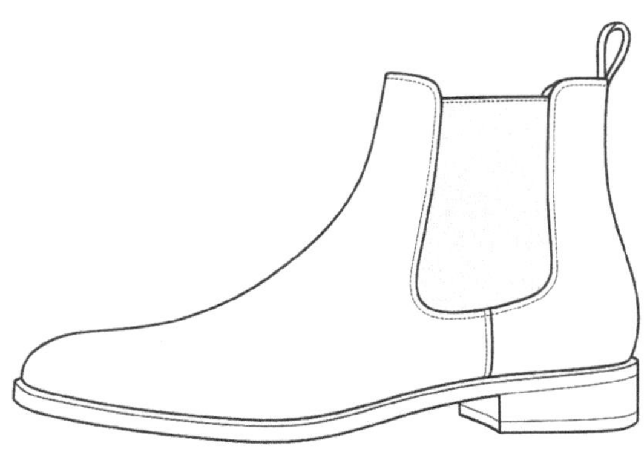

The Jodhpur boot is an over-the-ankle low boot. This footwear originated in Jodhpur, India, where Indian polo players starter to wear this in the early 20th century. When they played in England, the English makers quickly adopted the style and started to make their own versions. Technically the design is fairly simple, not considering the crimping process. Precise pattern construction is necessary to have a well-fitted opening at the top.

Moccasins are soft, flexible shoes built from leather pieces that wrap closely around the foot, creating a comfortable, glove-like fit. They typically feature a stitched apron on the front, giving them their characteristic moccasin appearance. Whether constructed in a traditional soft form or in a more structured Norwegian-inspired style, moccasins are valued for their natural feel and timeless appearance.

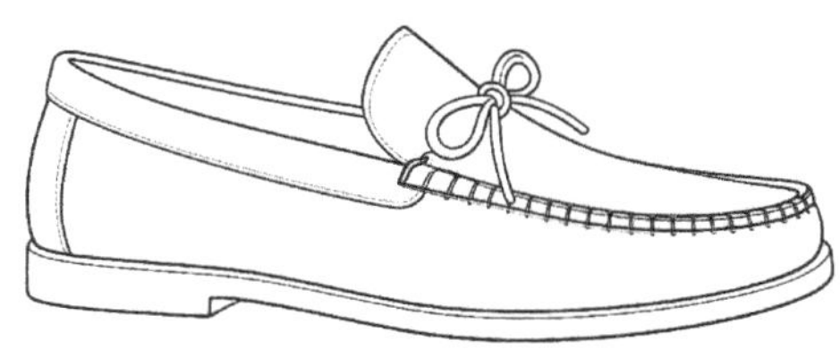

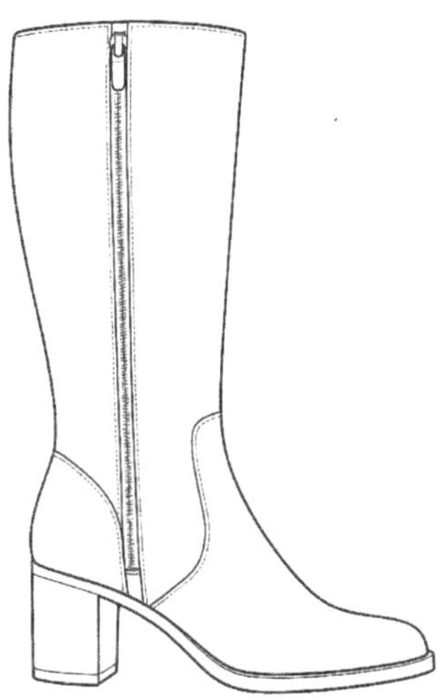

Knee high zipper boots evolved from traditional riding and military boots, where tall shafts offered leg protection and stability. With the introduction of reliable zippers in the early 20th century, these boots became far easier to put on and remove, making the style practical for daily wear rather than only specialized use. Today, they remain popular for their sleek silhouette, supportive structure, and the balance they offer between function and refined appearance.

Classic **Equestrian boots** might not meant for long walks, but purpose-built riding boots designed to provide stability, protection, and precise control in the saddle. Functional variations such as polo, dressage, and jumping boots each reflect the specific demands of their discipline – whether it is freedom of movement, maximum leg contact, or reinforced protection. Their rigid and tall shafts, structured vamps, and secure closures help the rider maintain proper leg position.

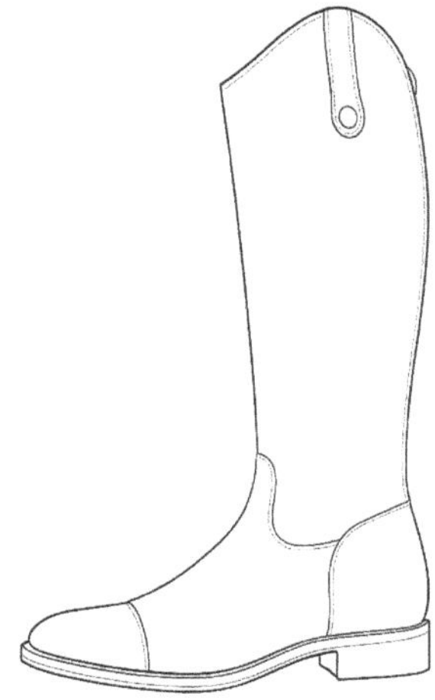

Engineer boots are rugged, pull-on boots originally developed for workers who needed durable footwear that could withstand heavy use and while protecting the foot and ankle. Characterized by their high shafts, minimal stitching, and adjustable instep and top straps, they offer both functional support and a distinctive, utilitarian aesthetic. Today they remain popular in motorcycle culture and heritage workwear across the USA.

Western boots have their roots in European riding and military footwear, brought to North America by settlers and adapted for frontier life. After the Civil War, soldiers' cavalry boots provided the basic form that the frontier shoemakers adapted into a more functional design for long hours in the saddle. The style became culturally iconic in the 1930s, when Hollywood Western movies romanticized the cowboy image and propelled these boots from practical working gear into mainstream fashion.

Fashion sandals are open, lightweight footwear designed to showcase the foot while offering some protection in warm weather. Their uppers range from minimal straps to more structured designs, allowing for endless stylistic variations. Materials, hardware, and decorative elements play a major role, making sandals as much a visual statement as a functional choice for everyday wear.

Slippers are typically soft indoor shoes with flexible soles, cushioned interiors, and open or closed uppers that allow the foot to slip in easily. They can range from minimalist designs to more structured models and may be intended for warm or cold conditions. The range of possible styles is broad. Many slippers use plush fabrics or felted materials to enhance comfort and insulation, while others incorporate more durable outsoles that permit short trips outdoors.

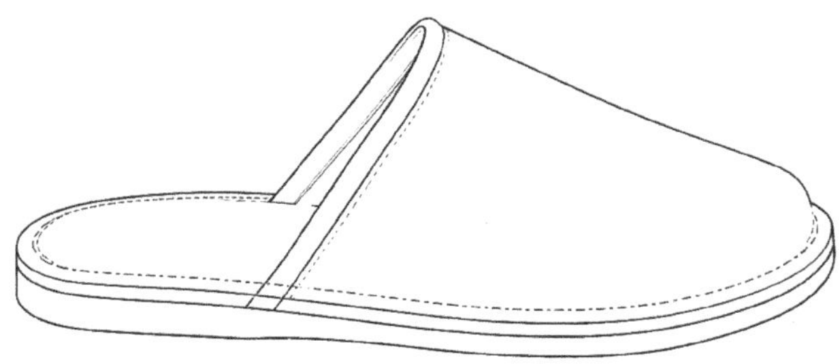

Workboots are durable, highly structured footwear designed to protect the foot in demanding environments. They have rugged outsoles that provide traction on a variety of surfaces. Materials such as full-grain leather, padded linings, and heavy-duty stitching contribute to long-lasting performance. Many models also incorporate safety features like steel or composite toe caps, puncture-resistant soles. Today, workboots have stepped beyond their functional roots, to become a staple in casual wear.

Hiking boots are designed to protect and support the foot over uneven, extreme terrains. They are rigid and solid, protecting the foot from outside impacts while supporting the ankle. The model is traditionally made from one vamp piece with an extension of a wider backstrap, and sometimes features top collar cushioning. The closure is a wider closed lace, featuring hooks and speed lace solutions. These boots almost always have bellows tongue against moisture.

Ballet Shoes are lightweight, flexible footwear designed to allow a dancer to move with precision, balance, and grace. Made from soft leather, canvas, or satin, they fit closely to the foot, enhancing articulation. Traditional ballet shoes are fully symmetrical, allowing them to be worn on either foot and gradually shaped by the dancer through use. Pointe shoes, the more advanced form, include reinforced boxes that enable dancers to rise onto the tips of their toes.

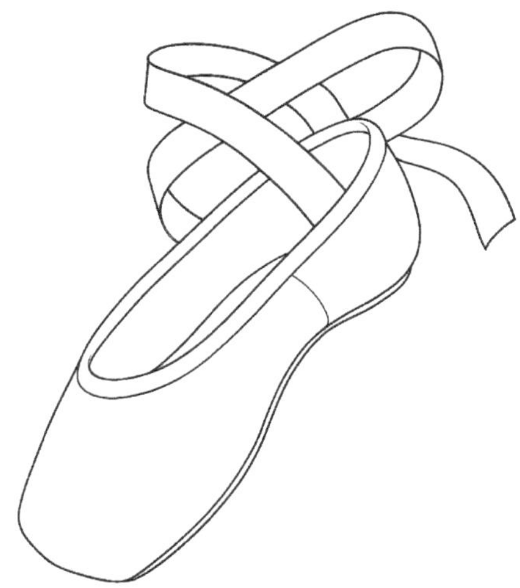

The rise of popularity of the **athletic shoes** synchronous with the ever broadening spectrum of synthetic materials used in the footwear industry. Athletic shoes are designed to support the foot during dynamic movement, combining cushioning, flexibility, and stability. The same time these footwear have reached new heights in performance in their functionality, they are the de-facto choice of footwear for consumers across the globe for a slew of occasions. Popularized by the celebrity culture fashion street wear athletic shoes are now an acceptable choice for high-end events.

Last Fundamentals

For pattern making, a thorough understanding of last terminology and geometry is essential. Like any other discipline within the footwear industry, pattern making and upper making cannot exist in isolation. All related trades – most notably last design and last manufacturing – have a direct and significant influence on technical upper design.

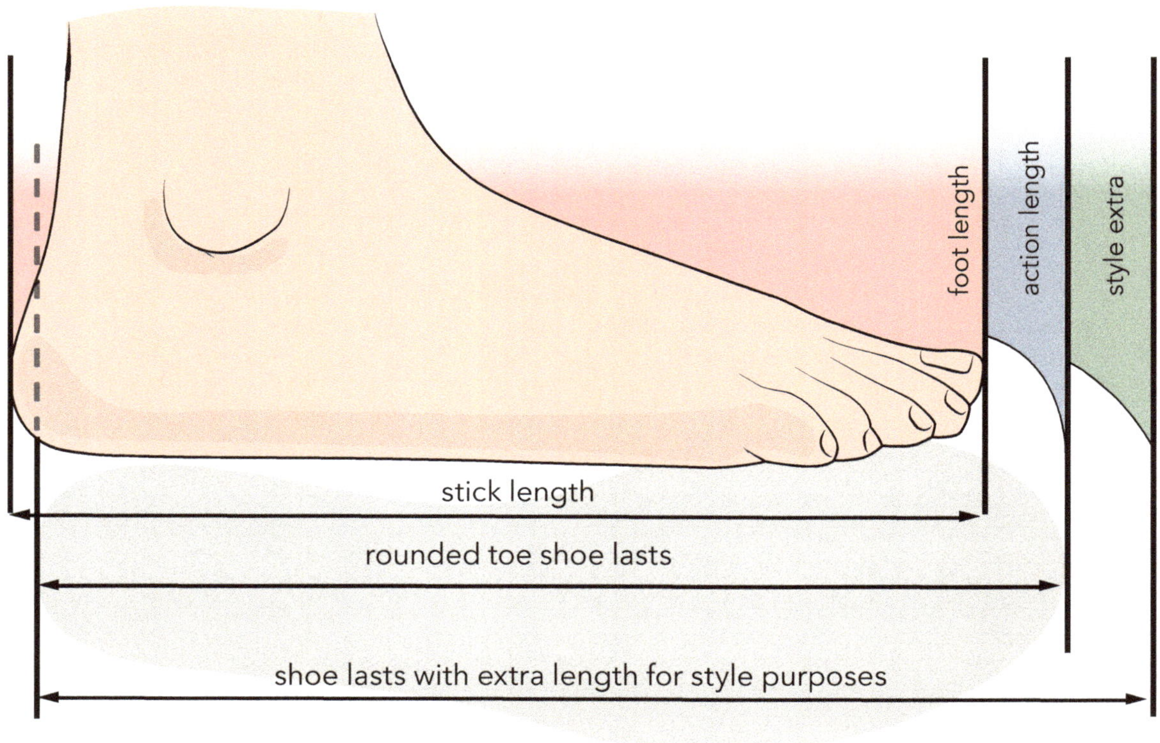

A **shoe last** is an abstract representation of the foot. This means that the last is not the shape of the foot, what's more, it's surprisingly not the size of the foot either, but a form that can be used to create a fitting footwear. The abstraction depends on the purpose of the footwear (some special ones need special fittings or even shapes). Generally speaking the footwear last is bigger than the foot to accommodate its movements during walking, running. In case of a regular footwear, that is about 2-3 sizes (20 mm, 3 EU sizes, 3/4"). A last length, which is determined by this 20 mm extra is called "**standard last length".** Regardless of the modifications due to style ("style extras"), we use this number to determine the last size.

Sandal lasts are substantially smaller, only a couple of millimeters longer than the foot itself since there is no restricting toe cap above the toes.

Longer toe styles requires additional elongation above the regular action length. Extreme pointy toe pumps or western boots are good example for these.

It is also important to note that the lasts are measured where the heel hits the ground, not from the farthest point back.

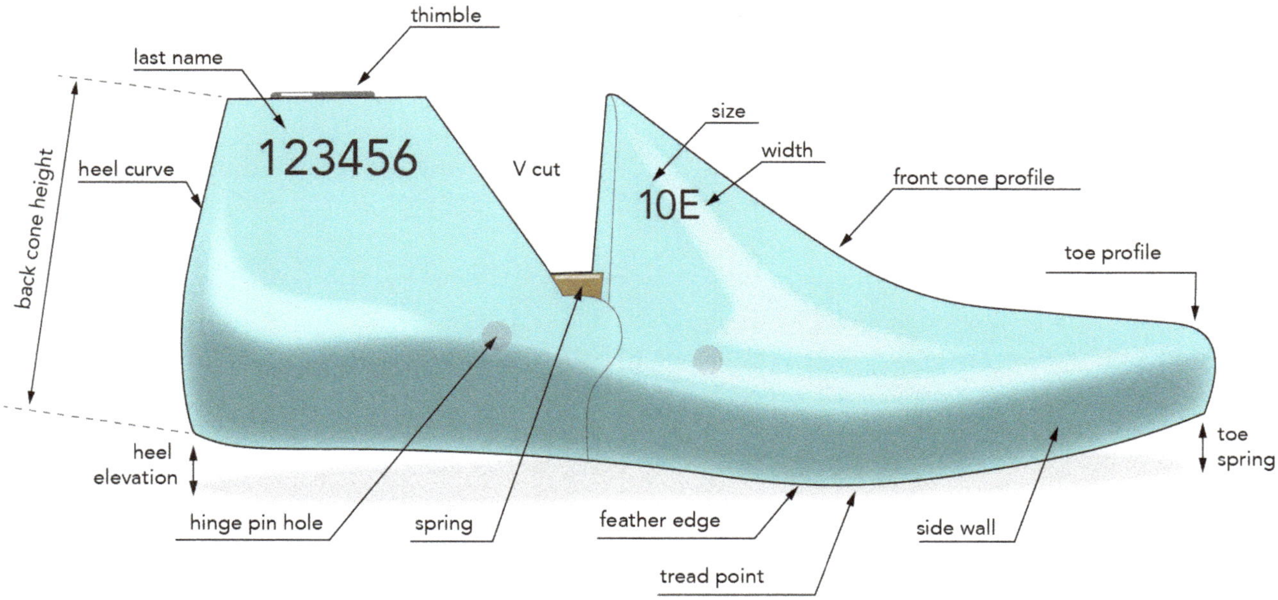

The **heel angle** is defined as the angle between the heel feather-line plane and the ground. This angle is primarily determined by the heel elevation rather than the nominal heel height. Heel height alone can be misleading, as the presence of a platform increases the measured height while leaving the actual heel angle unchanged. In a practical scenario, two identically matching 10 cm high heel can have different heel angles if one is designed for a platform footwear.

The **heel seat** is roughly 1/5 length of the last on the bottom. It is usually only slightly curved on modern last, because of technical limitations, older or bespoke lasts tend to have closer to the natural, curved shape of the human foot.

Hinge pin holes and the **spring** are structural parts of the last without much relevance in pattern making. This can be different for other hinge types or entirely missing.

The **front cone** is the front half of the last, while the **back cone** is obviously the back half. Some lasts are divided differently or not divided at all.

The **feather edge** is a sharp line running all around the last (for most constructions). Not all last have this and some don't have it around the medial shank. Some special lasts are made without any feather edges or it is missing only in certain sections, which is very common for sleek, dress shoe lasts.

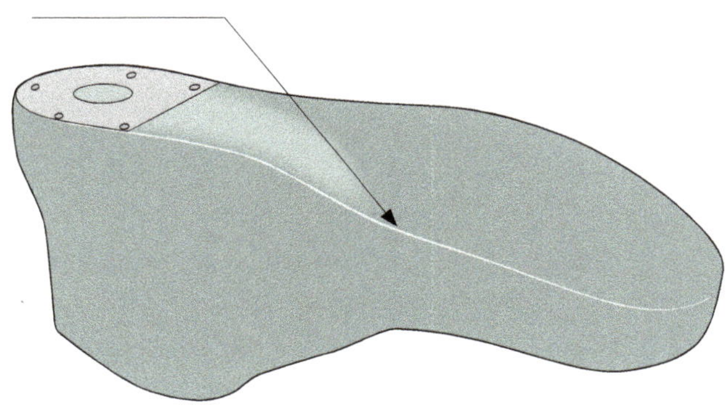

Last bottom or **plantar surface** is the underside of the last, shaping the insole contour and bottom curvature. In pattern making, pattern pieces for the bottom are just as integral part of the design process.

Side wall is the roughly vertical surface of the last that runs along the outer edge of the bottom, connecting the feather-line (where the upper turns under) to the plantar surface of the last at the front. It is a very characteristic element of the last, especially made for cupsoles and extended shoe cavity. The presence of an excessive side wall is a good indication that the last is made for cupsoles so the special side wall stitchers machines can reach it easily.

The **toe spring** is the upward curvature of the toe area of a shoe or last, measured from the ground to the highest point of the lifted toe tip. It is designed to support the natural rolling motion of the foot during gait. Some footwear need zero or even negative toe spring.

The **front cone profile** is the shape of the last looking from the side view. The shape of the front cone profile is a good indication if a last is made for boots, shoes or loafers.

The **back cone profile or back curve** is the shape of the back portion of the back cone. This is also a good indication what kind of footwear meant to be made on that last.

The **thimble** is a reinforced hole in the last, this plays a very important role during the making/ manufacturing process – many of the industrial machines utilize these hole to hold the last firmly. It also helps then remove the last with a help of a lasting pin in it.

The **toe shape** is a form of the toe from the top view, it is one of the most defining characteristic of a footwear. The toe shape of the last is mildly different from the ready made footwear, as there will be additional layers, such as lining, stiffeners, reinforcements and upper layers added. The toe shape influences the comfort of the footwear. It can be: rounded, square, pointy, almond, chisel and so on. Toe shapes influence the design itself and this is one of the reason modern pattern making cannot rely on geometrical methods solely.

The **toe profile** is the shape of the toe from the side view.

The **tread point** is the anatomical pivot point under the ball of the foot–usually located beneath the first metatarsal head – where the foot transitions from weight-bearing to toe-off. In footwear development, it marks the key reference point for flex line placement, sole bending behavior, and toe spring geometry.

Laser etched lasts with quality control lines.

The **etched marks** on the last – stamped or laser etched – are important information for the design team and factory. These identify the name (or number), size and the width size. The production date, last factory name also can be indicated here, as well as the counter and the vamp heights in some cases. If the last is made for a specific product (or a specific type of product), the last manufacturer also can add marks for making and quality control reasons. In our illustration we can see a laser marks for a topline for a pump. This will help the factory to control the topline position and spot a faulty product in time.

The Bottom of the Lasts

The bottom of the last – or the "**bottom paper**", when we create a copy of it – is worth discussing.

The last is can be dissected into three parts from the heel to the toe:

1. The symmetrical heel length. This is roughly 1/3 of the entire last length.
2. The waist area to the sole section of the ball girth. This line defines the medial and the lateral ball points on the feather edge. The medial ball point distance from the heel is a bit longer.
3. The toe area. This is where most of the last design is happening, as this remains the most visible part of the footwear.

The bottom paper has two major axes: the first, the **main axis** (marked with stacked red line here). This is where we measure the length of the last.

The black stacked line indicates the **heel axis**. This divides the front of the heel (the heel width) accurately two equal portions. This is an important line to measure the heel length and design bottom features of the footwear, such as sole, sockliner, etc.

The grey line around the sole shape signifies the **feather edge**, but as mentioned before, not all last has feather edges, or at least not always all the way around.

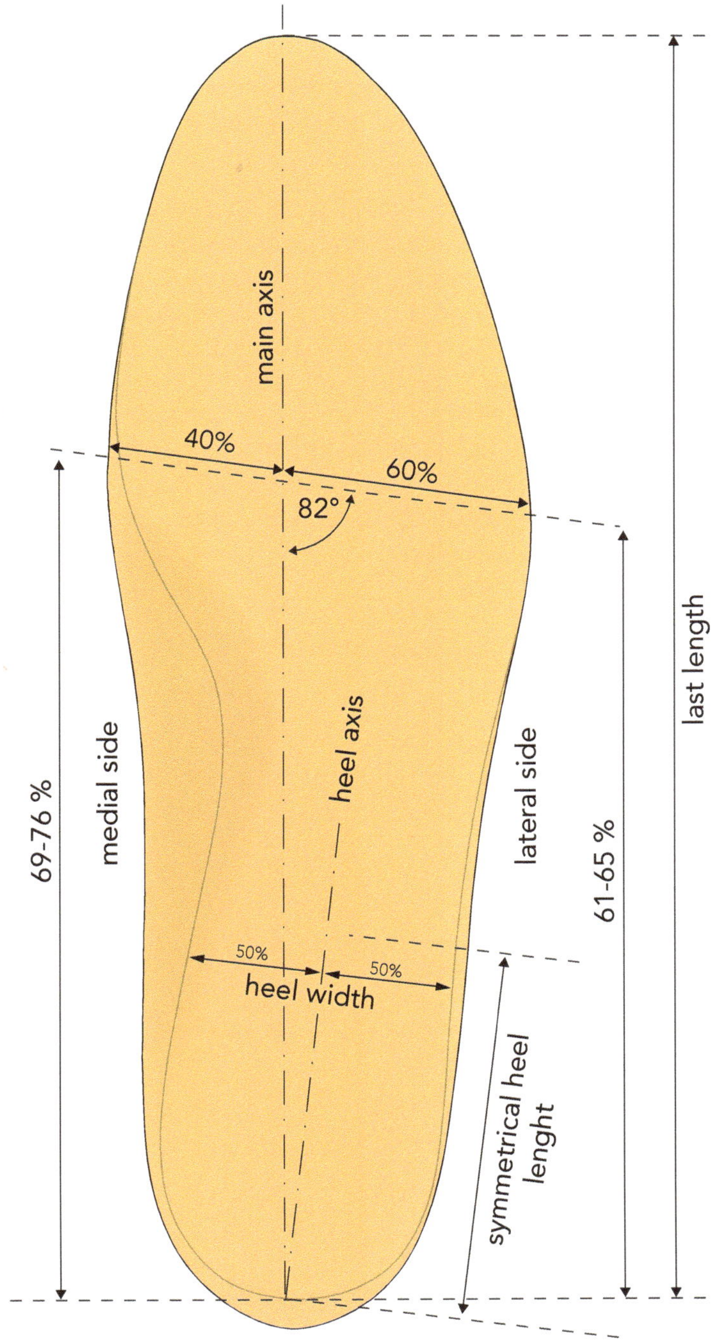

Heel Elevation and Heel Height

All footwear developers learn this sooner or later: heel elevation must not be confused with heel height. This distinction is equally important in pattern construction, even though pattern makers may not always concern themselves with heel angles or platforms, despite the fact that these elements also require proper pattern development.

Heel elevation: the height of the heel when the tread point touches the ground and the last is held in the proper position.

Heel elevation is defined by the spatial relationship between the ground plane on which the last is positioned – when held in its correct working orientation (the tread point on the ground). In pattern making, this value is fundamental: it determines critical angles and serves as the basis for several subsequent calculations – especially in taller boots, where even a small angular deviation can accumulate and cause the boot shaft to lean noticeably forward or backward.

Heel height: the actual height of the heel, heel elevation plus the elevation of the tread point (for example platform or just a thicker sole).

The height refers to the vertical distance from the ground plane to the back of the heel of the last. This measurement includes not only the heel itself, but also the cumulative thickness of all materials built up in this area, such as the insole, heel counter, upper, midsole, outsole, and heel lifts.

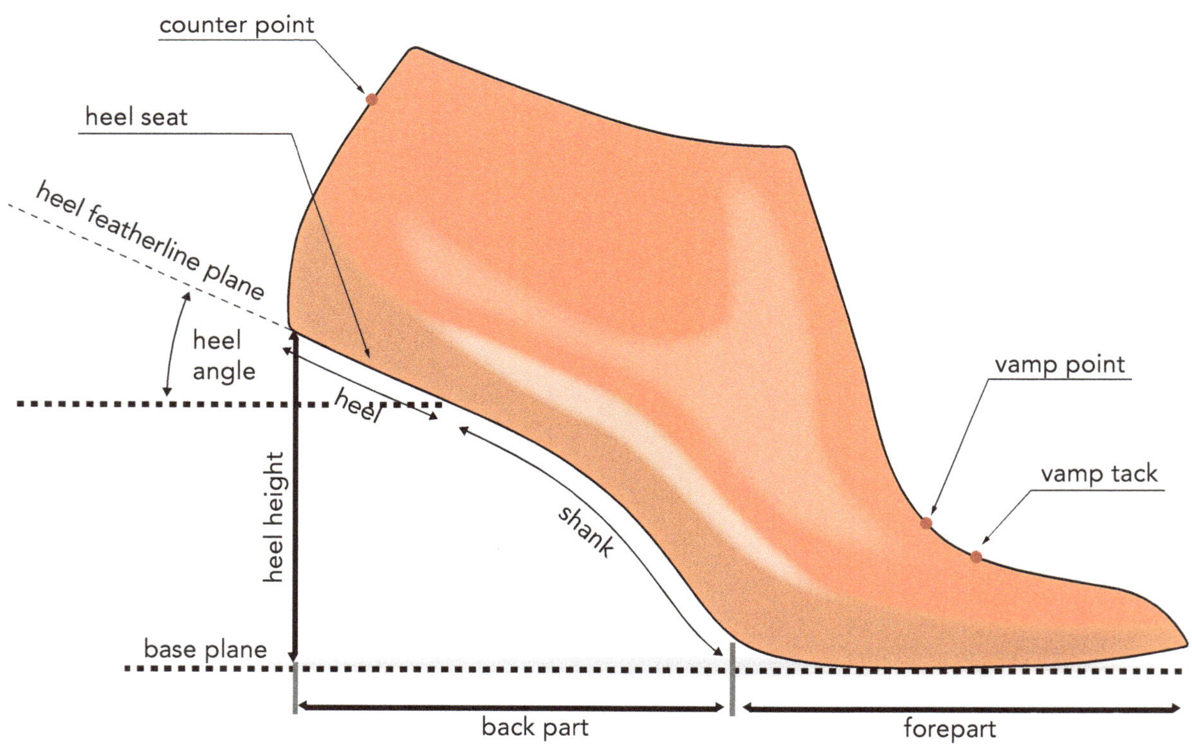

The same heel angle can belong to very different heel heights. Ignoring the heel angle and only focusing on the height number results in faulty products.

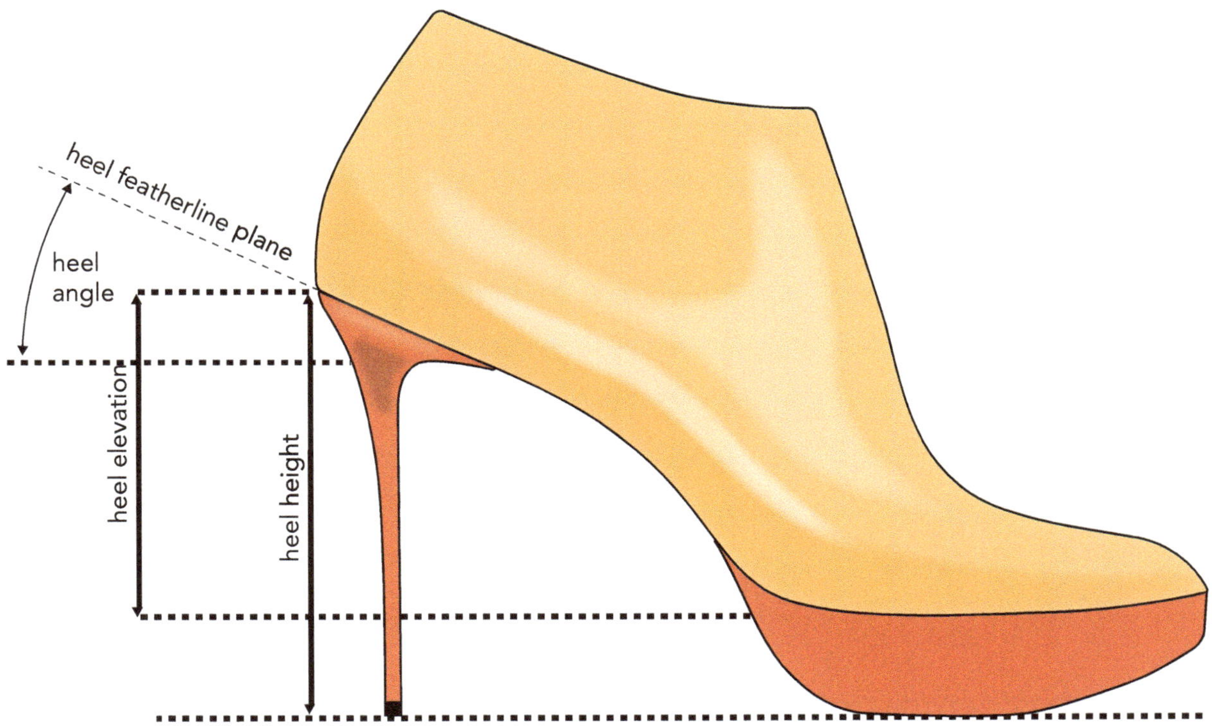

The difference between these two concepts is clearly illustrated by a platform shoe with a higher heel. Let's say the heel height measures 120 mm, using this value for pattern calculations would be a serious error. When determining the correct last orientation, the platform must be completely disregarded: while it raises the wearer higher off the ground, the foot itself remains in the same functional position relative to the last.

A platform directly affects heel height, but not heel elevation or angle. Technically, a platform represents a parallel shift of the heel, adding height without changing its angle. A very high heel can therefore have the same angle as a relatively low one, and confusing these concepts can result in a heel being attached at an incorrect angle.

In pattern making, heel elevation is closely related to shaft angle. While this relationship may be insignificant in low-heel models, in higher shaft designs even a few degrees of difference can produce a serious and clearly visible change in angle.

Last Types

Lasts can be classified by material, hinge or joint type, and specific fit or functional features. Not every last is suitable for every type of footwear: shoe, boot, and sandal lasts differ to meet distinct construction and fitting requirements.

There are also specialized lasts for specific technologies, such as metal lasts used in molded or injection-based manufacturing, as well as finishing lasts, which are now rare. Symmetrical lasts are used only in limited cases, mainly in ballet footwear.

The **scoop block lasts** represent the old fashion last making standards. Back in the day this was the most common type, because of its simplicity. They were used in most footwear, even the removing process was more challenging than the modern versions.

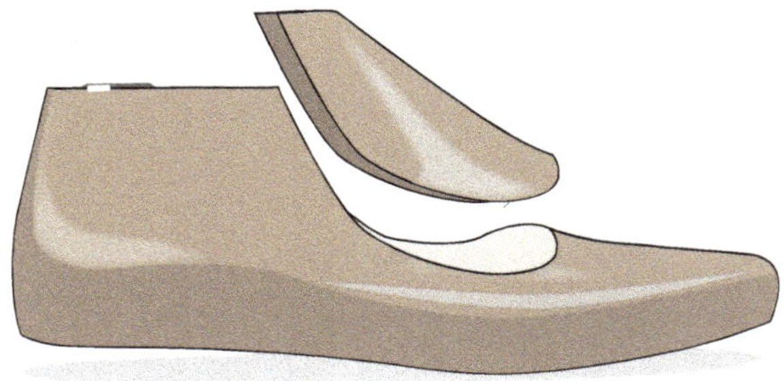

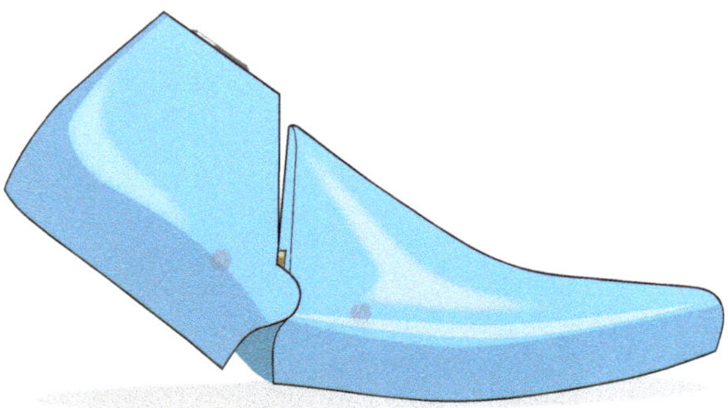

The **V-hinged lasts** have an extremely strong spring between the cones which makes it stable when it is used in its original form, but possible to hinge it closed and back up. These are the most common lasts in the industry. They are relatively easy to remover it makes them perfect for low cut shoes, derbies, oxfords, wholecuts, etc. and some boots as well.

The **reverse-hinged lasts** are also easy to remove, but most importantly they don't stretch the topline during the process. This last feature makes this a perfect choice for any technology, where the last has to be placed back to the shoe or slip lasted constructions. This is not a good choice for boots where the backline is somewhat rigid, as there will not be space to move the back cone.

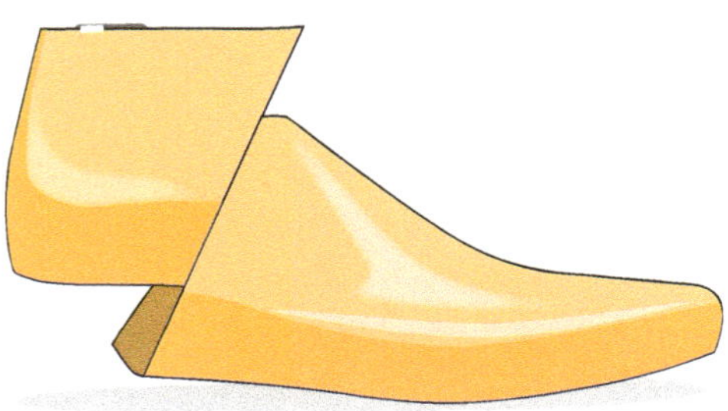

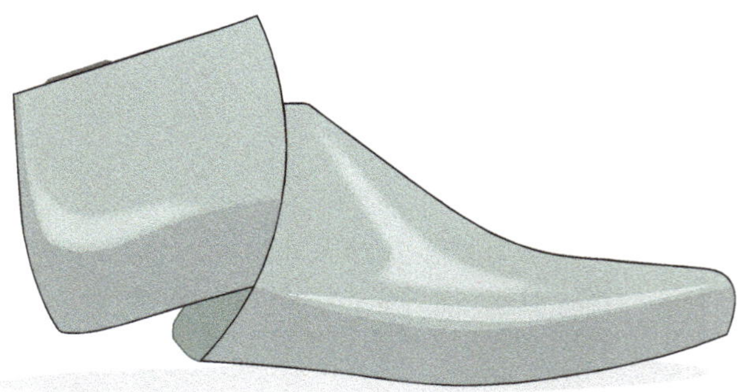

The **sliding lasts** have a back cone sliding upwards towards the toe. It has different versions for the mechanism, depending on the usage and technology. It is a good choice for re-lasting, forced lasting and some types of boots as well.

The **three part scoop-hinged lasts** is a special last for boots. Boot lasts has two distinguished function for specifically for boots. First of all the backline and the front cone profile is a bit different, because of the fitting requirements. The other important feature is - how they get removed and that is very dependent of the upper model. A regular shoe last can actual function as a boot last in some cases, but the other way around it is problematic.

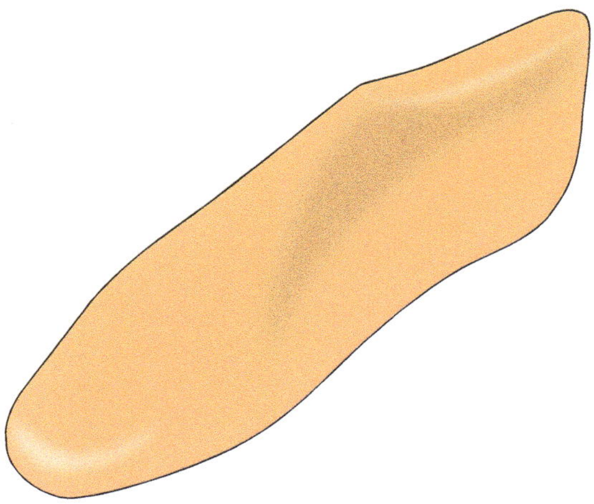

Solid lasts can be used for footwear with a very long and wide opening, but they are also common in baby shoes, where the size is too small to accommodate hinges. The generous opening allows the finished shoe to be removed easily from the last. Ballet lasts–whether for pointe shoes or soft slippers–represent a special category of solid lasts. They are traditionally made symmetrical, reflecting the historical method of manufacturing ballet footwear. Sandal lasts are likewise typically produced as a solid block.

Sandal lasts – especially those designed for open-toe constructions – are typically shorter than shoe or boot lasts, because the open forepart eliminates the need for full action length (sandals still have action length, but only 6-7 mm). They are usually manufactured as solid lasts, since sandals do not require a hinged mechanism to release the finished product from the last.

Some sandal lasts, like the example shown here, extend above the anklebone. This additional height allows the use of ankle straps or higher, boot-like pattern elements while still retaining the open, lightweight character of a sandal last.

The Volume of the Last

The shape and size of the last–or its overall mass–define the internal volume of the footwear, which is intended to accommodate the foot, but not exclusively. An obvious example is socks, which also occupy space within the shoe. There are, however, less apparent elements to consider.

The sockliner also takes up space in the internal cavity, and although it may appear insignificant, it rarely is. In some cases, the materials involved are substantial, such as the thick, padded linings used in athletic shoe footbeds. Additional foams and cushioning may be present at the topline, tongue, sides, and other areas.

In these illustrations, the cross-sections clearly show how last manufacturers account for these additional materials in their designs.

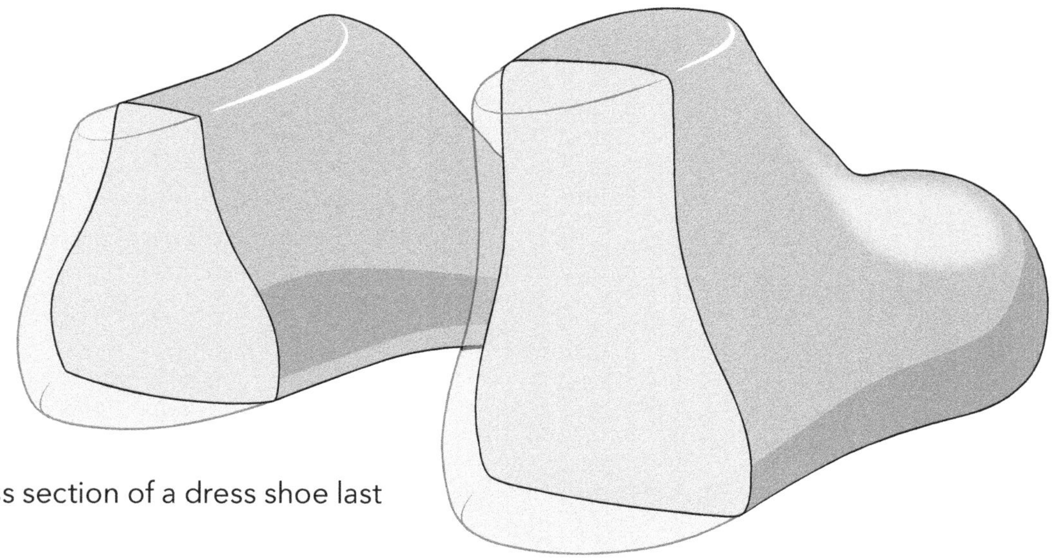

cross section of a dress shoe last

cross section of an athletic shoe last

A regular dress-shoe last has a relatively narrow top portion. There are two reasons for this: first, a snug fit at the topline is expected; second, the ankle bone sits above the topline and can therefore be disregarded. The bottom of the last corresponds to the insole plane, and the sockliner typically occupies only a minimal amount of space.

In some cases – such as ballerinas, where a tight topline is essential – the last can be even narrower at the top.

The athletic shoe last generally has more volume that adds to its "chunkier" appearance. This is to provide the internal space required for substantial padding and other materials. The topline of athletic footwear extends above the ankle bone, so its exact anatomical shape is less critical, while the required volume is essential. The bottom of the last has less pronounced curvature, as the structure of the footbed provides much of the final shape.

Last Manufacturing

Traditional bespoke last carving in the Koronya workshop

Last design is an integral part of the complex footwear development process and forms the foundation of pattern making.

Not long ago, last making was an extremely labor-intensive process that was impossible to be executed at scale, and achieving consistent results was highly challenging. In the bespoke trade, hand-carved lasts are still part of the tradition; however, industrial production has long since shifted to mechanized processes.

The first machines to enter the industry were copy-lathe systems. These machines mechanically "read" a physical master last and transferred the information to another section of the machine, where the new last was carved from a solid wooden blank. Within certain limits, these machines were even capable of modifying specific parameters of the last, such as turning a right last into a left one or changing the size.

CNC technology was the next step. It became possible to design a last in a 3D computer program and translate the information into a footwear last. The 3D design data can be used for insole and outsole design, or even upper CAD applications, keeping the entire design process digital and opening the door to collaboration between distant parties, data transfer across countries, and reduced development time.

Modern last-making machines can handle multiple pairs at the same time. Using pre-made last blanks, multi-axis CNC machines carve lasts from standardized blanks with extreme precision, ensuring exact control over toe shape, heel curvature, spring, and sectional geometry. This level of accuracy delivers unmatched consistency across sizes and repeated production runs.

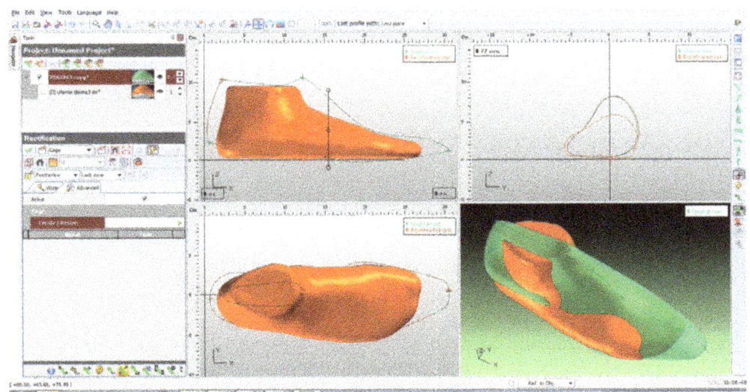

Foot scanners capture the three-dimensional geometry of the customer's foot with high accuracy

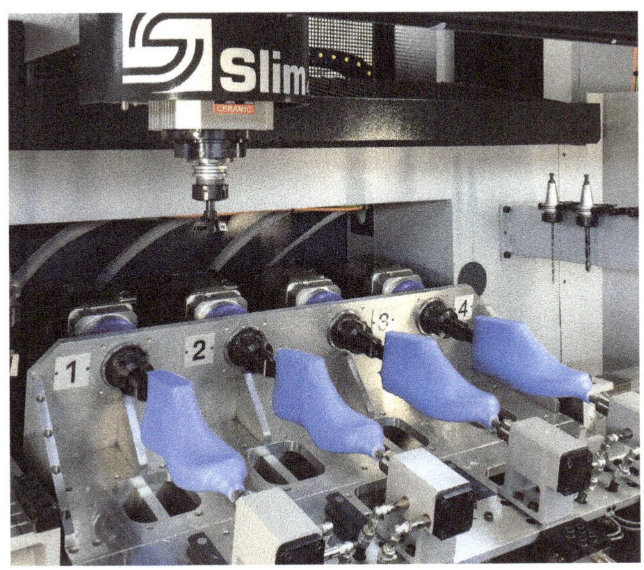

High end machines such as this one by NewLast is capable of making several lasts at the same time

The Size of the Footwear

Assigning standardized numerical sizes to footwear was not necessary as long as customers could buy directly from the maker. Although ready-to-wear footwear already existed, sizing was of limited importance under these conditions. When this direct connection disappeared with the dawn of mass production, some form of standardization became necessary. Different parts of the world responded to this demand in different ways; as a result, currently there is no global standard of sizing. Instead, several distinct systems are in use today, based on different units and even different zero points.

The existing systems can be categorized into two fundamentally different approaches:

Indirect sizing systems

These systems consider the rounded-toe shoe last as the size reference (indirect sizing) and indicate the corresponding footwear based on that predetermined size. The actual footwear length can vary significantly: sandal length may be well under the last length, while extremely pointy-toe boots may extend far beyond it. On the other hand, this approach helps customers find the correct size in shoes and boots without being distracted by toe shape or footwear type. In other words, customers can rely on the same size regardless if the style is a sandal or a dressy pointed-toe shoe.

Direct sizing systems

Direct footwear sizing systems, such as the Mondopoint system, measure the actual length and sometimes the width of the foot in millimeters, providing a precise and standardized approach to footwear fitting. Originally developed for military applications by the International Organization for Standardization (ISO), Mondopoint was designed to simplify logistics and ensure consistent sizing across different manufacturers and countries. Instead of relying on arbitrary size numbers, it corresponds directly to anatomical measurements; for example, a foot 270 mm long equals to a Mondopoint size 270. Some versions also include width in millimeters, allowing for finer fit adjustments. Because of its accuracy and universality, Mondopoint is still used in military, ski, and safety footwear, where precision and interchangeability are critical. This direct measurement approach has also influenced modern fit analysis and international size standardization by reducing inconsistencies found in traditional US, UK, and European systems.

Regardless of the fact that the toes sit in noticeably different positions and the toe shapes vary widely, all of these footwear examples are the same size because they share a standard last length. This principle is extremely useful, not only from a sales perspective but also for technical design.

If footwear were sized by actual overall length, as this example clearly demonstrates, the result would be both confusing and impractical. Customers would effectively need different sizes depending on toe shape or the amount of "style extra," which would make little sense from either a functional or a commercial standpoint.

In reality, the same foot size fits into all of these models, which means sizing must rely on the one consistent and meaningful datum available: the standard last length. This is the only logical and scalable way to define sizes across different footwear types and stylistic variations.

Interestingly, up to this specific point, all major sizing systems around the world agree on this principle.

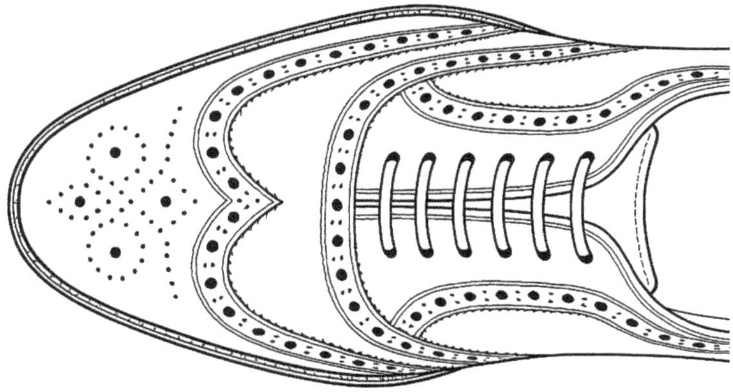

A rounded toe shape shoe is usually as long as the standard last length dictate

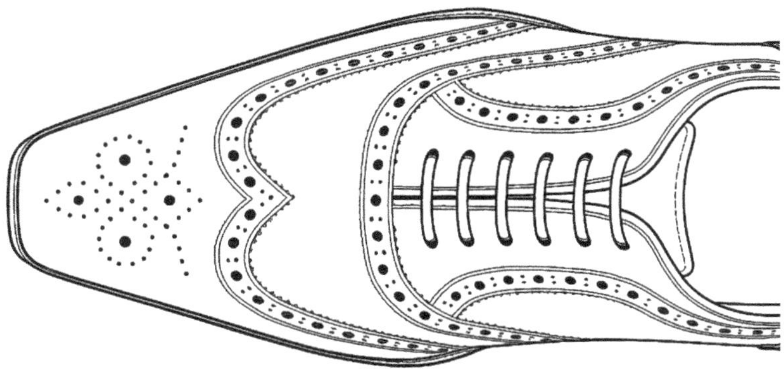

Pointy and square toe models use a "style extra" on the top of the "action length"

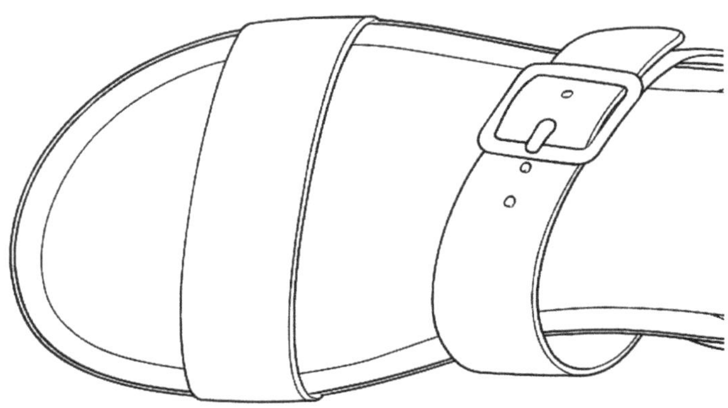

Sandals still use some extra length, but only a few millimeters

The Units of the Footwear Sizing Systems

Footwear sizing systems use different measurement units. The imperial systems are based on the inch. Both use 1/3 of an inch (written as 1/3"), known as the **barleycorn**.

Europe mostly uses the **Paris Point**, which equals 2/3 of a centimeter (written as 2/3 mm). The Mondopoint system uses metric units, expressed in centimeters or millimeters.

Zero Point

Defining the zero point is relatively straightforward in the Paris Point and Mondopoint systems: zero is zero, and this is where measurement begins. Technically, even a doll's foot can be measured as long as it has some length, regardless of its practicality. This is the reason these systems still make more sense for pattern making, because the numbers are proportional. A size 24 is truly half of a size 48, which allows the system to be used reliably in different formulas.

The UK and US systems are more complicated. Their zero point is not actually zero, but is determined by the smallest foot size, starting with a newborn. In the illustration, a last of a given size is compared with the foot of a newborn baby. Even though it is rarely necessary to use a last at this age, it remains an interesting historical detail.

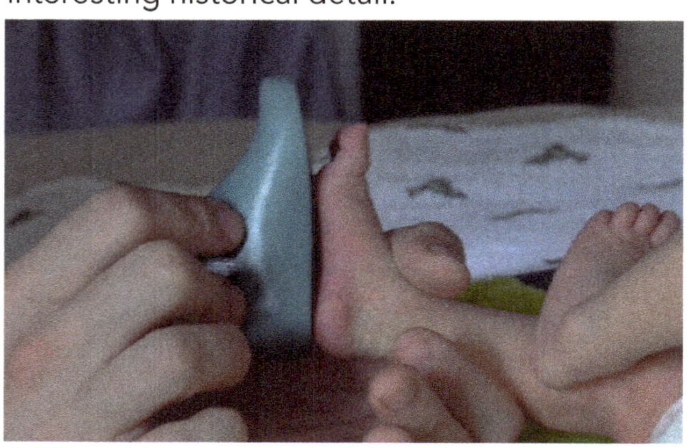

A US size 0 last compared to a foot of a newborn.

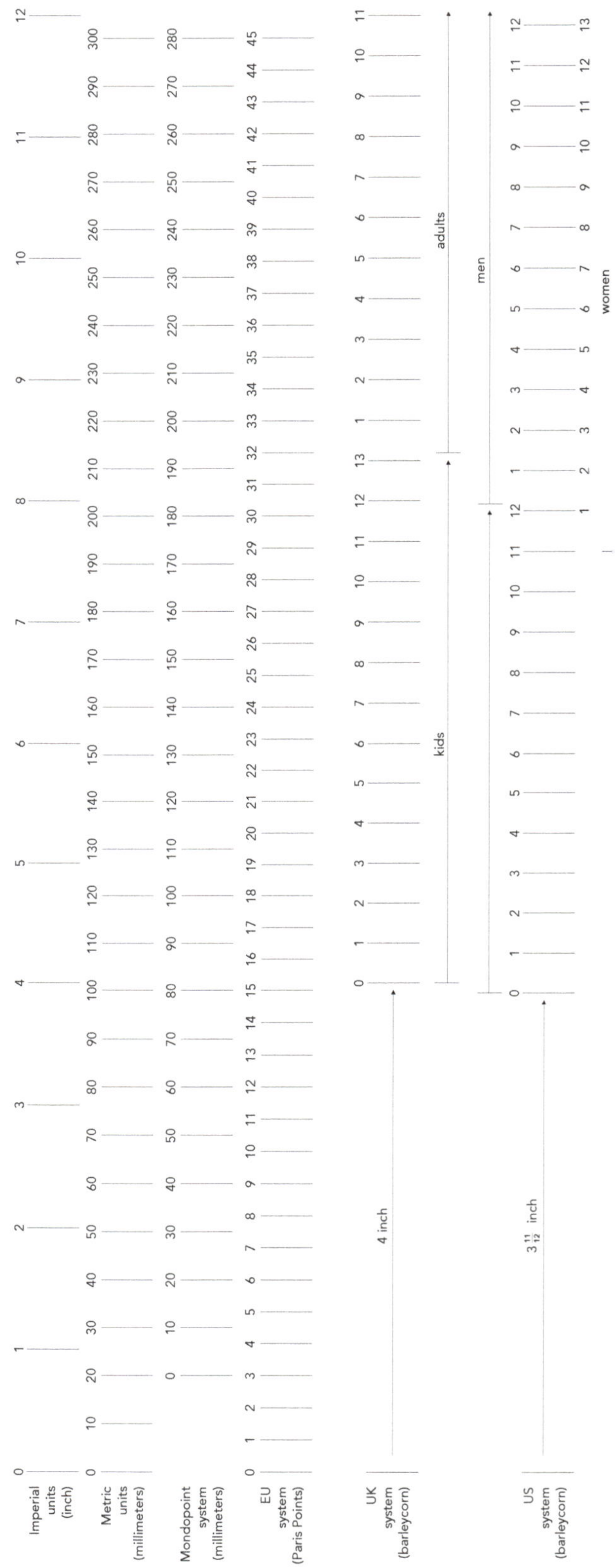

The **UK system** starts sizing at 4″ and increases by 1/3″, or one barleycorn, reaching size 13, after which it restarts at size 1. Technically, these numbers are proportional as long as the initial, disregarded portion is added back. For example, an adult size 10 equals the original 4″, or 12 barleycorns, plus the 13 children's sizes and another 10 sizes, resulting in a total of 35 barleycorns, which converts to around 11.6″.

The **US system** developed differently. Edwin Simpson, a New York businessman, is credited with introducing a detailed and formal sizing system around 1880, standardizing and modernizing the UK approach. This system was widely adopted by the US footwear industry in the late 1880s. Instead of a 4-inch minimum, the US system starts at 3 11/12″ and uses different numbering for men's and women's sizes, with a one and a half size offset between them.

Width Sizes

While length determines how the shoe aligns with the foot, width defines how much space is available across the ball and instep areas. Widths are typically represented by letters or letter combinations such as A, B, C, D, E, EE, and EEE, with A indicating narrow fits and D or E signifying standard or wider fits. Other systems may use extended sequences like 4A through 6E, or descriptive codes such as N (narrow), M or R (medium or regular), W (wide), and XW (extra wide).

These conventions vary by region and manufacturer. For example, a U.S. men's D width is considered standard, while a B width is standard for women. Some European makers use numerical girth codes or design around a single, assumed width.

The Brannock Device

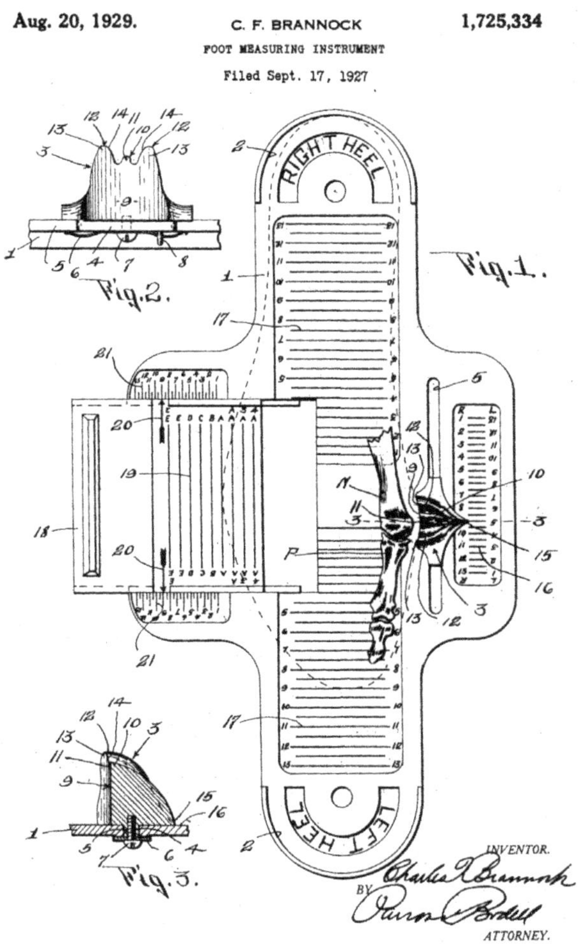

The Brannock Device is a mechanical foot-measuring instrument developed by Charles F. Brannock in 1927 to standardize the sizing process. It consists of a metal base plate with calibrated scales for three primary dimensions: overall foot length, width, and arch length (heel-to-ball). These measurements allow for a more precise correlation between the anatomical dimensions of the foot and the corresponding last or shoe size.

2 PATTERN MAKING BASICS

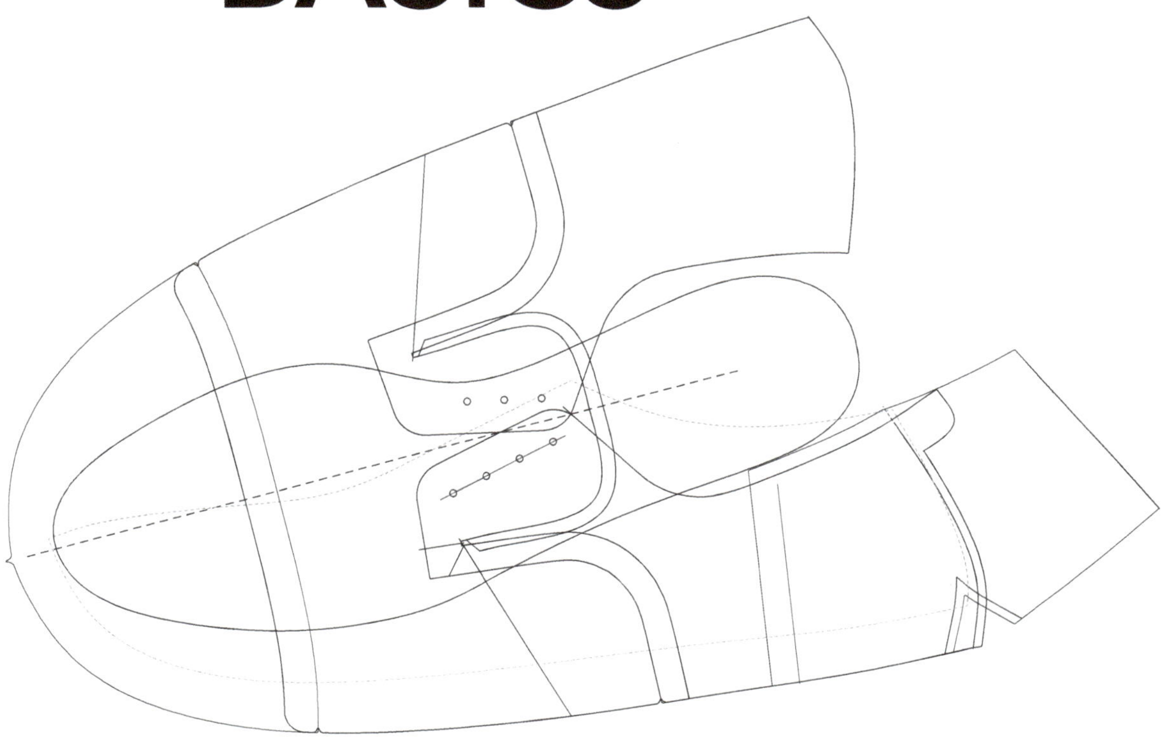

Building upon the foundations of the previous chapter, it's time to move onto the techniques that will bring those theoretical parts together. Generally speaking, the pattern is a 2D template that is the technical translation of a 3D design. It contains the necessary information to get to as close to the original vision as possible, which includes functional data such as seam allowances, alignment marks, and construction guidelines, etc. as well.

There are numerous ways to construct a pattern. A competent pattern maker will take into account the material, the scalability, machine-requirement and a myriad of other considerations that is pertinent to the production. The instructions shown in this book is what I've found the most effective and straightforward way to teach the basics of pattern making.

The Significance of the Geometric Method

The **last standard** functions as the structural framework of any footwear pattern. Most patterns are built on a base geometric layout. These reference lines provide guidance for style proportions, seam placement, toplines, and pivoting, all calibrated to foot size, heel height, and key measurements. These guidelines are not absolute rules. Even older pattern books that are otherwise quite rigid in their formulas and proportions, acknowledge that individual styles, specific foot measurements, and deformities can influence the placement of these lines. At the same time, the relevance of classic proportions remains significant. What is commonly perceived as "classic" is largely the result of long-term visual imprinting derived from these proportions and placements, which are practical as well as aesthetic. Although the method may appear complex at first glance, it is logical and can be learned relatively quickly.

Robert Knöfel (1834–1884)

It is essential that any serious shoemaking book mention and acknowledge Robert Knöfel. His geometric system laid the foundation for modern footwear pattern making. Knöfel approached shoemaking as both a craft and a science, applying mathematical precision, and proportion to create repeatable and logical methods for pattern construction. His work helped elevate shoemaking from a purely empirical trade to a disciplined technical field and influenced generations of craftsmen and educators.

Robert Knöfel, shown on the "Denkmalfonds - Robert Knöfel" commemorative stamp. Knöfel lived during a transitional period when photography was becoming more common, the image is not based on a photograph, but on a lithographed portrait created after his death.

Robert Knöfel was born on January 5, 1834, in Wilsdorf, Germany. His father was a master shoemaker, and as the eldest of eight children, Knöfel was expected to continue the family trade. From an early age, he demonstrated exceptional intelligence and curiosity. Encouraged by his father, he developed a strong interest in the sciences while completing his apprenticeship in the family workshop.

As a young apprentice, he studied geometry and anatomy with great dedication and later taught these subjects to fellow apprentices. During this formative period, he developed his geometric system of footwear pattern making—a structured, analytical approach that would later influence generations of shoemakers.

In 1860, Knöfel passed his master shoemaker's examination in Dresden. He founded the Workers' Self-Education Club, through which he promoted intellectual and technical advancement among tradesmen. He also authored a textbook on footwear that combined scientific reasoning with practical craftsmanship.

Throughout his career, Knöfel traveled extensively across Germany, lecturing to shoemakers and apprentices. His teaching addressed not only pattern construction, but also materials and cost calculation, reflecting a comprehensive understanding of the trade.

The Standard Pattern Ruler

Even in the early stages of the geometric method's development, shoemakers sought ways to simplify the process, avoiding laborious calculations. It helps that many of the calculations used are derived from proportional relationships – such as halves, thirds, and quarters – as well as from specific angles and established standards. The Standard Pattern Ruler is my contribution to help accelerate the work. This specialized tool is the culmination of years of teaching experience and has gone through multiple versions and iterations. It helps the pattern maker to work efficiently with only a pencil and this instrument in any hand patterning process, but especially to create a pattern from scratch with the geometric method. Apart from the most commonly used units, classic curves and angles, it also has template for many shoe-specific details. It's transparent in order to see the existing lines, and it's flexible enough to be able to use on the last directly.

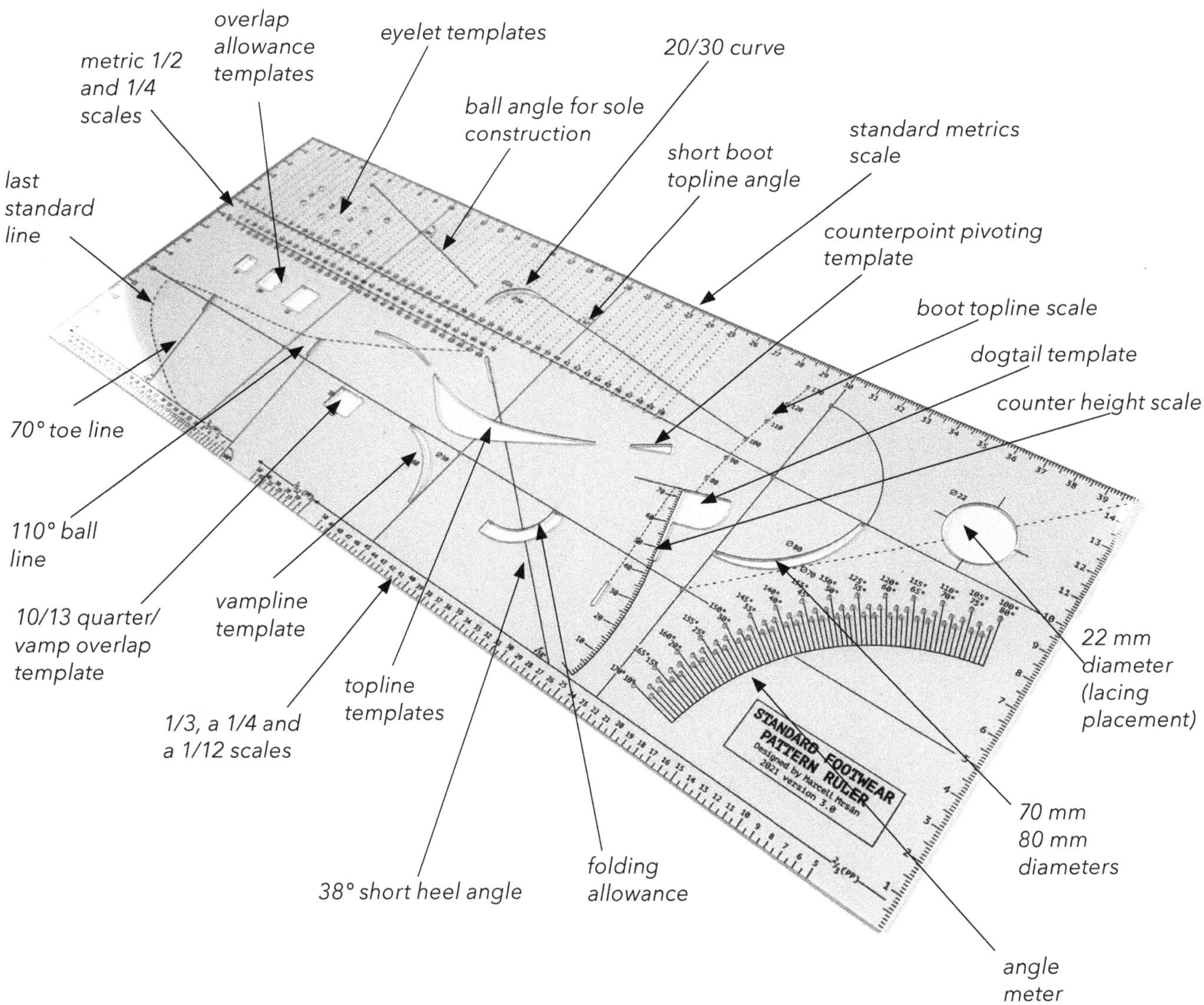

Pattern vocabulary

In the following chapters, specific terminology will be used. This vocabulary defines the fundamental terms used in footwear pattern making. Establishing a shared technical language helps avoid ambiguity and ensures that methods, reference lines, and operations are interpreted consistently. The vocabulary presented here forms the foundation for the techniques and processes introduced in the chapters that follow.

Ball girth line
The ball girth line is a reference line on a footwear pattern marking the widest circumference of the forefoot at the ball, used to control fit, volume distribution, and pattern proportions. Ball girth can be calculated based on shoe size and width, or it can be measured directly on the foot. For bespoke footwear, direct measurement is clearly the preferred approach.

Ball line
The ball line represents the ball girth position on the last copy. It is slightly under one third of it.

Baseline
Baseline is the reference line on a footwear pattern representing the floor contact plane of the last, used to establish vertical positioning of boot shafts, angles for girth lines, and proportions of pattern elements.

Excess, allowance
For certain seams and processes, an extra amount of material is required at an edge. Specific seam types, edge decorations, and technologies require clearly defined allowances.

Folding axis - main folding axis
The analogue counterpart of the folding axis is the edge of a folded piece of paper. This is the line along which a pattern piece is unfolded. The main folding axis is the line along which the entire pattern can be unfolded and is typically used for the vamp.

Heel width

Heel width is used for insole patterns and is usually measured as one fifth of the short heel measurement.

Last copy
Last copy is a two-dimensional representation taken directly from the last, capturing its actual shape and measurements for use in pattern development.

Last standard
Last standard is the geometric reference framework derived from the geometric method, defining the primary reference lines, angles, and proportions used for footwear pattern construction.

Lasting allowance
Lasting allowance is a special type of allowance. It is generally much larger than other allowances and depends on the lasting technology. Some technologies require a generous amount, while others need very little or none.

Master pattern
A master pattern is the complete reference pattern that contains all essential lines, allowances, and construction information, from which all working and derivative patterns are developed.

Parallel shifting
Parallel shifting is a pattern-making operation in which a line or pattern element is moved in a constant direction by a fixed distance while maintaining its original orientation and geometry. If the line is curved–as is the case in most situations–the operation is referred to as offsetting.

Pivoting
Pivoting is a pattern-making operation in which a line or pattern section is rotated around a fixed pivot point to redistribute length or shape while maintaining continuity with adjacent elements.

Rotation
Rotation is a geometric operation in which a line, curve, or pattern element is turned around a fixed pivot point by a specified angle while maintaining its original shape and dimensions.

Short heel angle
The short heel angle indicates the position of the short heel measurement on the last copy (38° + heel height). For example, a 2 cm heel corresponds to 40°, while a 10 cm heel corresponds to 48°.

Short heel measurement
The short heel measurement is the distance measured around the back of the foot and heel, used in footwear pattern making to control heel fit, topline position, and rear volume. It is especially critical for models where this line cannot be opened, such as tall boots.

Style lines
Style lines are design-defined lines on a footwear pattern that determine the visual layout of the upper, including seam placement, panel division, and decorative elements.

Toe line
The toe line is a reference line on a footwear pattern indicating the position of the toe area; in the geometric method, it determines the amount of pivoting that can or should be applied during pattern development.

Unfolding
Unfolding a pattern piece is the process of opening a folded pattern piece along its folding axis to obtain the complete, flat pattern geometry.

Allowances

The values below represent average allowances for each specific type. In certain cases, different values may be required. Some allowances are determined using specific mathematical formulas to calculate the appropriate amount.

Lap Seam Allowance

$$3.5\,T + 2.5 = \text{allowance in mm}$$

T represents the material thickness in mm.

So, a 2 mm thick material should have:

$$3.5 \times 2 + 2.5 = 9.5 \text{ mm}$$

Realistically we rarely calculate with formulas, a material test on some scrap leather gives an equally good result for the trained eye.

Suggested allowance widths	
Folding	4 mm
Trimming	3-5 mm
Reverse seam	3-5 mm
Lap seam	8-12 mm
Lap seam for under brogueing	15 mm
Turnseam top side	5-10 mm
Turnseam bottom side	0-3 mm
Tacking	10 mm
U-binding	0 mm
French/Italian binding	0 mm
Gimping	2-5 mm

The Last Standard

Here's how to create a last standard for a size 42 (280 mm), low heeled classic men's shoes, using the standard lats measurements (on the following page) .

1. Draw two perpendicular lines. Point "A" is their intersection.
2. Measure 2/3 of the SLL to the base line. In our example this is 280*(2/3)= 187 mm. This is point "B".
3. From "B" point measure the leftover 1/3 SLL to the base line. This will be point "L".
4. Measure an additional 10 mm from "L" and this will be point "P".
5. Measure the heel height from "A" upwards. This is point "H"
6. Connect "H" and "B" with a line
7. Measure 38°+2 (the heel height in cm) from "HB" line. In our example the heel height is 2 cm, therefore we use 40°
8. Measure the half of the short heel measurement on this new line, this is point "I"
9. Measure 100° from "B" point on the "BH" line
10. Measure the 1/3 of the ball girth on this line. In our example this will be 240/3=80 mm, which becomes point "V"
11. Find the half point on the "BV" line, which is point "M"
12. Measure 110° from "L" on the base line. Measure the half of "BV" line + 10 mm. In our example: 80/2 + 10 = 50 mm. This is point "T"
13. Connect "I" and "V" and "V" and "T" and elongate this line with 25 mm. This is point "S"
14. Connect "P" with "S"
15. Measure an additional 20 mm from "H" and call this point "C"
16. Measure the counter height from "H", this is point "R"
17. Draw a perpendicular line from "R" and measure 20 mm, which makes point "E"
18. Connect "M" with "E"

With this last step the base form is ready.

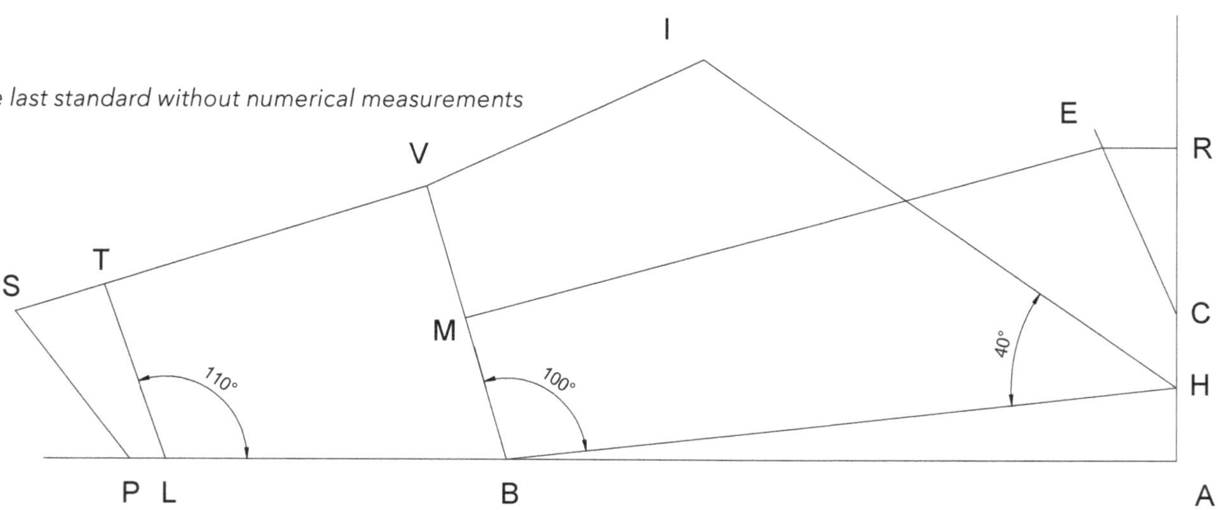

The last standard without numerical measurements

Standard Last Measurements

When an actual last is used for pattern making – which is the case in most professional workflows – these measurements primarily serve as inputs for formulas or as a means of double-checking conventional size references. It is important to note that such "standards" should be treated as broad guidelines, as both standardization principles and foot anatomy varies from region to region. All numbers (except the header row, which uses Paris Point) are given in millimeters in the spreadsheet below.

EU size	36	37	38	39	40	41	42	43	44	45	46	47
Standard Last Length	240	247	253	260	267	273	280	287	293	300	307	313
Ball girth	210	215	220	225	230	235	240	245	250	255	260	265
Waist girth	220	225	230	235	240	245	250	255	260	265	270	275
Short heel measurement	290	297	303	310	317	323	330	337	343	350	357	363
Standard low boot height	100	102.5	105	107.5	110	112.5	115	117.5	120	122.5	125	127.5
Counter height	59	60	61	62	63	64	65	66	67	68	69	70

The Simplified Last Standard

An experienced pattern maker can use the physical last to establish all primary pattern lines. For later adjustments, proportion checks, and pivoting operations, a simplified last standard – essentially an angle-based reference system – can be used. This standard must be derived from the correct last length; otherwise, a full last copy already contains all necessary girth measurements and line placement information.

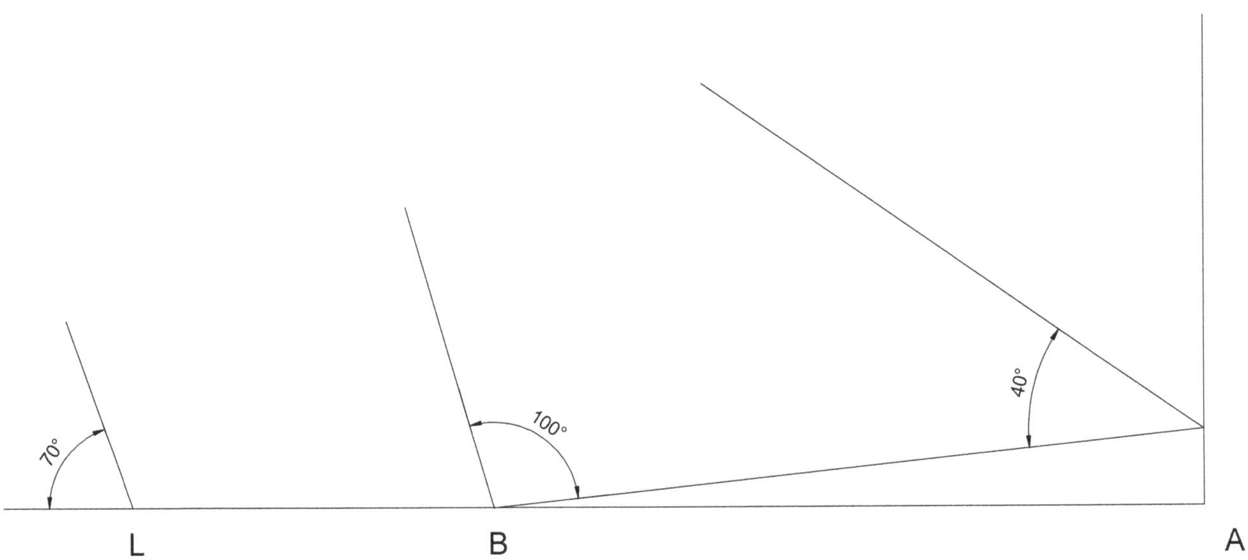

STANDARD
PATTERN
2021 version

Creating a Last Standard with the Pattern Ruler

Creating the last standard with this ruler follows a similar process, but involves very little calculation. Only a few basic standards need to be kept in mind.

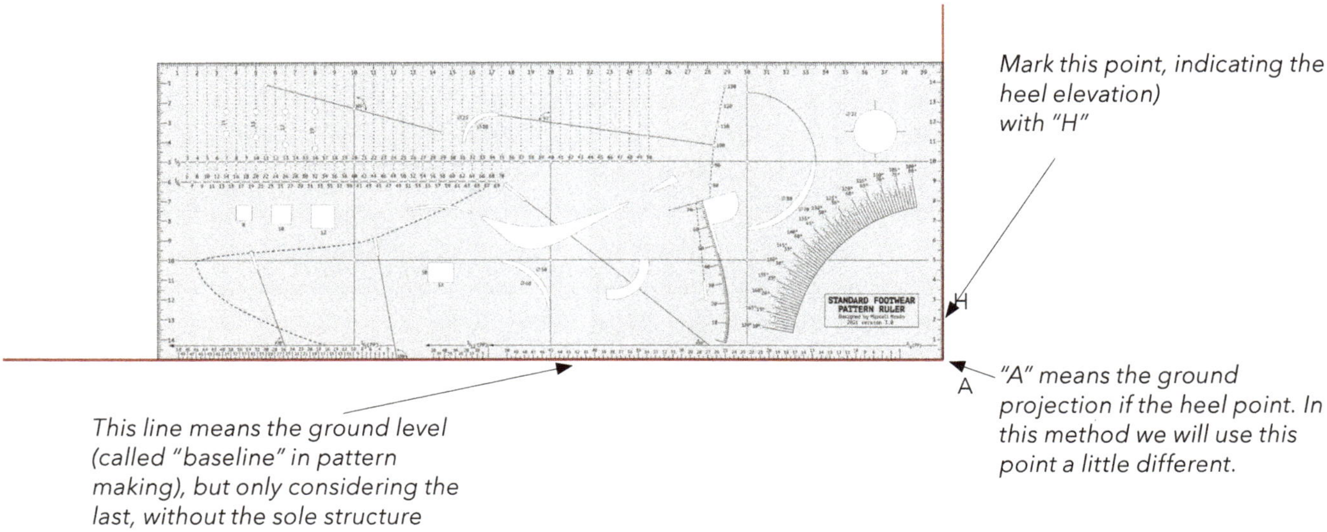

The correct two-thirds proportion of the SLL can be measured directly on the heel elevation line, without calculations or additional tools. Align the bottom-right corner of the ruler with point "H" and rotate the ruler until the desired size meets the ground line.

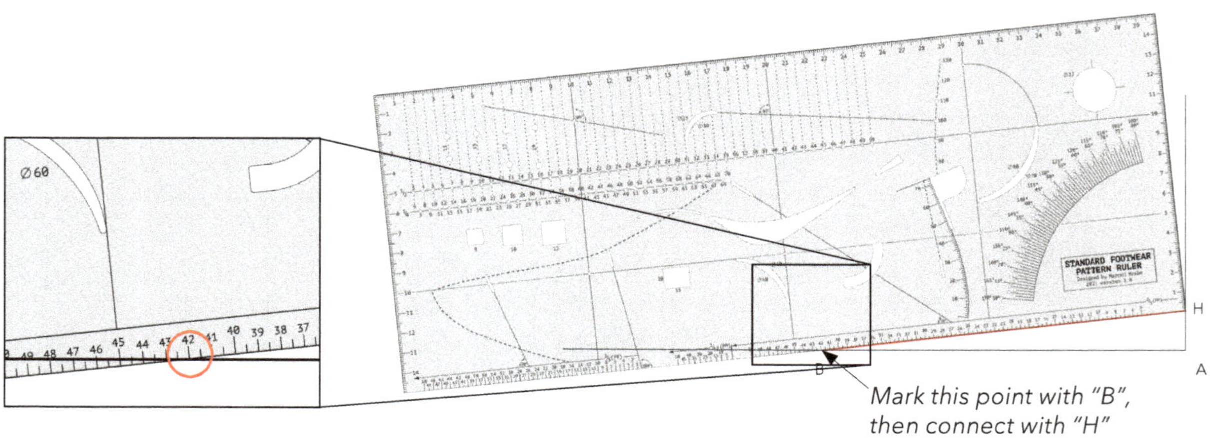

The next step is to mark the one-third distance from point "B" toward the toe. The ruler provides this scale accurately. Mark the appropriate size (in this case, 42) and label the new point as "L."

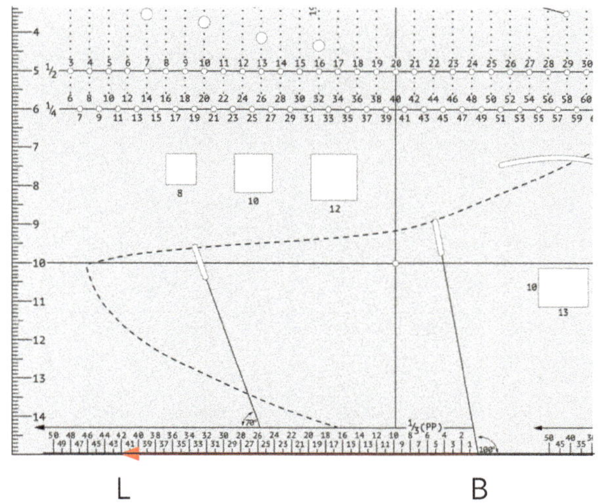

This distance represents the shorter portion of the classic 1/3–2/3 ratio of the last. As discussed in the sizing chapter, UK and US sizing systems are not suitable for proportional measurements, and differences between genders and adult-child sizing make them even less practical. For this reason, this part of the ruler uses EU sizing, which is proportional and has a defined starting point.

The Short Heel Angle

Measure 40° (38° plus heel height in centimeters–in this example, 38 + 2 = 40) from point "H." Some references use 38° for all low-heel models, and in practice this is not incorrect; a 2° difference has little effect on the topline.

If using 38°, the ruler includes a fixed angle for this purpose, indicated from the symbolic last shape.

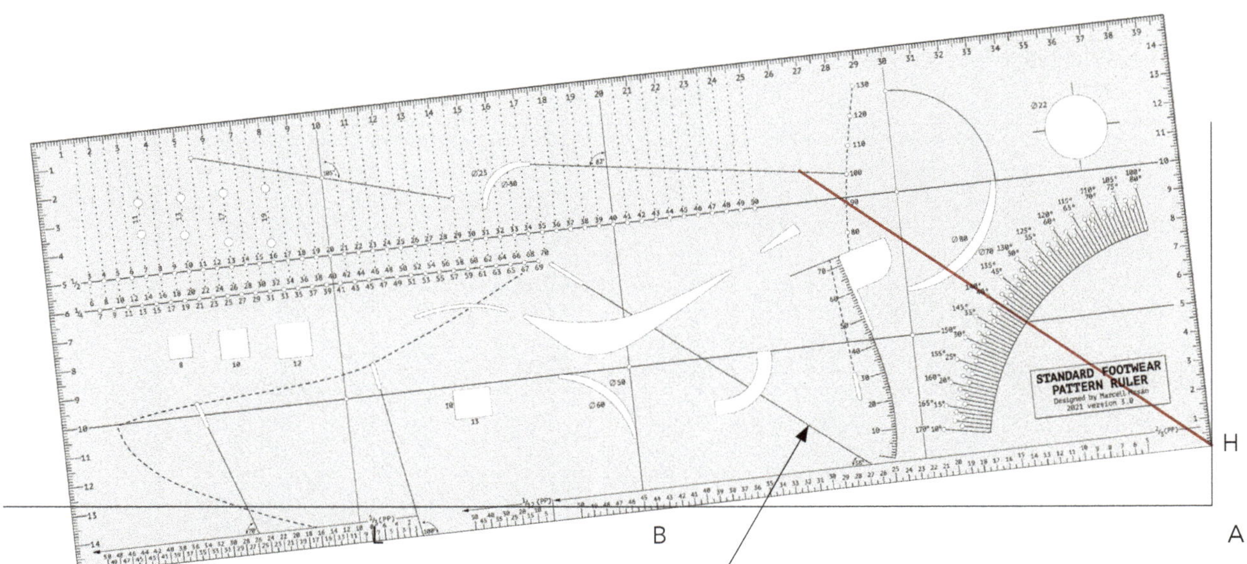

This line represents the short heel of low-heel footwear and can also be used. It will result in a slightly lower lace line.

Draw the ball line. This line is one of the most important elements of the last standard for several reasons. Its angle is not measured from the baseline, but from the "BH" line. As heel height increases, this line rotates progressively forward. Use the preset line on the pattern ruler, which is set at 100° from the "BH" line.

Note that the ruler is not aligned with the ground line, but with the heel elevation. This is critical for achieving the correct angle and is why this angle remains 100° across different heel heights.

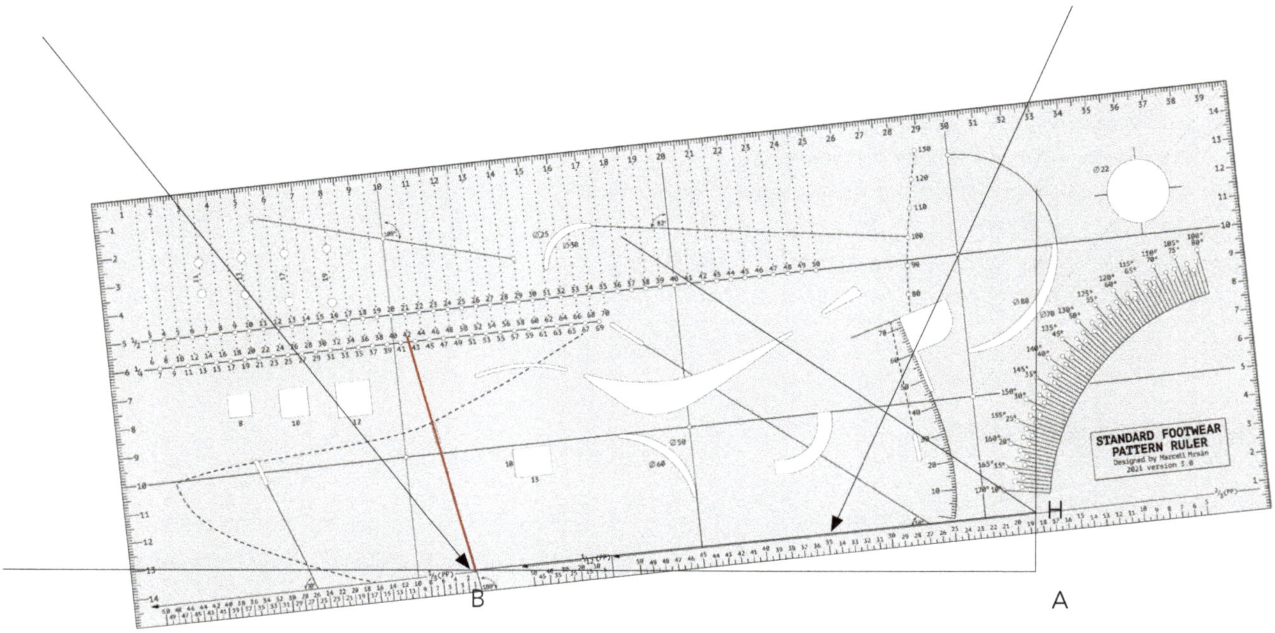

For the toe line, use the 70° angle, as shown in the illustration below. This angle is standard for all models and heel heights.

The function of this line, in addition to defining toe length, is to help the pattern maker determine the necessary pivoting.

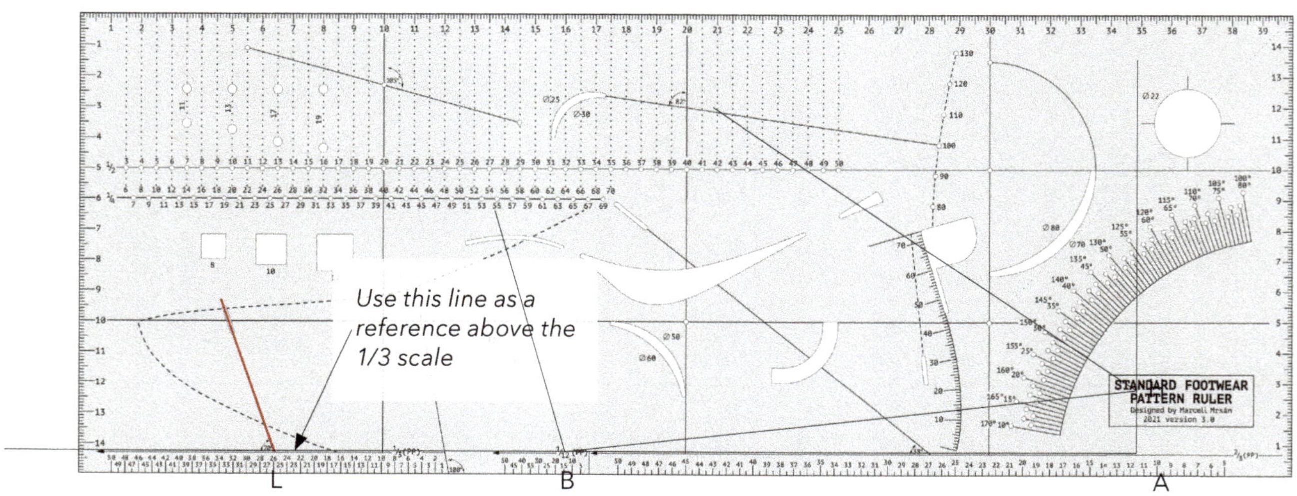

The next step is to measure the correct values along these lines. Use either the actual foot measurements or the data from the "Standard Last Measurements" spreadsheet. The three lines are defined by three simple formulas. As an alternative to formulas, the actual foot measurements of the customer or model can be used. These measurements form the basis for different styles and are essential for proper fit.

The short heel line also serves as a reference for certain boot styles.

The Geometric Standard

H - I

The short heel measurement/2. In our example it is 165 mm

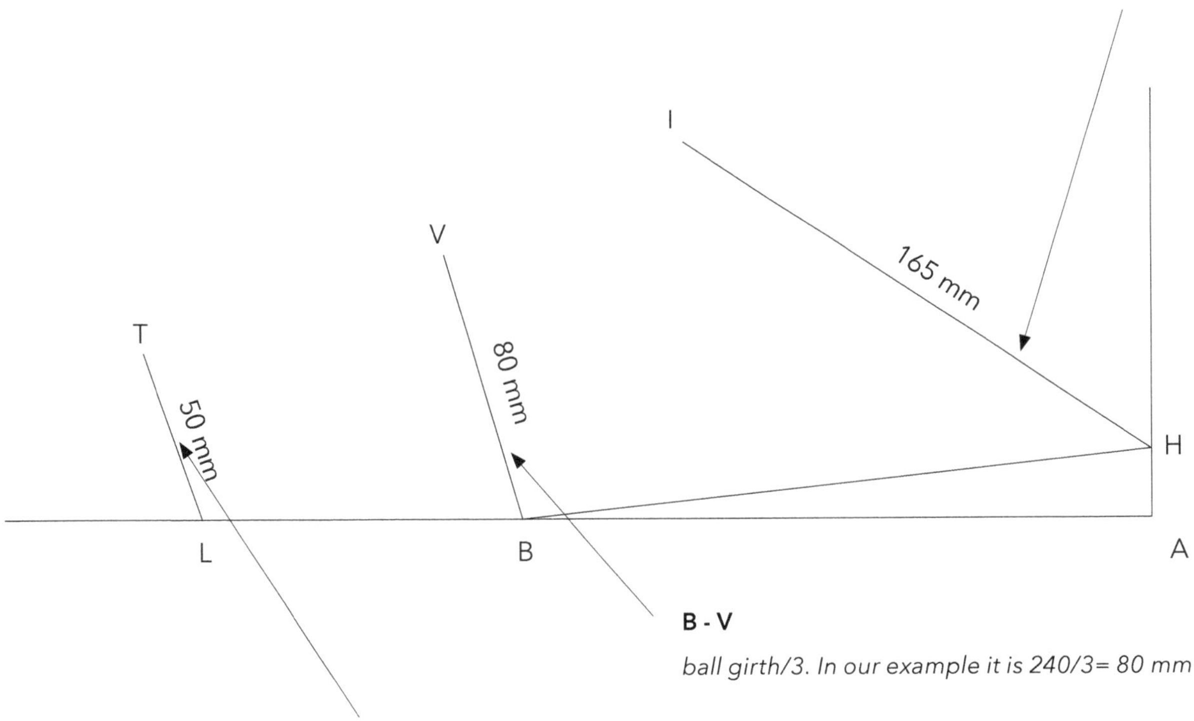

B - V

ball girth/3. In our example it is 240/3= 80 mm

L - T

BV/2 + 10, in this example: 80/2+10 = 50 mm

Some pattern books use a slightly different formula, which is the following: take 10 mm from the girth measurement, considering that the sole portion requires this allowance. Use the remaining value to take the third. Thus, (240 – 10) / 3 = 230 / 3 = 76.67 mm. The difference is not very significant.

Connect these new point and extend the line from "T".

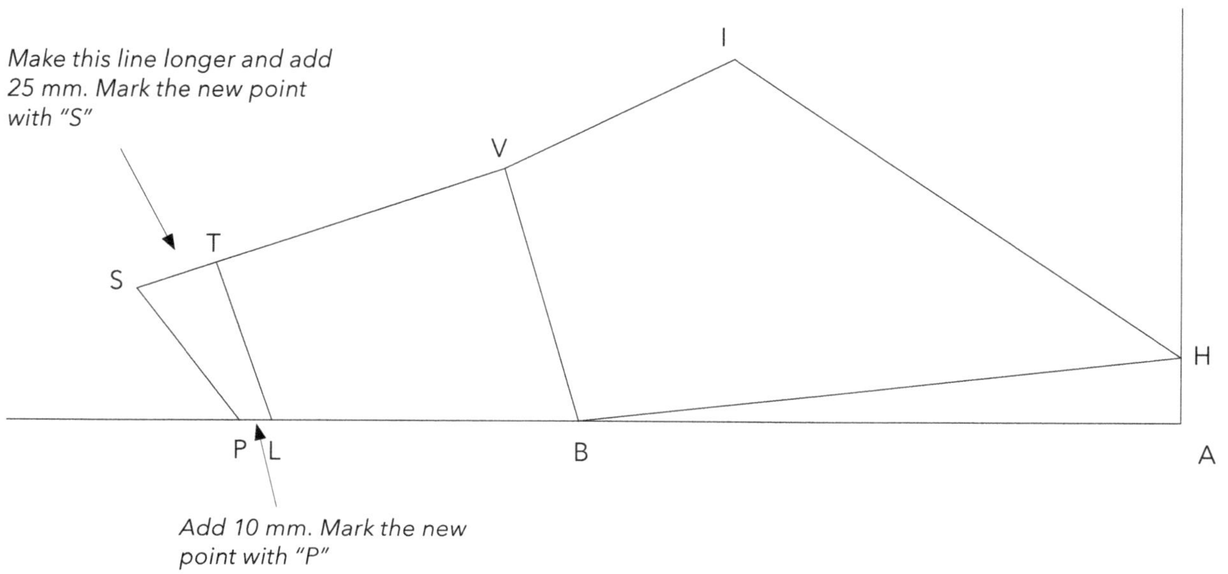

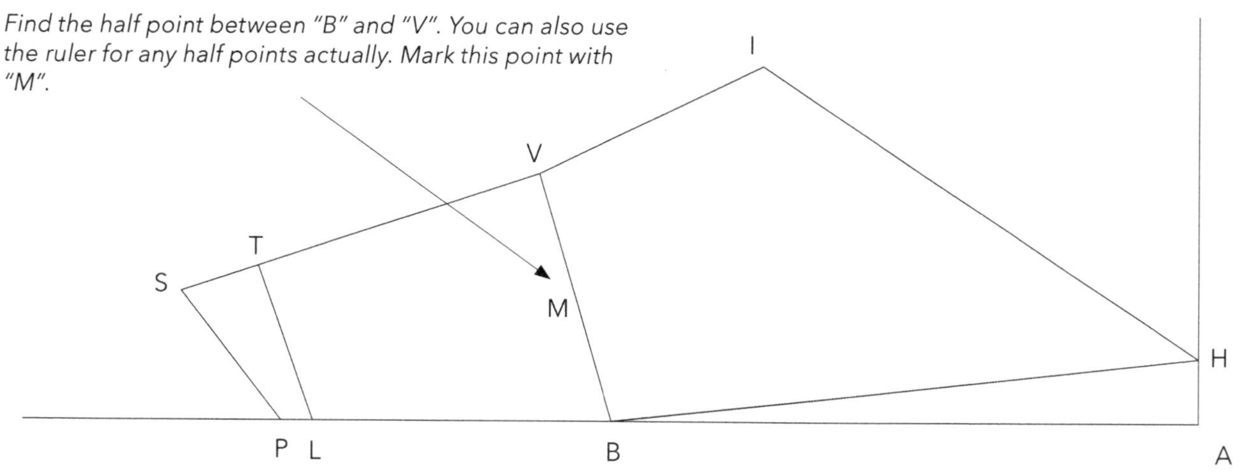

Finding Proportions with the Ruler

The Pattern Ruler includes a function for quickly finding half (or quarter) points. The process is simple:

Measure a distance on the pattern and note the numeric value – for example, 17.

Locate this number on the half-scale.

Marking this point gives the half distance. This can be done on the same line or on a different one. If used on the same line, the dotted guide can be followed up to the top-edge scale.

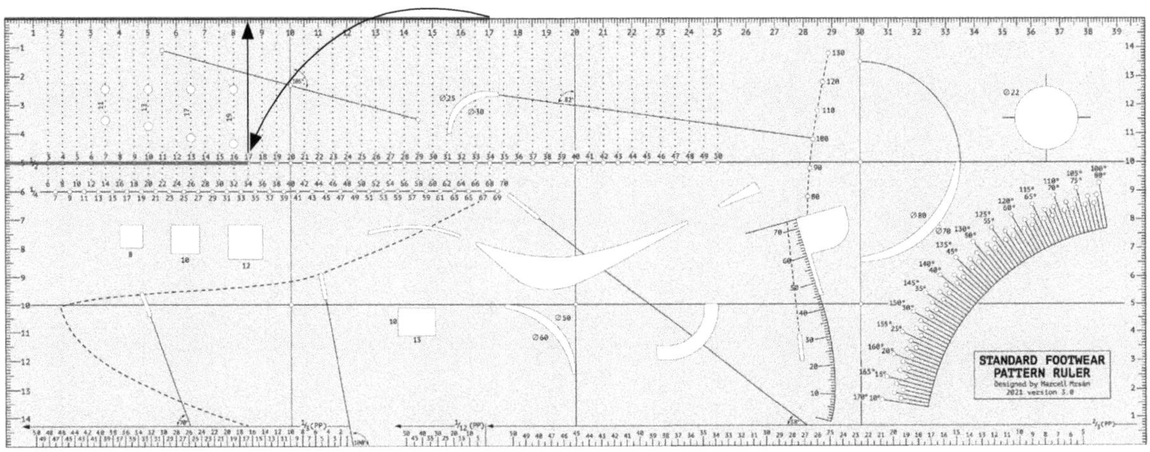

Using the same method, one-quarter of a length can also be found by using the quarter scale, located just below the half scale. This function is particularly important for side-seamed boot patterns.

These scales also work in reverse. For example, to find double a length of 13, locate 13 on the scale, follow the dotted line, and read the corresponding value on the original line. The same principle applies for quadrupling a value using the quarter scale.

It is also possible to measure proportions such as 2/3, 1/6, 1/8, 1/12 (which has a dedicated scale on the bottom edge), as well as 4/9 and 2/9.

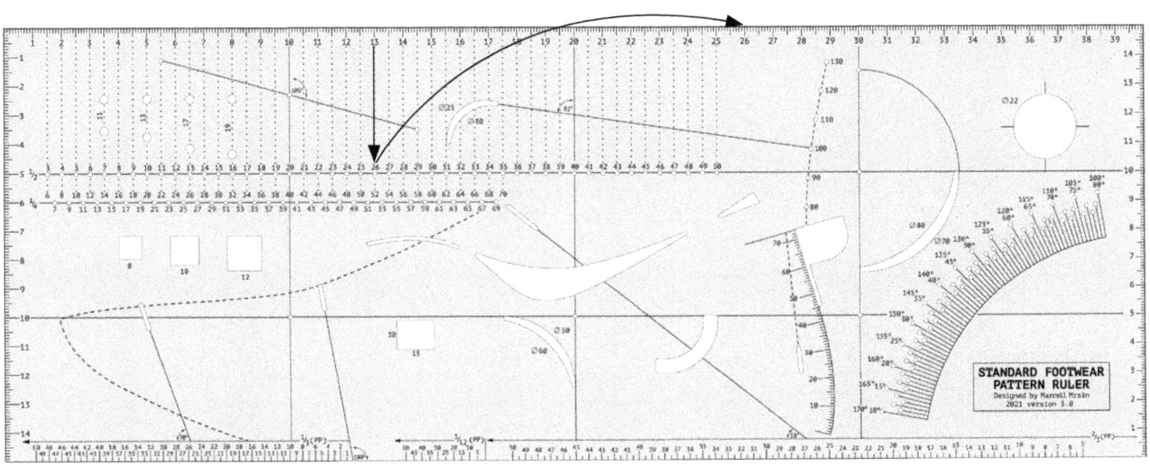

The Heel Line

Draw a line parallel to the vertical one, 20 mm away. Using the parallel guides on the top edge of the ruler makes this easy.

Use the top left corner of the ruler to shift the vertical line with the desired 20 mm.

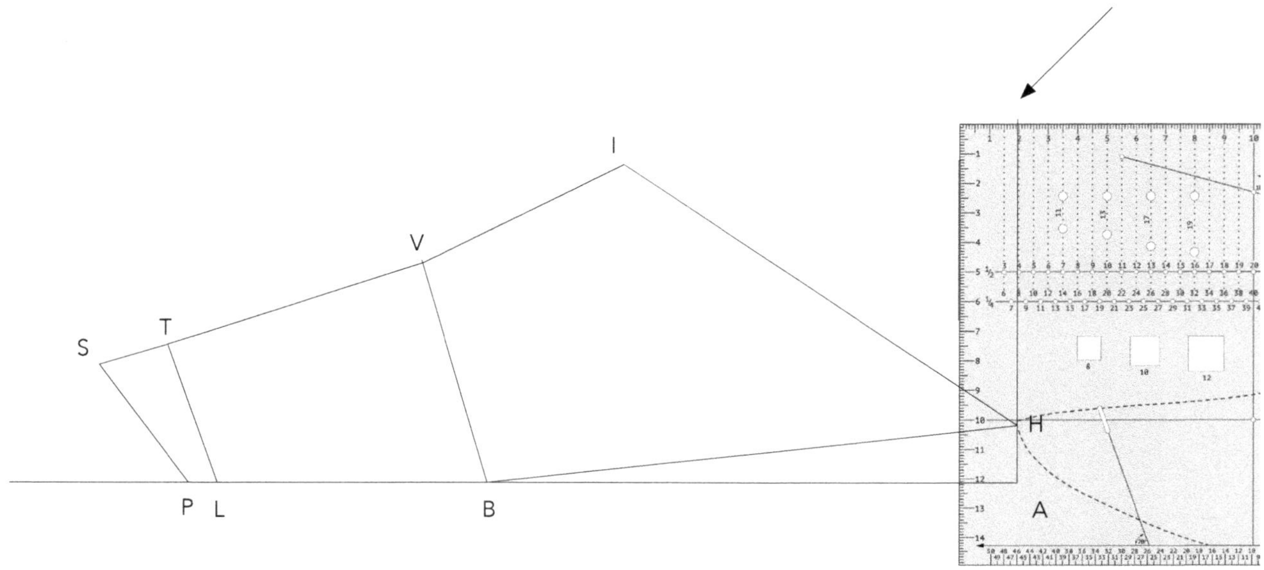

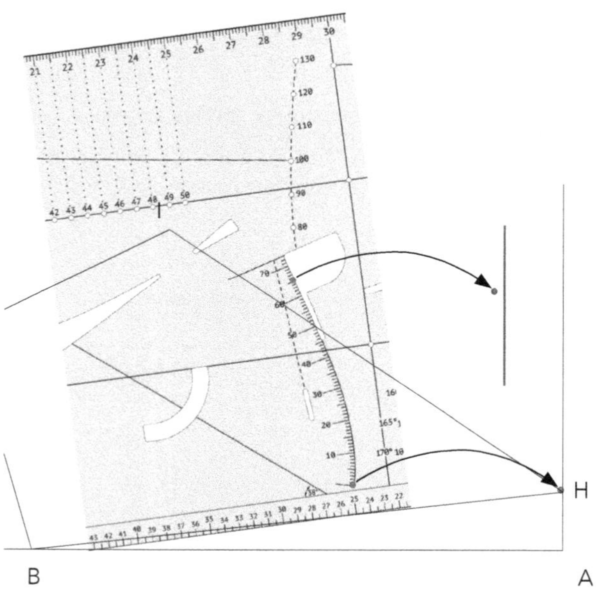

The next step is to create the backline and determine the counter point. This can be done using the geometric method, but there is also a simple and practical approach. Find the counter height in the spreadsheet; for size 42, this value is 65 mm (for certain models, this number may be reduced). The formula is straightforward:

SLL (EU) + 23 = 65.

This is the value to locate on the counter-height scale. Align the bottom-right corner of the ruler and match this value on the scale to the appropriate vertical line on the pattern.

Draw a straight line here, following the template on the ruler.

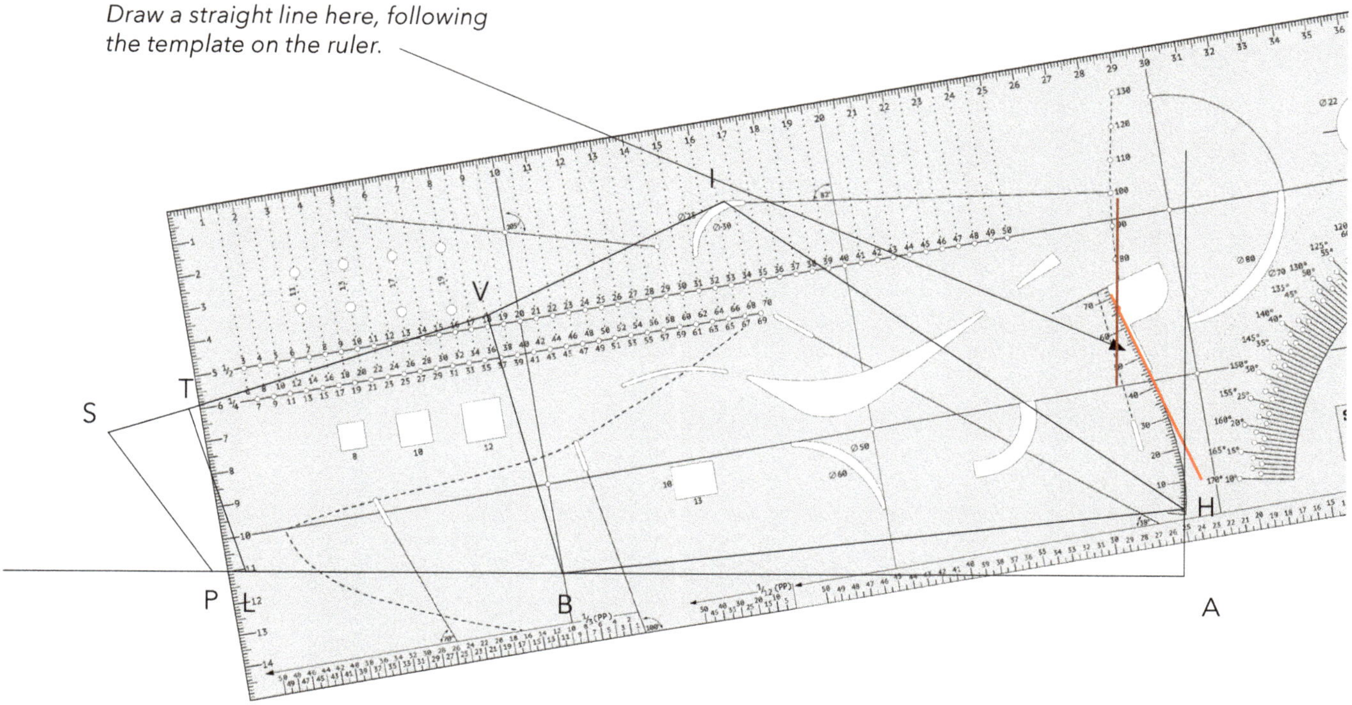

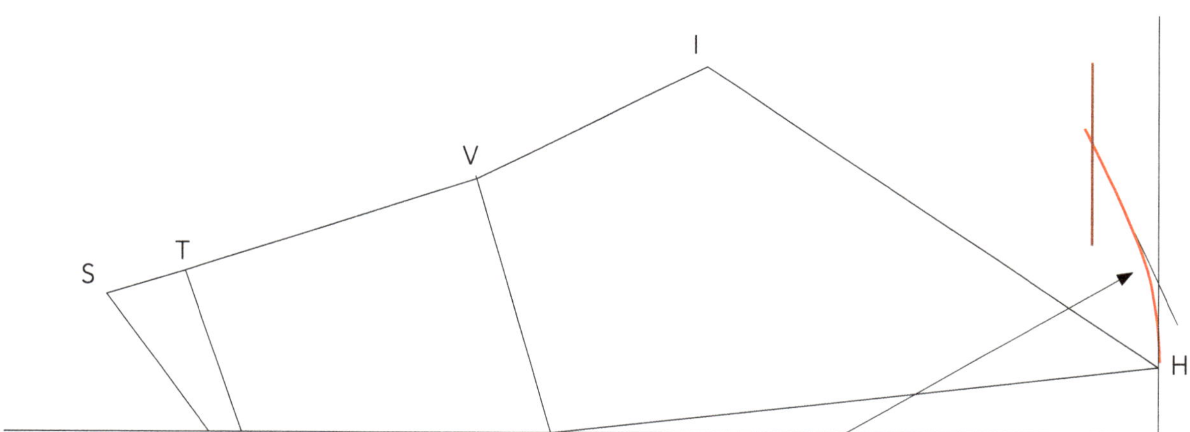

With these points, a smooth backline curve can be created. This step is optional; if the next step is adding a last copy, it can be skipped.

Different leathers, their thickness, stretch, and even different counter closures may require different heel curves. For example, with very stretchy leather and a long topline, it may be necessary to pivot the curve inward later.

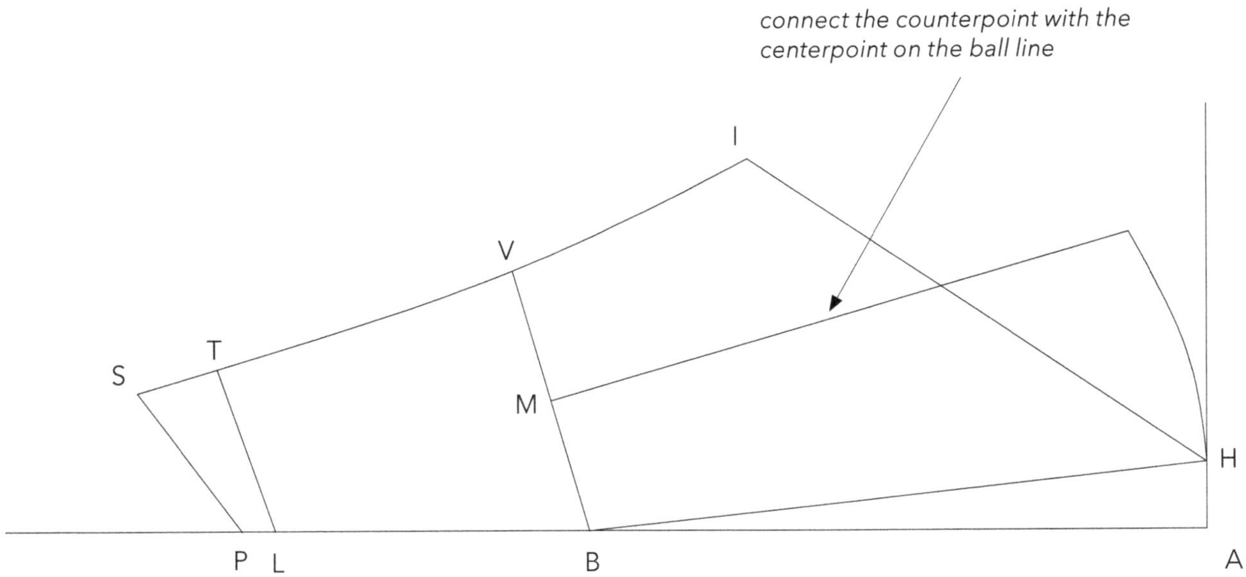

Placing an actual last copy is informative. If the last is close to a standard shape – without an elongated toe or unusual width – the result should look similar to this. One notable difference will appear at the toe. While the last copy shows a more curved form, the geometric method produces a more turned-down shape.

The reason is straightforward: many classic models require a pivoted last shape to establish a proper main folding axis. A wholecut is a good example of this. To determine such shapes, a pivoted last copy is a valuable reference.

The Finished Last Standard

The last standard is now prepared for further work, and most of the patterns discussed in this book are built on these reference lines. It is common for more experienced pattern makers to work without a formal last standard, developing designs directly on the last or on a flat last copy. This does not mean, however, that these reference lines are ignored or lack relevance.

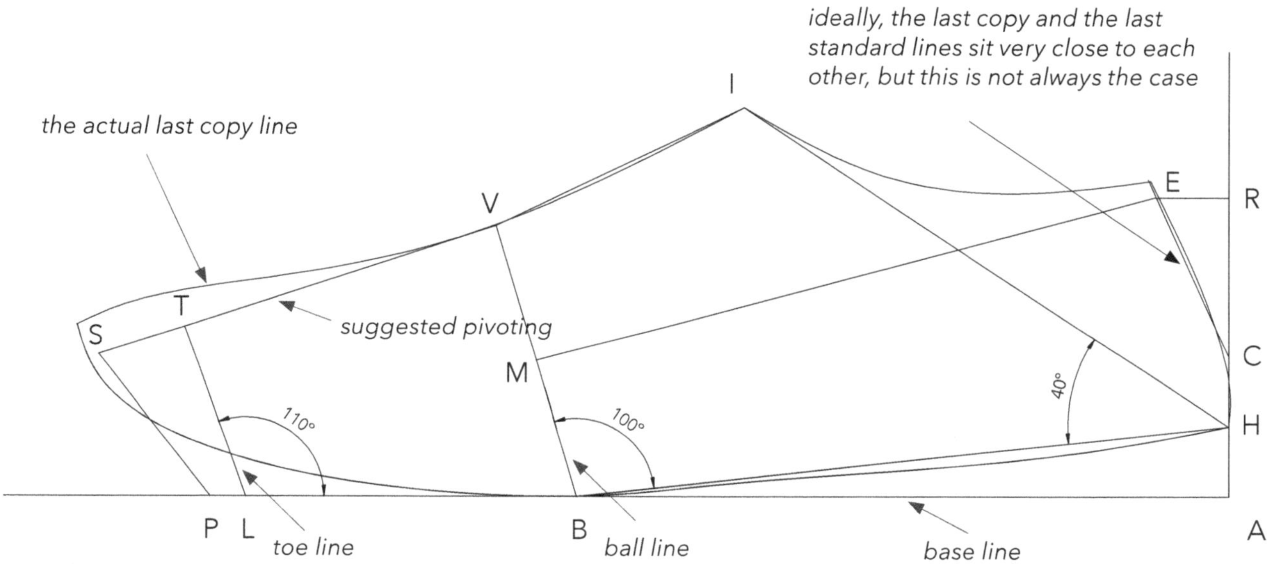

The last standard is now prepared for further work, and most of the patterns discussed in this book are built on these reference lines. It is common for more experienced pattern makers to work without a formal last standard, developing designs directly on the last or on a flat last copy. This does not mean, however, that these reference lines are ignored or lack relevance.

Placing critical lines–such as the topline or the vamp line–in an incorrect position is not only an aesthetic issue; it also affects wearability and durability.

Using a last copy together with the last standard may appear excessive, but in practice it is an effective way to reduce preparation time, as all measurements are already accurate. In this case, the geometric standard alone is sufficient.

If the last copy and the last standard lines differ significantly, the most likely reason is highly non-standard foot measurements. In such cases, the last copy should take precedence, and the geometric standard serves only as a general guideline for proportions rather than a strict reference.

These lines and proportions also can be used in CAD software. The idea is to create a "parametric" design, where a few lines guides the whole structure and a designer can create a last standard in second, just adjusting the heel height and the size. This can sped up the work and help creating patterns very effectively, even for orthopedic purposes.

Sockliner Patterns

No pattern book would be complete without addressing sockliner design, even if this topic is not, strictly speaking, part of upper design – it still falls within the pattern maker's responsibilities and is therefore worth discussing at least briefly.

In most cases, sockliners are attached to the insole after the footwear is completed and the last is removed. There are shoes in which the sockliner is part of the insole structure and the footwear is lasted with the sockliner in place. As these cases are more closely related to product development, the focus here is on the first case.

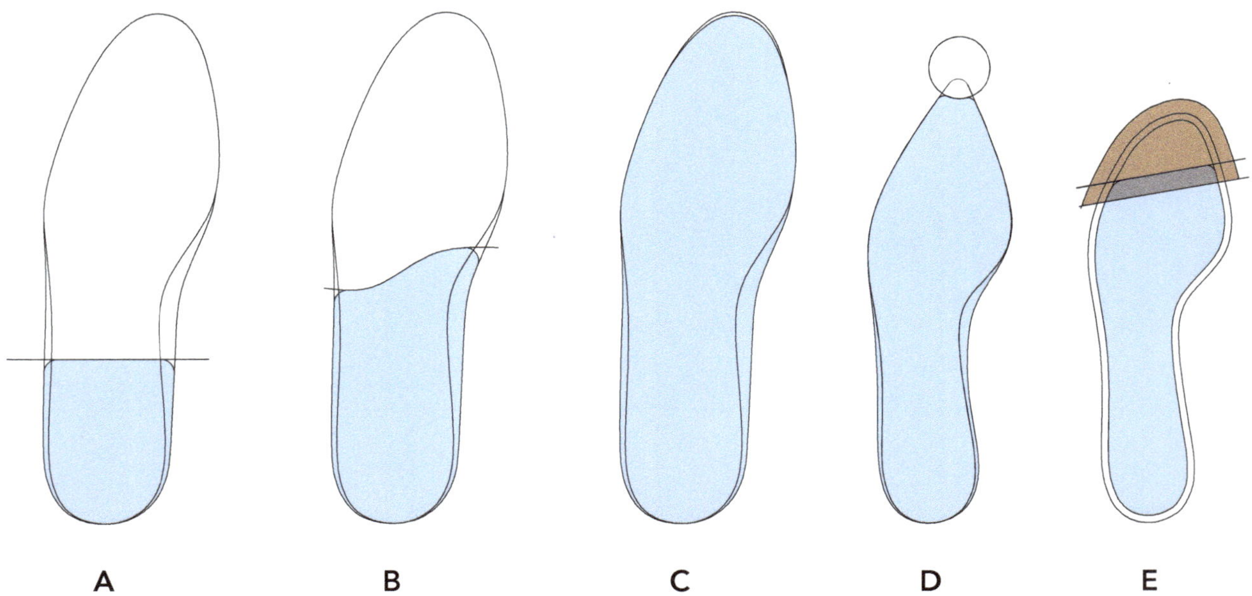

A. The short sockliner is typically used for higher boots in which the insole is made from vegetable-tanned leather, so its function is only to cover the nails and pegs. Since it ends at the symmetrical heel length, it is symmetrical, and one size is used for several boot sizes.

B. A slightly longer version can also be used for more than one size. The front is slightly asymmetrical. It can be used for dress shoes as well.

C. Full sockliner. The shape is widened around the midsection so it covers the insole perfectly. If the sole shape calls for it, the length is slightly shortened because of the orange peel effect, avoiding the leather bunching up in this area.

D. While high heels might follow the same structure shown in "C", going all the way to the toe, taller, pointed-heel boot sockliners do not need the pointed end, as the toes do not extend into this area and this point would only cause technical difficulties.

E. This is a sandal sockliner. It is slightly narrower than the insole shape, as the edge will be covered with a wide strip of leather and should remain visible. The front has a larger, heavily thinned upper leather extension so it can cover the insole and be folded under.

Pivoting

Pivoting is a pattern-making process in which the pattern maker rotates a section of a piece around one or more points to change its shape. It is commonly used to straighten elements such as a last copy, a tongue, or a backline. The purpose of this method is usually to create a folding axis from a curved line – such as the front profile or the backline in previous examples – but it can also be used to translate a curved line into a straight one, for example when matching a binding piece. When pivoting, the pattern maker needs to consider distortion and how it can be balanced later in the process to counter the loss of information, should that be the case.

Example 1: Pivoting the toe

The most common practice is putting the pivot point at the top of the shape; therefore, the length of the front line remains unchanged, and only the bottom line is shortened. In this case, the last copy is straightened.

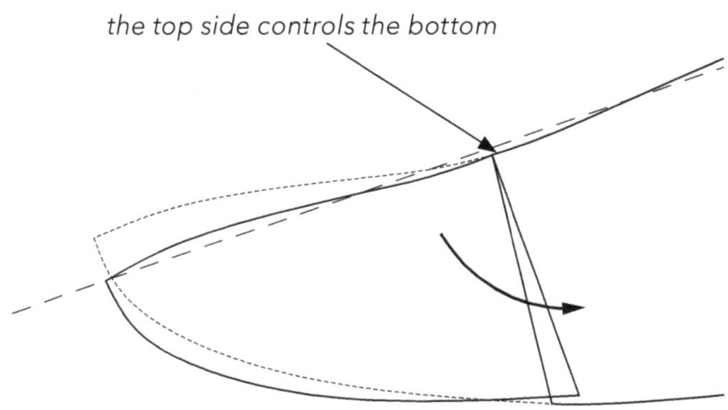

Choosing the rotation center closer to the middle of the line produces a very different result. The bottom still becomes shorter, but not as much as in the previous example; the front line, however, changes. Even though this may appear minor, it can cause serious fit issues. Lengthening any front decoration or pattern element – such as a medallion, apron front, or wingtip – by this amount would make it appear disproportionate and far longer than intended for its position.

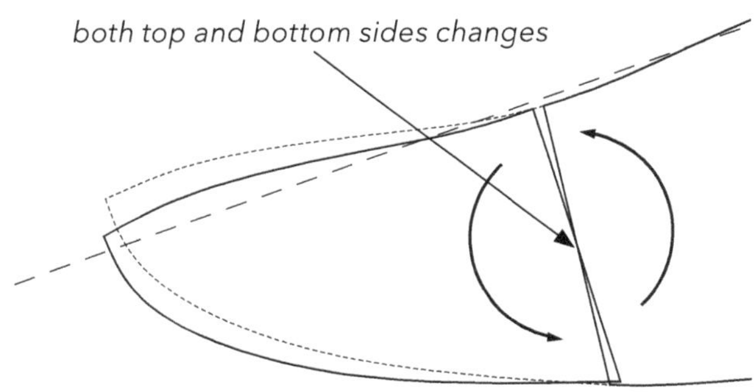

Choosing the pivot point on the bottom line is rarely used, as it would distort the top edge and elongate the entire pattern. While uncommon, this approach can occasionally be useful.

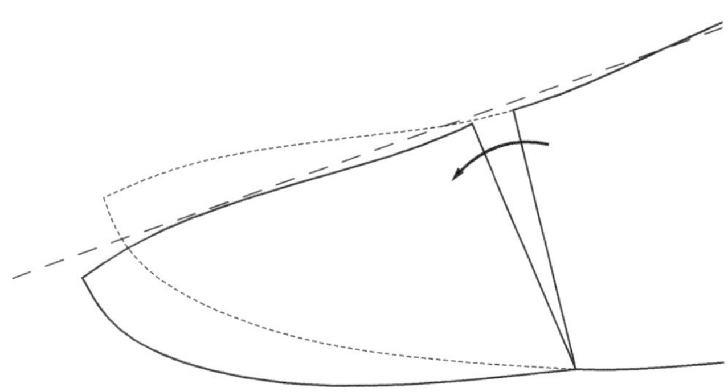

Example 2: Pivoting the tongue

The tongue is a good example where a centerline is the correct choice for pivoting. Instead of using a single pivot point, a series of small pivots is applied until the curved line aligns with the straight axis. This will be discussed in more detail in the Balmoral chapter.

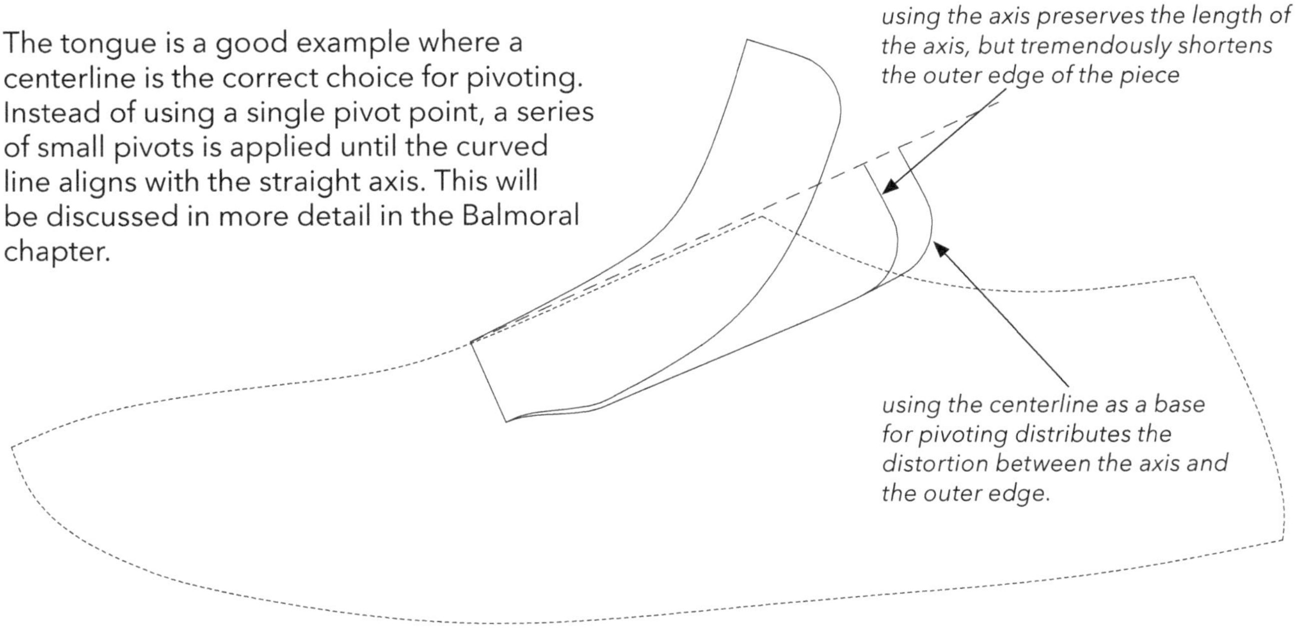

Example 3: Pivoting the backstrap

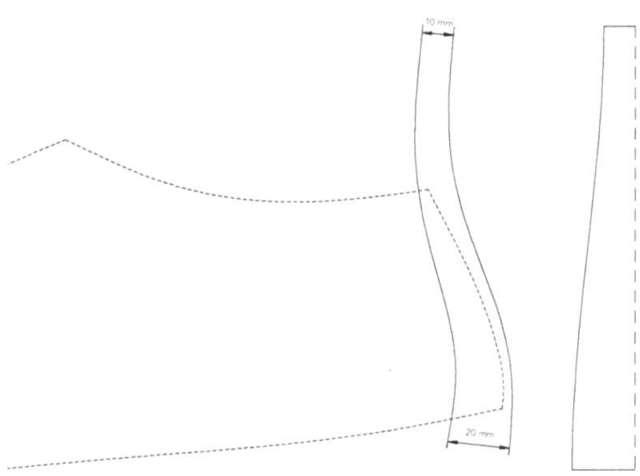

In some cases, very little information is sufficient to straighten a piece, and the narrow backstrap is a good example. Only the axis length and the offset at both ends are required (or additional points if the backstrap has a special shape). With these three pieces of information, the curved backstrap from the master pattern can be converted into a straight piece.

Example 4: Pivoting the boot vamp

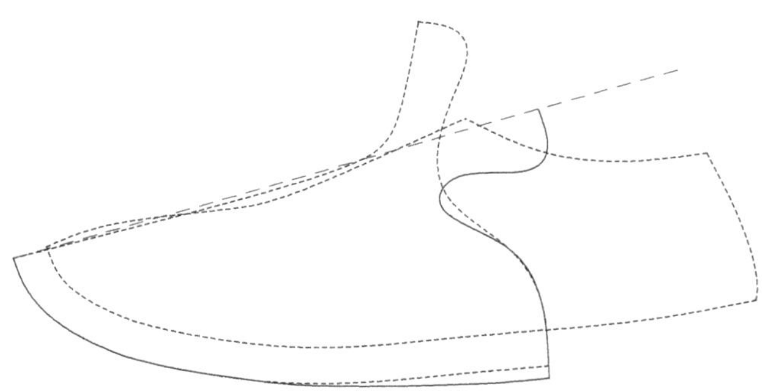

One of the most common cases for pivoting a pattern is a boot vamp. This is a special situation where one round of pivoting may not produce the correct result, and more than one prototype may be required. As an alternative, the pattern can be roughly approximated then traced accurately once the front profile has been established. This process is known as "re-cut" and is used in many complex pattern constructions.

The Last Copy

Over the past century, last design has progressed far beyond simple changes in toe shape and heel height. Modern footwear fashion and industrial production demand millimeter-level accuracy in patterns and grading, which means that an average last standard – such as the one produced by a purely geometric method – is often insufficient. For this reason, pattern makers have long used various techniques to capture the actual two-dimensional surface of a last in order to produce accurate patterns.

Different types of footwear require different last-copy methods, and no matter how precise the process, some information is inevitably lost during "flattening," or, in more technical terms, dimensional reduction. In the case of shoe uppers, a certain level of tolerance exists because leather (or its alternatives) is a forgiving material. The challenge for the pattern maker is to create a surface projection in which critical data – such as girths, heel height, and heel curvature – are preserved, while information that can be recreated later (for example, straight lines) or is nonessential (such as curves that will be transformed into axes) is allowed to be lost.

Designing on the Last

Certain lasts and footwear models require this type of design to function correctly. High sidewalls with sharp edges, apron-front constructions, anchor toes, wingtips, and most athletic shoes.

While geometric pattern making remains important, designing directly on the last is also fully justified. One key reason is that modern lasts are no longer as standardized as they once were. Pattern making today requires greater precision, with specific elements expected to sit in exact positions on the last – for example, a formal sole line that must continue accurately into the upper. Contemporary manufacturing technologies demand this level of accuracy.

Geometric principles applied to a three-dimensional surface function in the same way as on a flat sheet of paper. In addition, many modern lasts clearly indicate key reference points such as ball positions, counter heights, and other construction markings.

Some pattern features are best developed directly on the last. Typical examples include wingtips, apron fronts, moccasin constructions, anchor toes, mudguards, and medallion placement (the overall shape, not the perforation design). The last copy is then flattened and converted into a master pattern and, later, into final pattern pieces.

A single source copy can support multiple designs, resulting in time savings for the maker.

Alternatives

Masking tapes become widely available in the 1930s, but the concept of last copy is older than that. While the self adhesive tapes make the process very convenient and accurate - also provide a three dimensional sketching option, there are other, also effective techniques. First of all, it is totally possible to last a generous piece of leather on the last then remove it and use the information, it is possible with some textile as well.

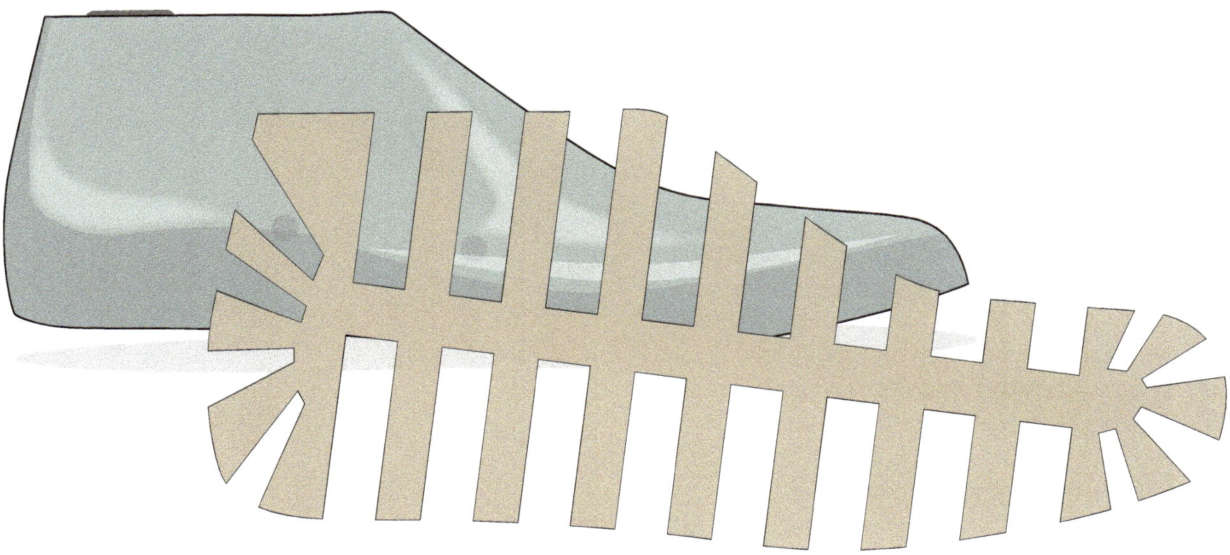

The cheapest and fastest option, however, is paper itself. A sheet of paper is cut generously larger than the last, and a rib-like structure is then cut into it. The pattern maker fixes this structure to the last with two tacks and can easily transfer the feather line as well as the centerlines at the front and back.

Admittedly, cutting this shape takes some time and it is not the most accurate method, as sections will be missing–ideally not in critical areas such as the heel point or the toe. Its major advantage is that this structure does not require any pivoting, as it remains a flat piece of paper. When used together with the geometric standard, this method can be sufficiently accurate for practical purposes.

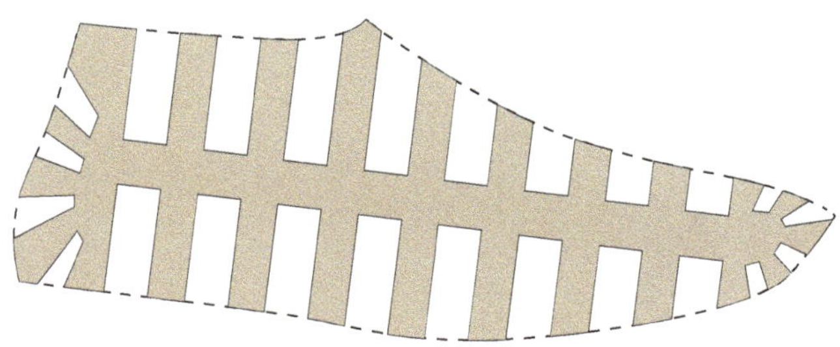

Reconstructing the missing sections is straightforward and requires minimal effort.

As the ball point is reached, the curvature changes, and it may be necessary to adjust the tape angle and/or the amount of overlap. The higher the heel, the more pronounced this change becomes. Flattening the tape can therefore be more challenging in this area. Make sure the tape overlaps the centerline.

One tape is placed along the heel curve, with a few additional strips applied across the last surface. There is no need to cover the entire surface; these strips serve only to allow the tape to be removed in one piece without accidental tearing. A 25 mm wide, non-stretch tape is preferred.

Once this is done, tapes can be placed from the top down to the feather edge. Make sure there is sufficient overlap between the strips – 10 mm is enough. Trim the ends cleanly at the edge. Doing this at this stage is straightforward and saves both tape and time.

After taping the lateral side (it may not be necessary to take last copies from both sides), use a bone folder to smooth the tape. This step is not critical, as a last copy with a less smooth surface can still be used, but drawing design lines is easier when the surface is reasonably smooth.

Drawing the centerline directly by hand is perfectly valid and works well. A cleaner and more elegant solution is putting down a straight line on a narrow strip of tape, then adhering the tape on the last. If the position is not correct, peel it off and adjust it until the result is satisfactory.

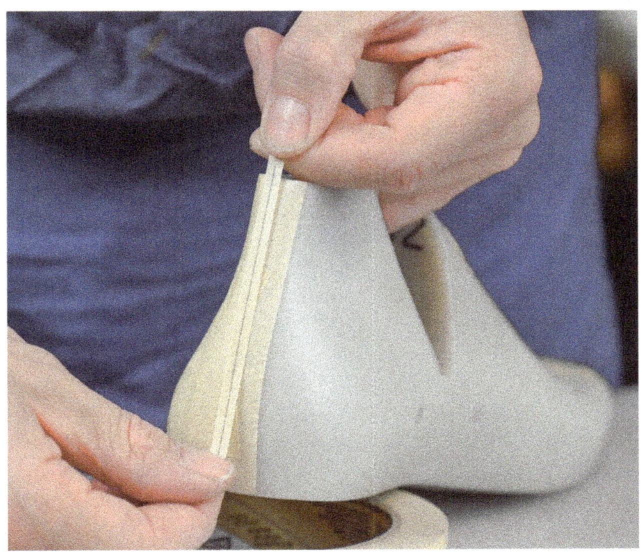

The same method works well on the backline. The symmetry of the tape can be checked visually and by observing where it meets the feather edge. Since the heel portion of the sole is symmetrical in this case, any misalignment is easy to detect. Trim the excess accurately at the bottom.

Peeling off the Last Copy

Removing the tape from the last is relatively easy, but a few points are worth keeping in mind. First, try to keep the tape structure intact. Minor tears will not significantly affect the pattern, but excessive damage should be avoided. Second, remember that the last copy is a three-dimensional shape. The removal process is not the time to flatten it, so preserve its 3D form as much as possible.

Finally, do not wait too long between taping the last and removing the tape. If the tape is left on for too long, the adhesive can begin to react with the surface of the last, making it difficult to remove in one piece. A useful trick is to lightly dust the tape before applying it. A small amount of dust goes a long way and makes separation faster and easier.

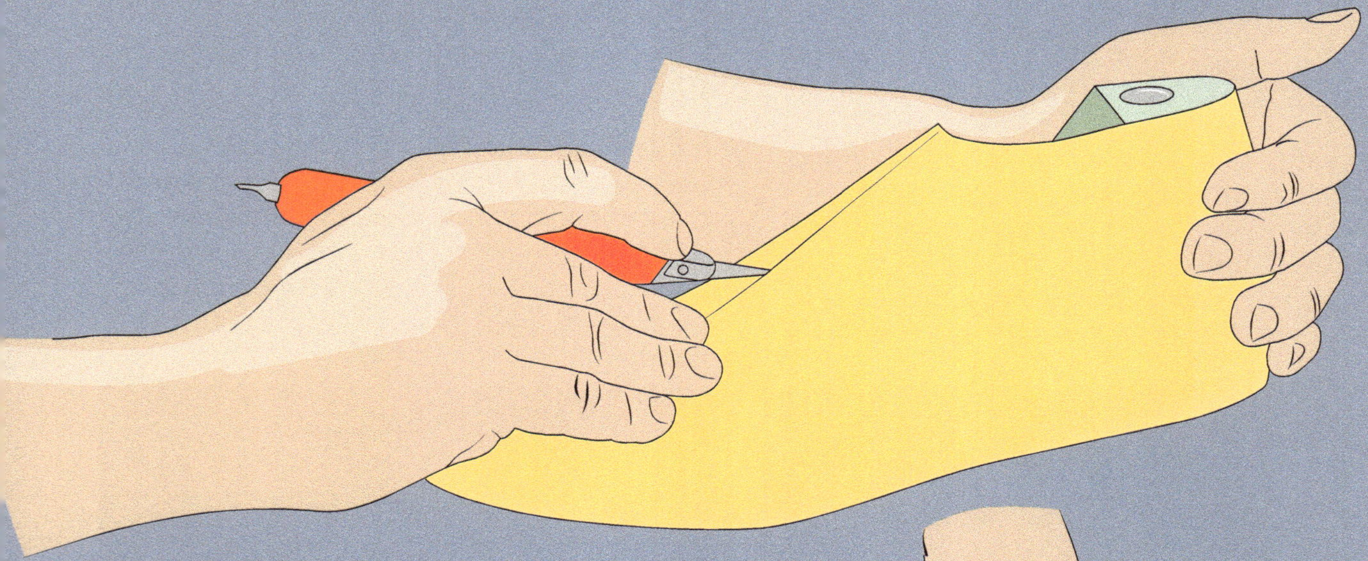

Make an accurate cut at the front and back of the last along the previously drawn or taped centerline. Precision here is essential. Even a slight deviation can cause problems later in the pattern. A small asymmetrical cut at the back will be obvious when comparing the two sides and is therefore easy to correct, but an error at the front may remain unnoticed until it is too late.

Safety first. Make short, controlled cuts and support the work with your fingers so the knife does not slip and cause injury. Do not cut with your hand unsupported in the air.

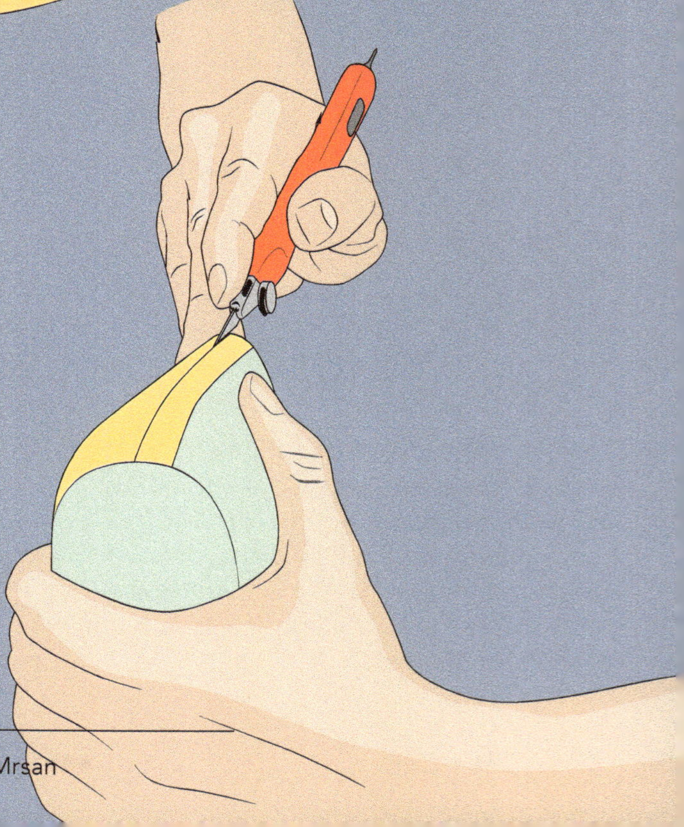

Remove the "half" (one side) of the last copy. Start from the back, from the direction in which the tapes were applied. Begin at the corners: the top and bottom of the heel curve. Work carefully and slowly.

When the instep is reached, peel off the corner at the very top of the front centerline. Continue little by little, preferably using both hands at the same time. If the tape tears, try to repair it and keep it together.

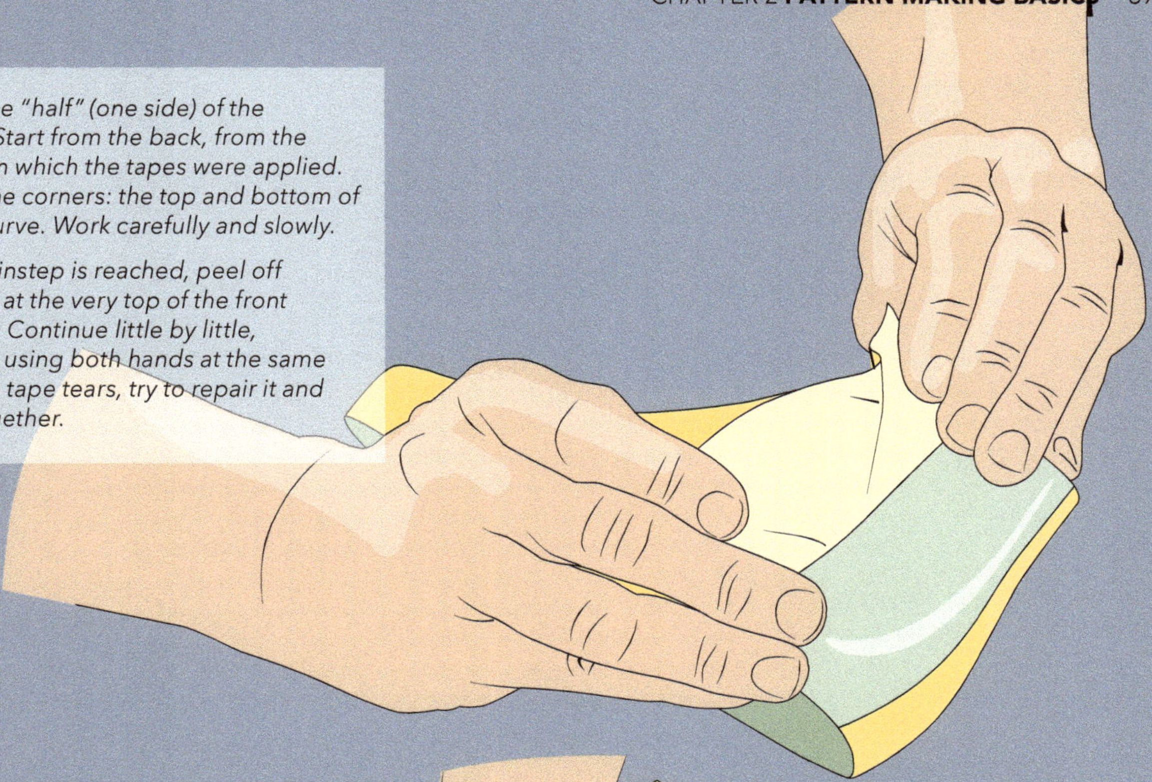

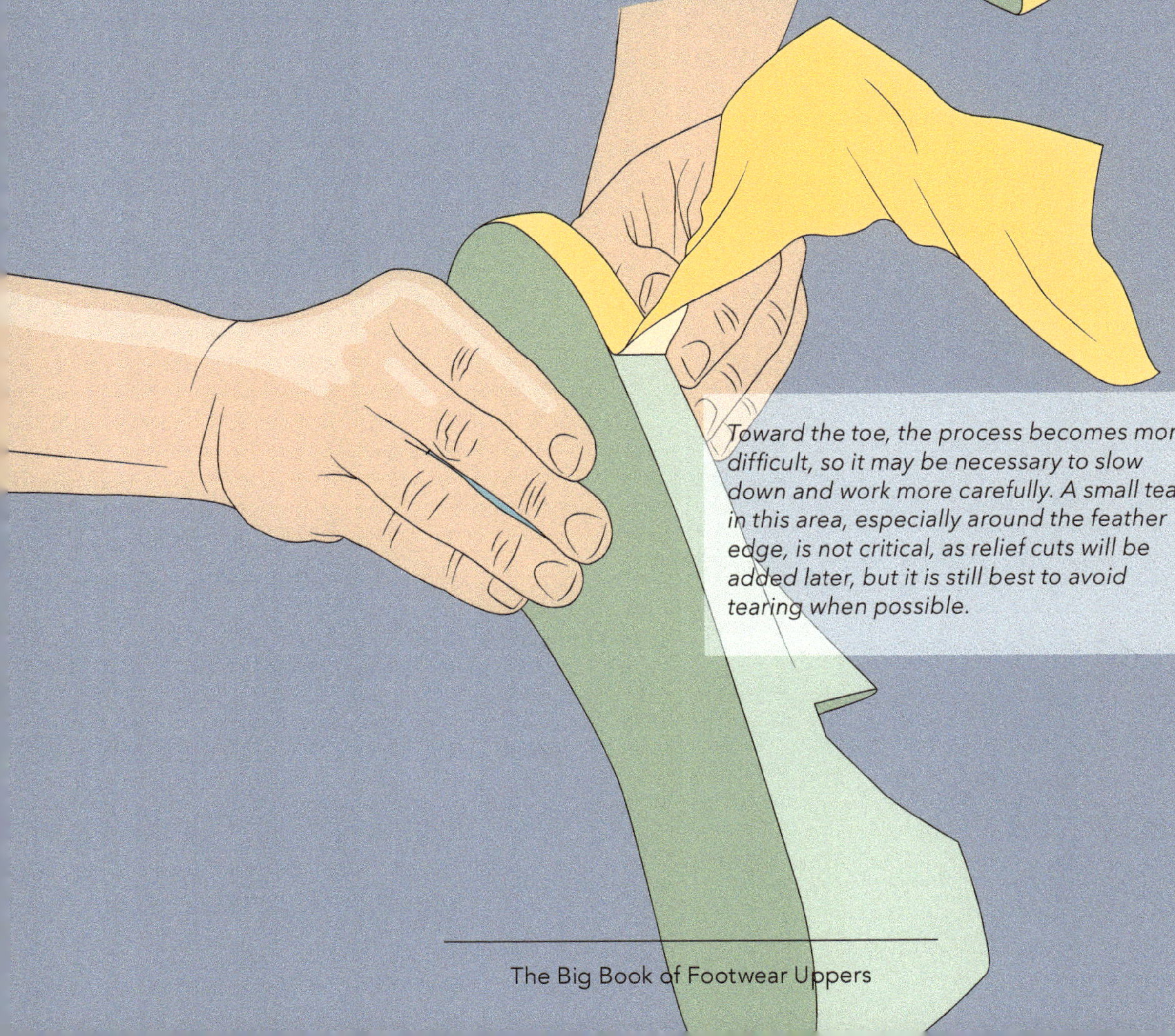

Toward the toe, the process becomes more difficult, so it may be necessary to slow down and work more carefully. A small tear in this area, especially around the feather edge, is not critical, as relief cuts will be added later, but it is still best to avoid tearing when possible.

Flattening the Last Copy

The last copy on the last is a three-dimensional object, just like the last itself. To transfer the necessary information for pattern making, this 3D shape must be "flattened" so a master pattern and individual pattern pieces can be created on paper. Flattening is the process of translating spatial information into a two-dimensional form.

As discussed earlier, the main goal is to preserve as much information as possible during flattening, or at least the information that is relevant to pattern construction. Some loss is inevitable, so the focus should be on retaining the most critical parameters.

The last copy is still sticky, so it should be pressed carefully onto the paper. Focus on the "lines of force," which are the main stretch lines that define the length of the upper and how accurately it matches the last surface. Follow this sequence to ensure proper flattening.

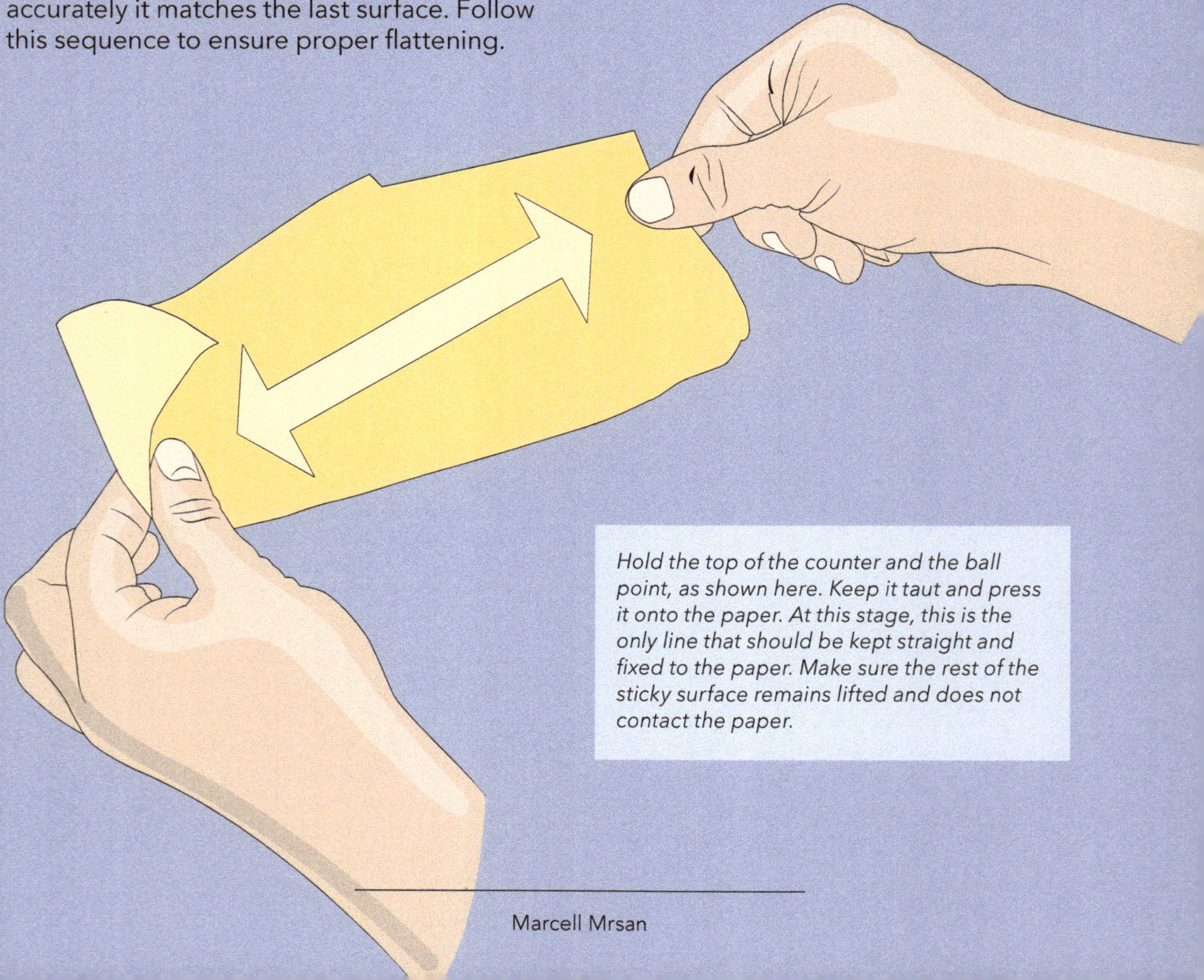

Hold the top of the counter and the ball point, as shown here. Keep it taut and press it onto the paper. At this stage, this is the only line that should be kept straight and fixed to the paper. Make sure the rest of the sticky surface remains lifted and does not contact the paper.

At this point, some parts of the last copy remain in the air, which is perfectly acceptable. The focus must remain on the "lines of force," indicated by the thick arrows, in order to preserve the last shape. The next line runs from the ball point toward the toe.

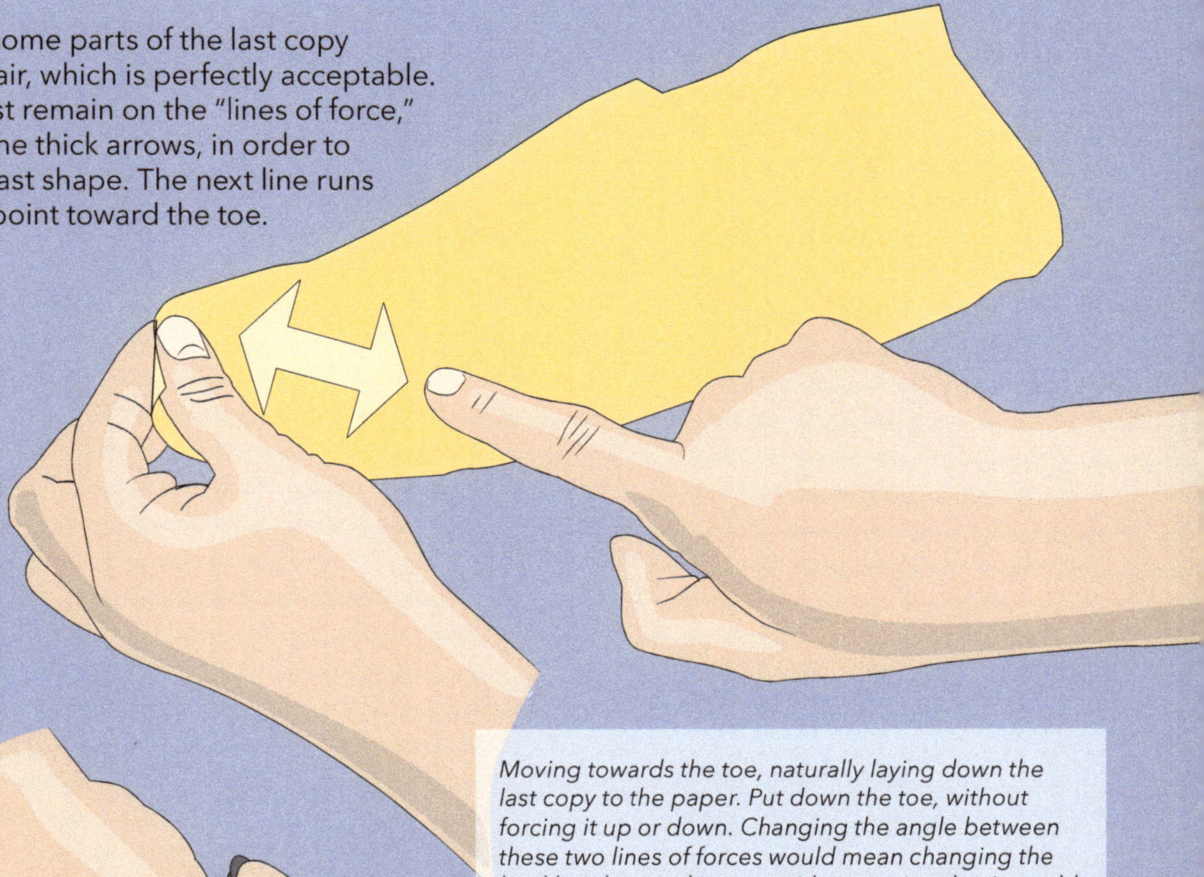

Moving towards the toe, naturally laying down the last copy to the paper. Put down the toe, without forcing it up or down. Changing the angle between these two lines of forces would mean changing the heel height – and it goes without saying that it would result in an inaccurate pattern. When this line is down, you are ready to make some relief cuts.

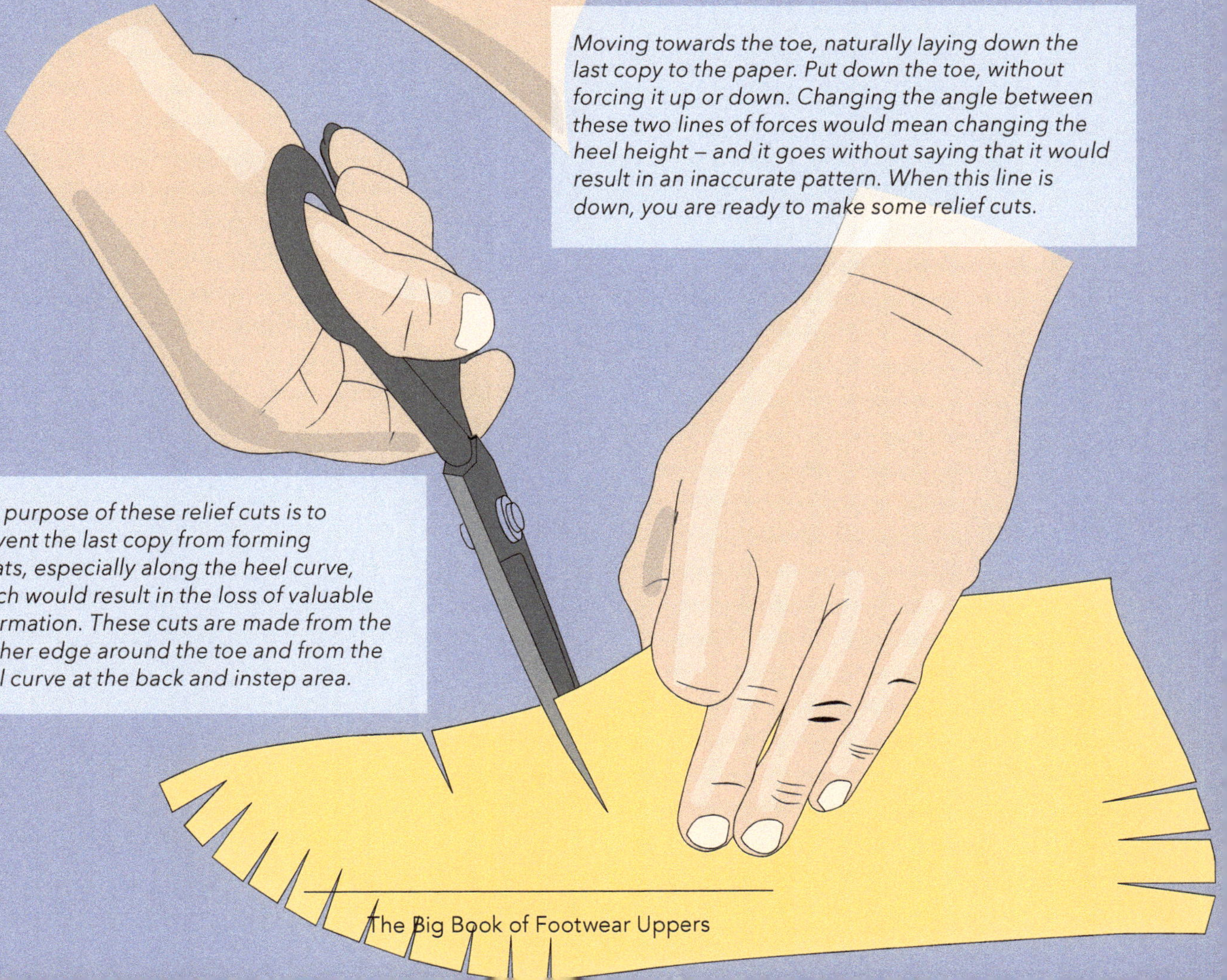

The purpose of these relief cuts is to prevent the last copy from forming pleats, especially along the heel curve, which would result in the loss of valuable information. These cuts are made from the feather edge around the toe and from the heel curve at the back and instep area.

Lines of Forces

The wide white arrows represent the two main lines of force. These lines, along with others, will be stretched during lasting. Flattening is always done away from these lines in order to preserve girth measurements and critical angles.

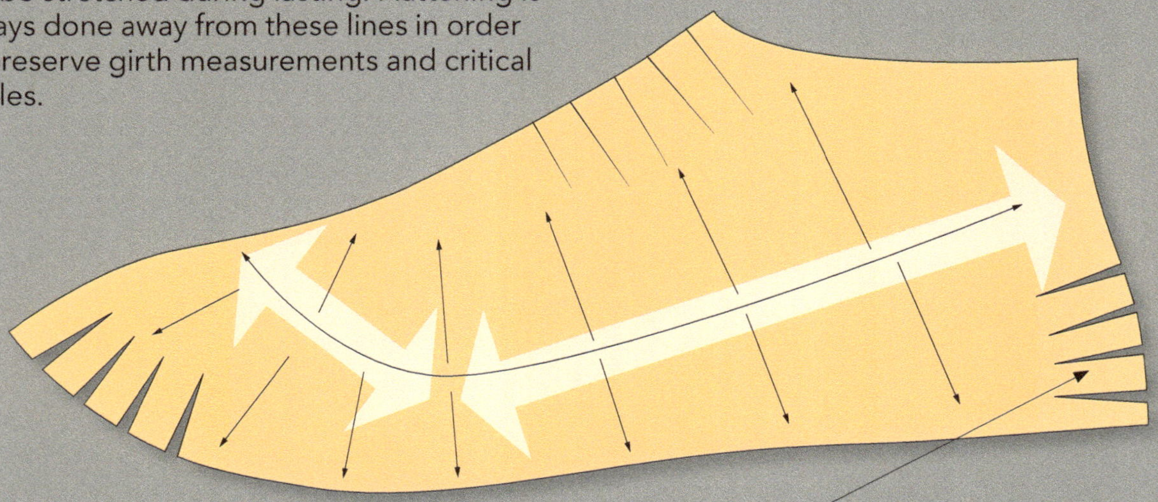

Cutting these relief cuts from the back curve introduces minor distortion at the bottom, which is largely irrelevant for the pattern. More importantly, the backline curve remains accurate.

Alternative Flattening

Relief cuts can be adapted as needed. For a mudguard model, it is more important to keep the profile edge close to straight and to preserve a bottom strip whose placement resembles the final shape, as this will be a double-axis model. Instead of making small cuts from the feather edge, a longer cut is made along the top edge of the mudguard. This preserves the length of the bottom and maintains the correct direction of the mudguard axes.

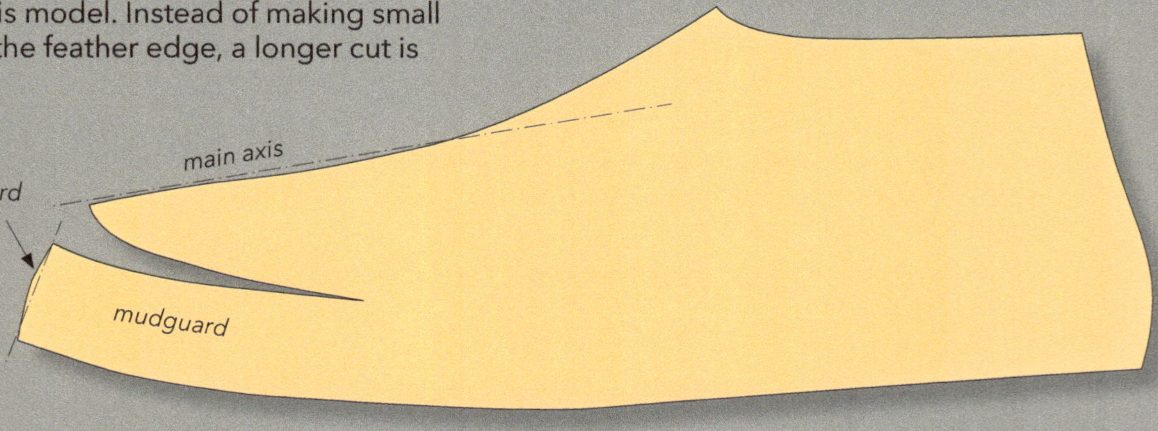

The Master Pattern

The master pattern is the foundation from which all other pattern pieces are derived. It may be created as a folded or unfolded pattern.

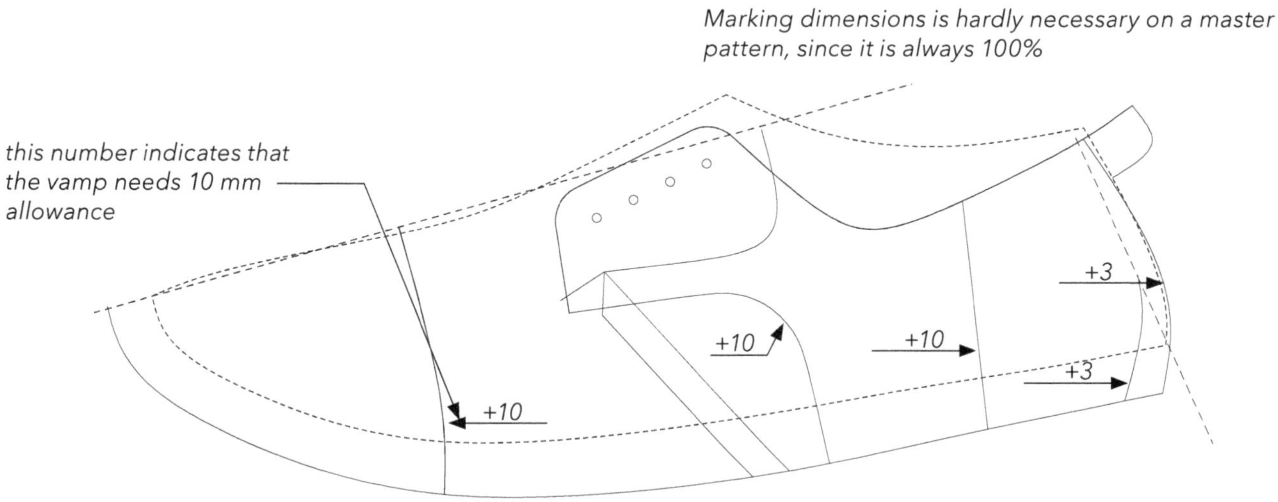

During product development, pattern lines often require adjustment and design changes must be applied. All such modifications should be made at the master pattern level to ensure consistency, repeatability, and technical traceability. For this reason, the master pattern serves as the single point of reference where corrections are validated and errors can be clearly identified.

Master patterns are typically developed in a mid-size, or in a specific size for custom footwear. Allowances may be marked on the master pattern when their function is not self-evident. In this example, the lap seam between the vamp lining and the quarter lining is defined by the pattern geometry itself, so additional marking is unnecessary.

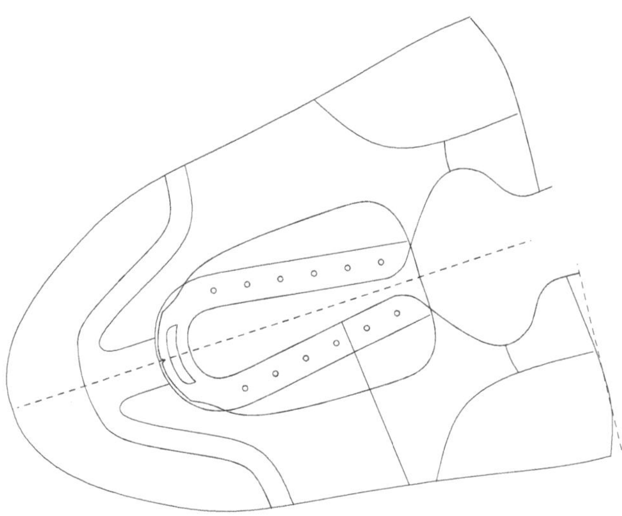

In the case of asymmetrical patterns, working with an **open** (unfolded) **master pattern** is often more practical. An open pattern allows each side to be defined clearly and independently, improving both accuracy and legibility.

Slip-lasted constructions – such as Strobel or stitchdown – are also frequently developed using unfolded patterns. In these cases, the open format simplifies repeated modifications during multiple development iterations.

The Manual Unfolding of a Pattern Piece

A process called "unfolding" is used to create pattern pieces involving a folding axis. The patterns discussed in later chapters all involve this step. The hand method goes as follows:

The first step is to fold a sheet of paper in half – large enough to accommodate the entire pattern piece. This fold must be precise and sharply defined. To achieve this, pattern makers typically use a scratch awl or another pointed tool to score a straight line before folding, ensuring a crisp crease.

Fold the pattern paper along the pre-scored line. Using medium-weight paper for patterning purposes will make both folding and subsequent cutting easier to control. The common oak tag paper is sufficient for making samples or custom, one-of-a-kind shoes.

Copy all relevant lines from the master pattern, including both edges and internal marking lines that belong to the piece. Follow any instructions provided with the master pattern.

Cut the outline of the piece. If the pattern piece is asymmetrical, always begin by cutting the exterior lines while the paper is still folded. After unfolding, re-cut the smaller side if necessary, and redraw any lines where the asymmetry is more complex.

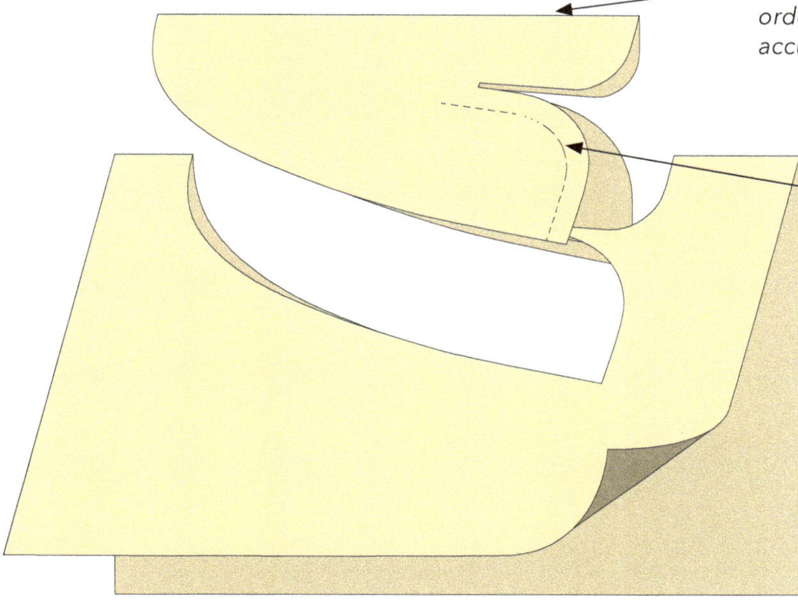

Pattern Types

Pattern pieces can serve different purposes. Custom and small-scale production may not require all of these, while industrial-scale manufacturing can justify the use of dedicated patterns for every operation.

Master pattern: all pieces combined

Clicking pattern: used for material cutting

Folding pattern: for accurate folding

Marking pattern: for stitch lines, piece placement, and hardware positions

Re-cut pattern: for example, after crimping

Special-purpose pattern: e.g. embroidery, tooling, or pattern holders

A marking pattern is used to mark stitch lines and the placement of a logo.

How to Read the Patterns Lines

This book aims to keep all content clean and user-friendly, yet some guidance is still required when working with the patterns. Unlike older pattern books, the patterns here are built up step by step, with each stage explained in detail. In certain cases, a pattern may repeat elements introduced earlier (for example, the geometric last standard). When this occurs, the relevant chapter is referenced rather than repeating the full explanation.

The patterns in this book are shown in this manner– folded in half. If the opposite side does not contain additional information, or if one side is sufficient to display all necessary details, there is no need to unfold the pattern to display the master pattern.

Simple lines refer to pattern construction lines. They can serve as the basis for other lines–such as allowances or mirrored elements–or be used on their own.

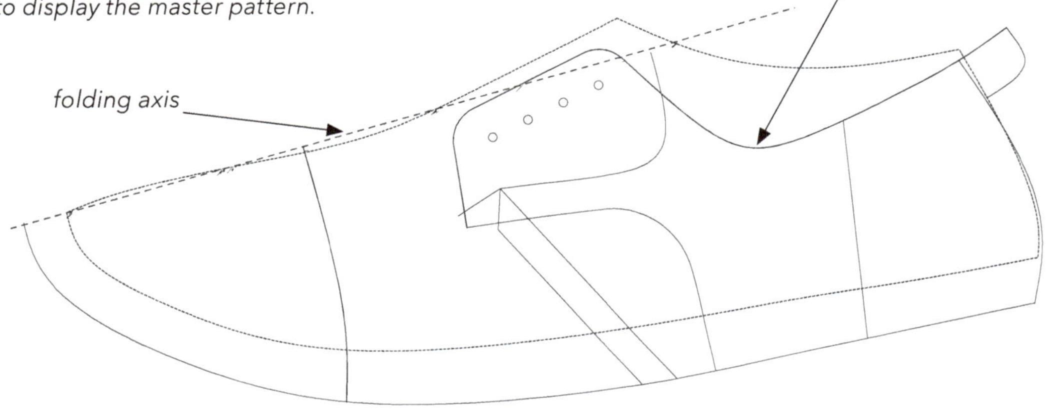

Size Codes on the Pattern

For obvious reasons, leather pieces cannot have the size written on them, although this would be useful. For this reason, the footwear industry uses a coding systems based on small notches to identify pattern pieces after cutting. There are different system, this is only one of them.

A large V represents 5, a small v represents 1. No notch means zero, and a simple straight cut represents one half. The rest follows basic arithmetic.

For example, V = 5 and v = 1. To indicate the number 8 – which may refer to 8, 18, 28, 38, and so on – the code would be V + vvv, or 5 + 3 = 8. It is unlikely that the same number would be used with different leading digits, but if necessary, this can also be encoded.

Additional information can be added using further codes, such as model number, variation number, or girth values. Pattern makers can apply these codes to digital pattern pieces or even to die-cutting tools.

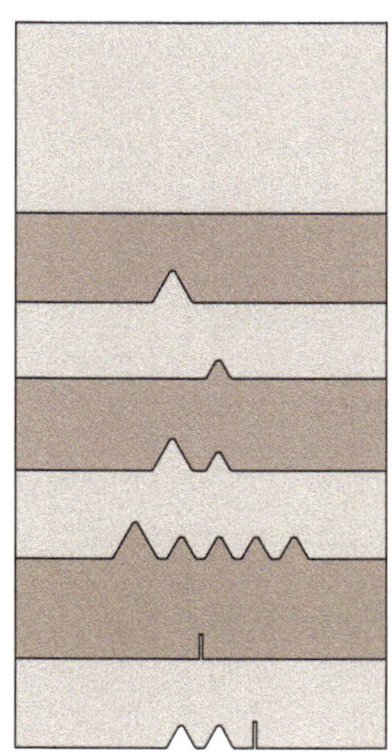

size 0, or 10, or 20, etc.

size 5, or 15, or 25, etc.

size 1, or 11, or 21, etc.

5+1=6: size 6, or 16, etc.

5+4=9: size 9, or 19, etc.

half

size 2.5, or 12.5, etc.

Colors and Markings

The patterns in the chapters are drawn with CAD software. As always, the pieces have to include all the pertinent information necessary for the clicker and/or upper maker's work. These are some of the things worth keeping in mind going forward.

lines inside pattern pieces there might be other lines to display the basis of allowances.

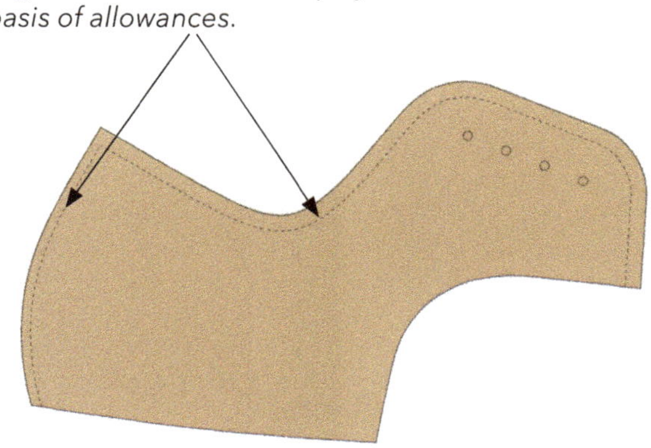

Light brown color on a pattern piece indicates an upper component. Note that the material itself and the function may not always be the same. For example, a piece cut from the same upper leather used for the exterior can function as an insert within the lining structure, or a lining leather may form part of the upper. The patterns are kept as simple as possible, so this overlap occurs only in the sock lining examples.

Light blue indicates lining. This does not refer only to lining leather, but to any non-upper material used as part of the lining structure.

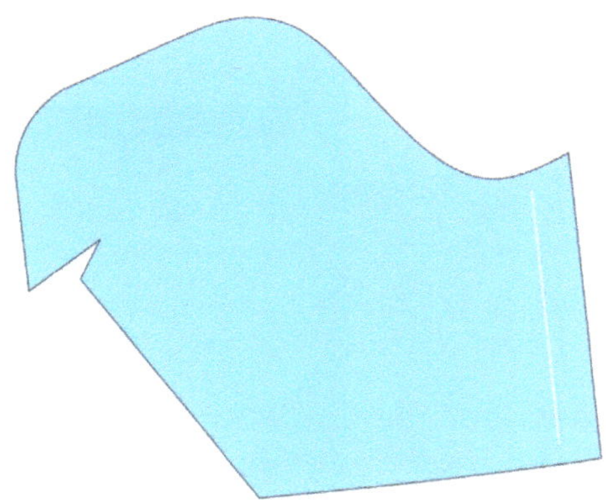

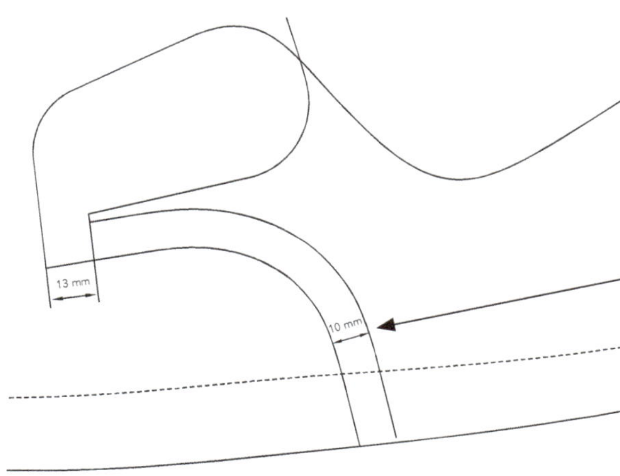

Dimension marks on the master pattern. These are marked, when these are relevant only.

Pattern CAD Systems

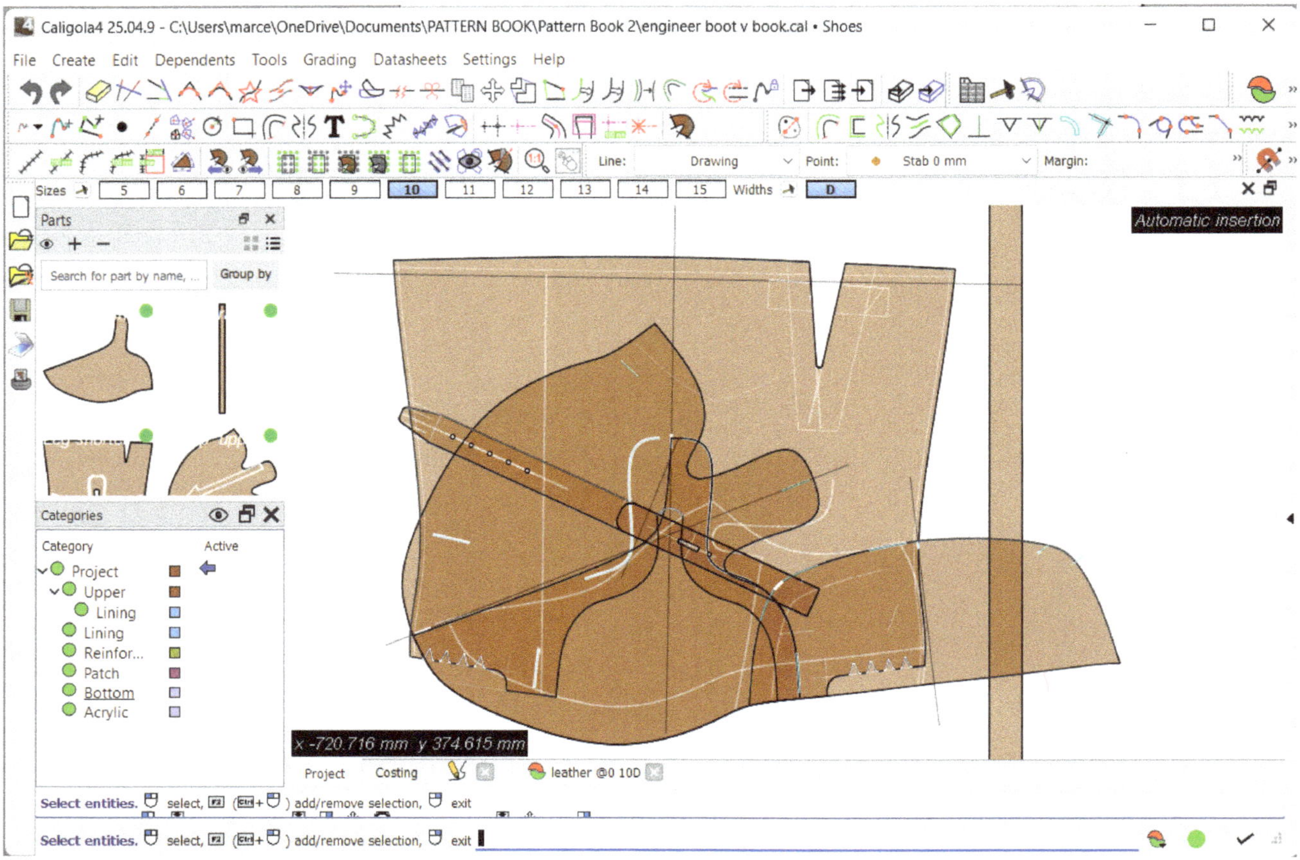

CAD systems can handle complex master patterns with ease. CAD software such as Caligola can manage hundreds of product variations (sizes, heights, girths, models) within a single file while remaining easy to handle.

In recent decades, computer software has largely replaced traditional pattern rooms. These systems provide significant support for pattern makers, either complementing hand methods or fully replacing them. CAD systems use digital vector graphics for lines, curves, points, holes, and other specialized objects. These elements are editable and repeatable, allowing operations such as moving, rotating, and mirroring. Positions can be controlled with sub-millimeter precision, which greatly improves accuracy.

CAD systems can handle highly complex pattern data. A single pattern may contain multiple lines associated with different pieces and materials. The system performs everything that analog pattern making can – and more – including pivoting, accounting for stretch directions and leather quality, calculating material consumption, weight, and production time, and generating specification sheets.

Product development is one of the key challenges in the trade, and software makes this process considerably easier. For example, modifying a main folding axis by hand can be time-consuming, as it affects nearly all pattern pieces, whereas such changes can be applied quickly and consistently in CAD.

Grading is another major advantage. Whether the output is paper patterns, die-cutting

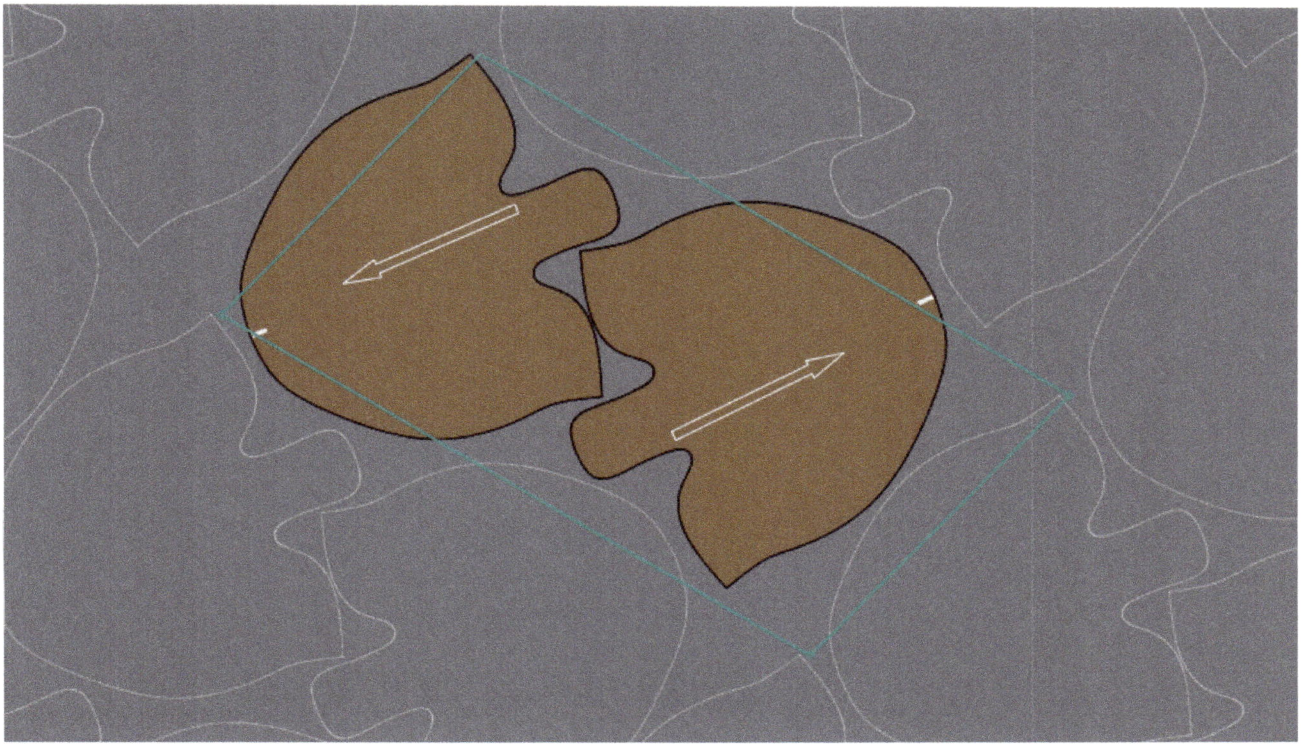

The parallelogram area is traditionally divided between the pattern piece or pieces and the surrounding material waste, making it a clear and classic example of yield evaluation. Modern CAD systems can extend this principle by introducing additional pattern pieces and even multiple sizes within the same layout to further optimize material utilization.

knives, tooling, templates, stitch patterns, logo placements, CNC cutting data, or a combination of these, CAD systems can manage it. In analog production, factories often avoid creating die-cutting knives for every size due to high costs. Components such as backstraps, tongues, or binding strips are good examples where size-specific patterns are unnecessary. This approach is known as group grading, where one size serves multiple sizes. For instance, a counter may cover three to four sizes, and a heel patch lining may fit half of the size range. All such information must be documented in specification sheets and considered in material consumption, nesting, and production planning.

Material consumption is one of the greatest challenges in footwear manufacturing, and the shape of the pattern pieces plays a major role. Software allows early, real-time evaluation of material usage, making it possible to determine whether a pattern meets waste targets or if further adjustments are required.

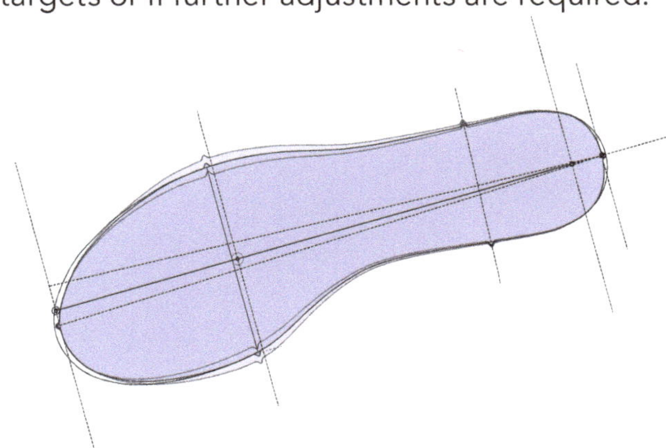

Caligola CAD allows sole grading and pattern grading to be performed within the same file, using separate reference numbers and grading ratios. In this illustration, width grading demonstrates that the operator has full control over how grading can be focused and concentrated specifically around the ball area.

3 UPPER MAKING BASICS

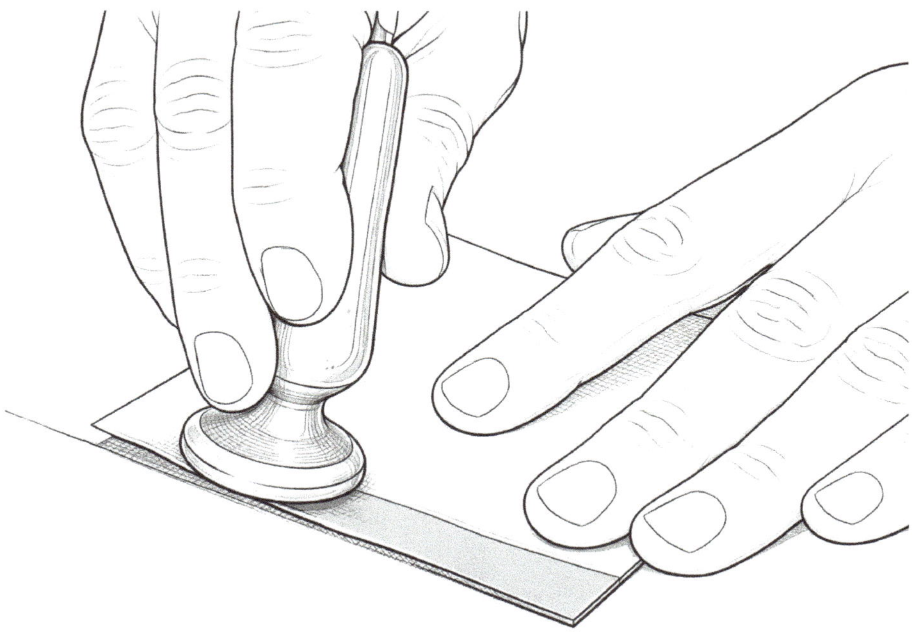

The uppers are the most visible part of the shoe at first glance. A professional shoemaker will judge the quality of the shoe by its sole and welt, but the majority of people will be first and foremost drawn to the look of the upper. What separates a high-quality upper from an average one is a series of small differences that stack upon each other, resulting in a visually much more pleasing work. In Section Two, each chapter discusses the steps of the construction of a specific upper in more detail. However, there are some overarching principles that applies to all uppers, which are laid out in this chapter. Hopefully, by the end of this book, you be able to know where to look to spot those tiny details that make a successful upper for what it is.

The Hand Tools of Upper Making

Unlike factory-scale shoe or upper manufacturing, small workshop upper making does not require an excessive number of tools or supplies. In most cases, processes commonly associated with machines can be executed effectively using hand tools. In fact, many machines were originally developed simply to speed up tasks that were first performed by hand.

The actual tool requirements depend heavily on the intended end product. In standard cemented or welted construction, the upper is typically completed by the upper maker, after which the remaining operations are carried out by a separate specialist. However, in the case of certain footwear types, the boundary between upper making and shoemaking – the latter referring to all processes carried out after the upper is completed – is not merely blurred, but effectively nonexistent.

As a result, when discussing tools for upper making, it must be understood that the scope can extend well beyond what is presented on this page, depending on the construction method and the degree to which upper making and shoemaking overlap.

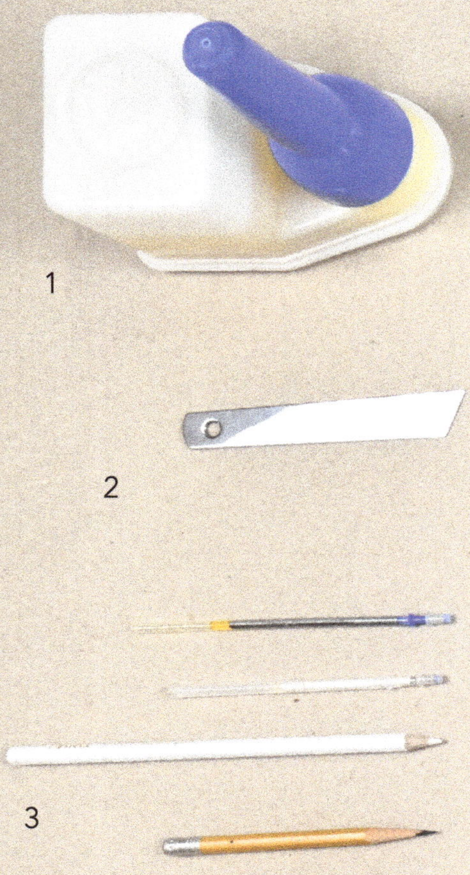

1. Adhesive
2. Glue spreader
3. Marking tools
4. Sharpening stone
5. Clicking knife
6. Scissors
7. Skiving knife
8. Knife for general use
9. Hole punches
10. Scratch awl
11. Hammer
12. Back-part shaper
13. Bone folder
14. Folding hammer
15. Thread burner
16. Lighter
17. Clicking weight

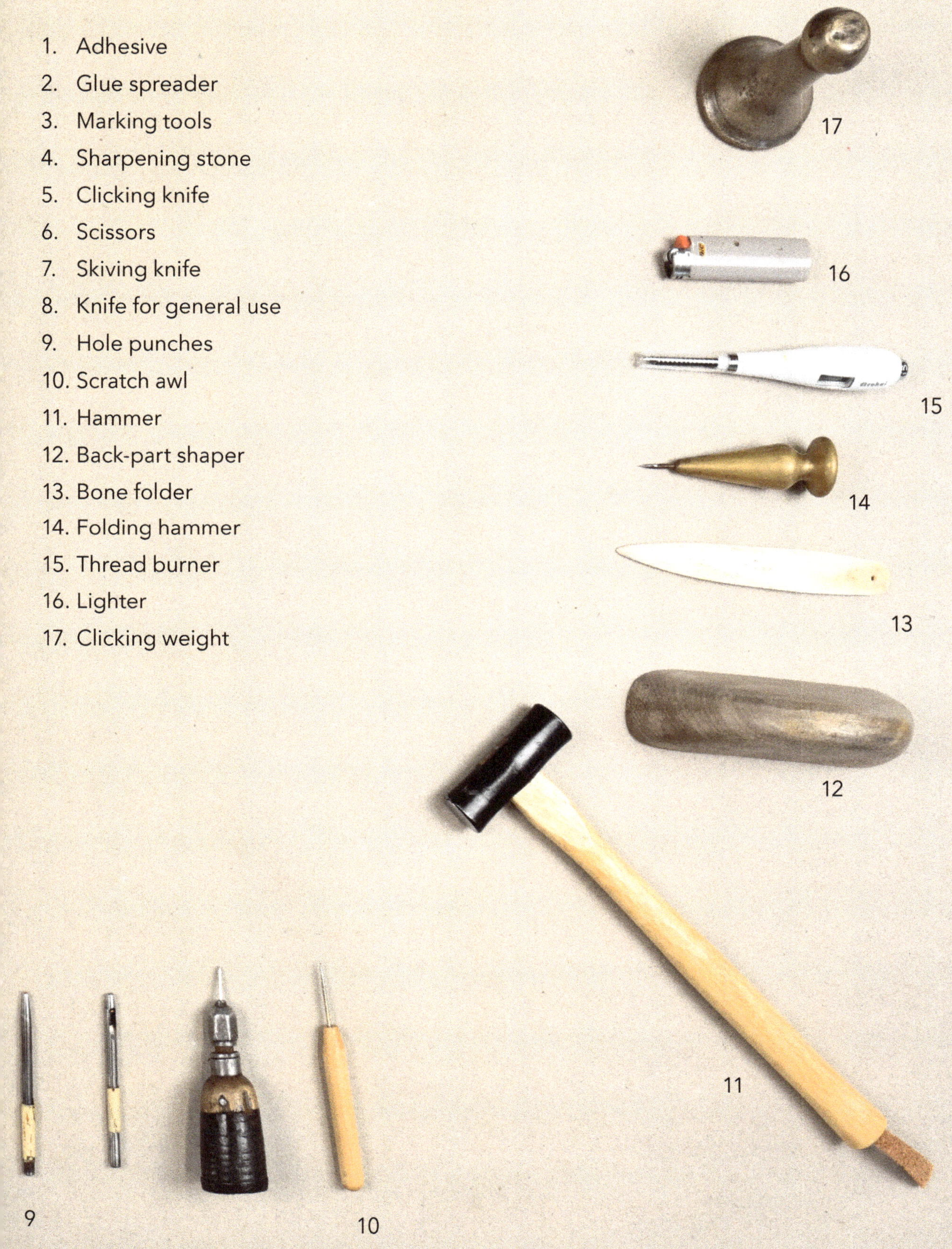

Clicking, Nesting

Early pattern sets were often edged or backed with steel to ensure accuracy and durability, and the repeated contact between knife and metal produced a sharp, audible click. This characteristic sound became closely associated with the cutting operation itself, giving rise to the term "**clicking**".

Clicking is the process by which upper components are cut from leather or other upper materials. It is one of the most critical stages of upper making and was traditionally carried out by the most experienced craftsman in the workshop. The clicker must evaluate material defects, stretch direction, visual appearance, fiber density, material yield, and the required quantity of components. A deep understanding of leather behavior is essential, as poor placement can result in defects, overstretched parts, reduced quality, or faulty products. For this reason, the clicker is often considered the worker who understands leather better than anyone else in the workshop.

In traditional workshops, clicking is performed using paper patterns. Although paper patterns wear quickly, they are sufficiently accurate for cutting a single pair of shoes and remain widely used in bespoke and small-scale production. Cutting is carried out with clicking knives, either traditional fixed-blade knives or modern versions with replaceable blades. High-density polyethylene cutting boards are used to support the leather during cutting and to protect the knife edge from damage.

Die cutting machines

Higher-volume production or more intricate pattern sets require industrial solutions. As shoemaking transitioned from small workshops to factory-based manufacturing, hand cutting with knives and paper or cardboard patterns became a production bottleneck. Steel cutting dies combined with mechanical presses provided a reliable alternative, allowing repeated and accurate cutting with reduced operator fatigue.

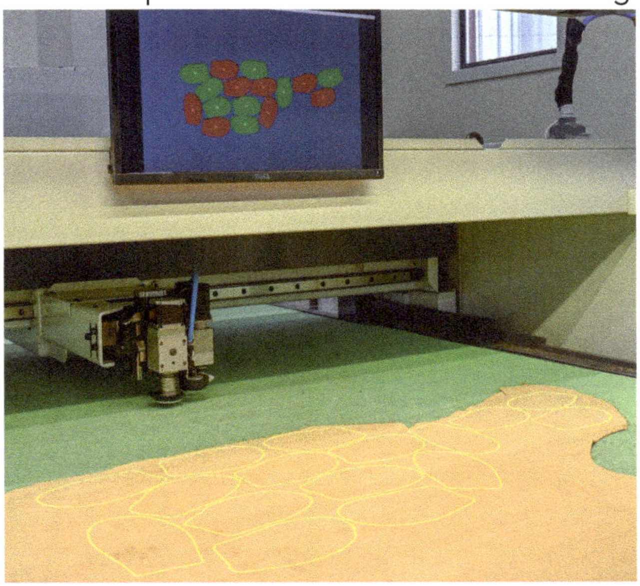

Modern CNC cutting machines such as this one by Caligola use laser or light projections to place the pieces

The earliest die clicking machines were mechanically driven presses, often powered by line shafts or flywheels in industrial workshops. Modern die clicking machines operate using hydraulic, pneumatic, or mechanical systems. Hydraulic machines are the most common today due to their smooth pressure build-up and precise force control, which is particularly important when cutting natural materials such as leather. Die-clicking machines are typically configured as swing-arm or traveling-head systems and are equipped with modern safety features, including photocells and guarded operating zones.

CNC cutting systems

Digital nesting and CNC cutting systems represent a major technological shift in the footwear industry. These systems began to appear in the late 20th century, with early adoption in the 1980s and 1990s, and became industry standards during the 2000s. CNC cutting machines use digitally controlled cutting heads capable of performing multiple functions, depending on the material and application.

The cutting process is integrated with digital nesting software, which optimizes pattern placement to maximize material yield while respecting grain direction and stretch constraints. Compared to manual or die-based cutting, this integration significantly reduces waste and increases consistency. The nesting process typically begins with material quality grading, which may be performed manually by an operator marking usable areas, automatically through defect-detection

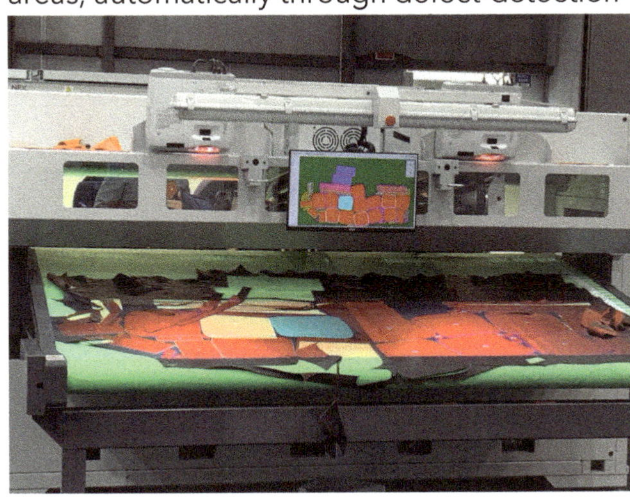

Nesting demonstration on the SIMAC shoe industry expo, Milan. A projector shows color coded pieces before cutting.

Skiving

As the upper consists of multiple parts and components, simply sewing them together would not only yield an aesthetically unimpressive result but it would also be unforgiving during wear. Skiving is the process, when the leather edge is getting thinned down to the width and thickness appropriate for the design and edge treatment. Depending on the size of the production the decision about skiving parameters can be the responsibility of the pattern maker/product developer, technician, upper maker.

Learning and using the skiving profiles is essential as it can ensure success both on the end user's side as well as the maker's side.

A - leather thickness

B - skived edge thickness

C - width of the skiving

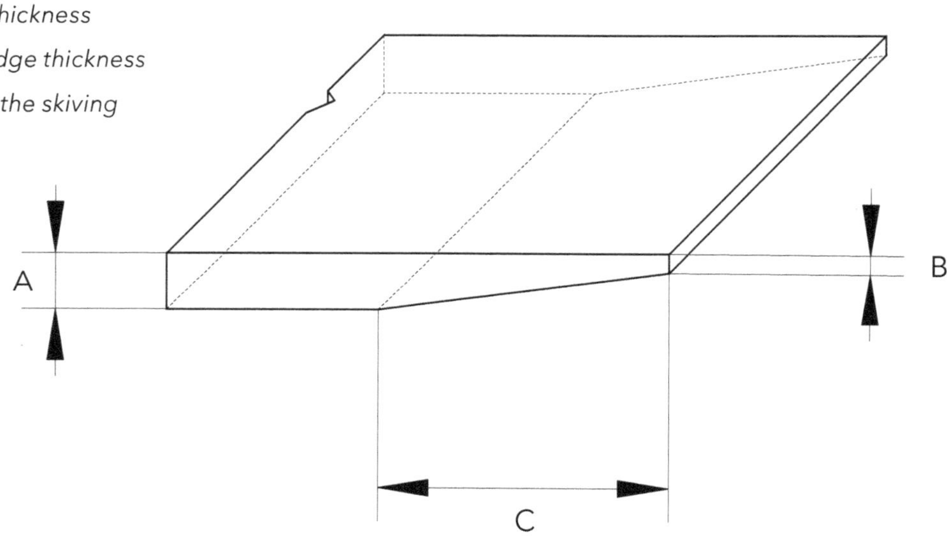

The skiving profile can be defined by only a few parameters: original leather's thickness, the skived thickness, and the width of the skiving. While hand skiving is possible, it takes years' worth of skill to be able to do it reliably. Hence a skiving machine is a worthwhile investment for even the smallest operations.

Older, analog skiving machines do not use numeric adjustments, the operator can rely on measuring the skived edge for feedback and evaluating the results. Modern digital machines provide very accurate feedback on the parameters used for skiving, such as skiving width, skive depth, and the angle of the presser foot, and can store these settings in program slots. These parameters, on both analog and digital machines, can produce very different results, and in fairness, some specialized skiving profiles are only achievable with specific setups or dedicated machine feet.

Skiving profiles can vary significantly depending on the presser foot design. A good example is channeled (grooved) shoulder skiving, which is typically used for folded edges. The score line helps create a smooth, consistent fold and provides a clear guide for applying reinforcement tape.

Most Common Skiving Profiles

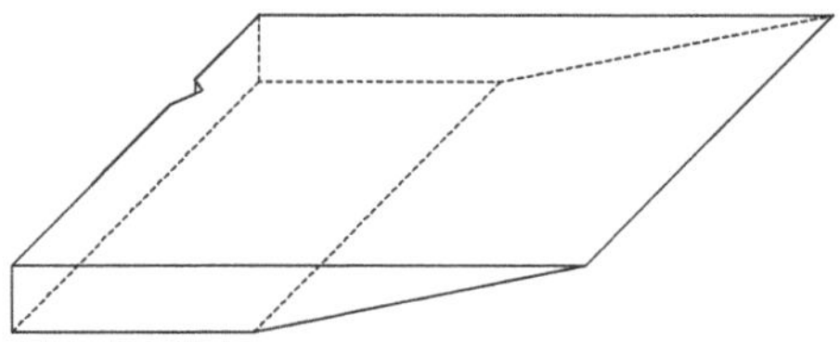

Wide skiving, zero edge thickness, perfect for overlap allowances, folding, turnseams, etc.

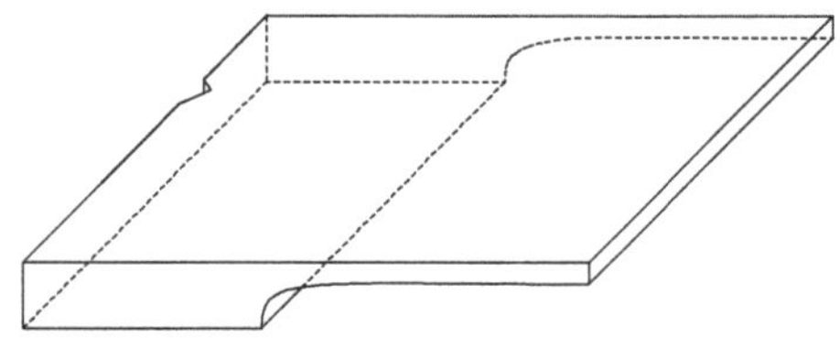

Wide shoulder skiving with some edge thickness. This can be done with any skiving machine with a regular foot by manipulating the foot angle.

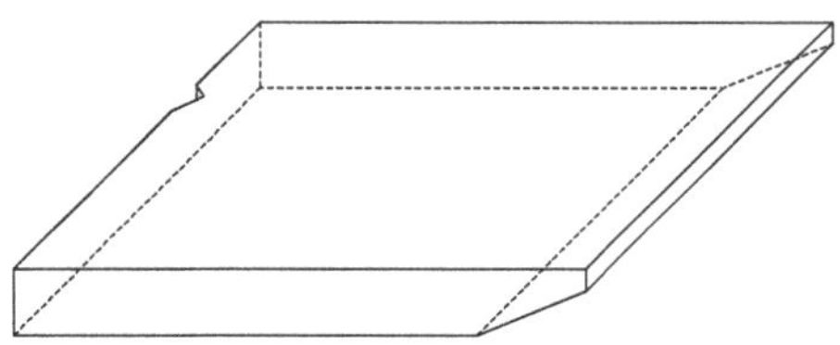

Short skiving, some edge thickness, used for reverse seams, overlapping edges, raw cut edges, turnseams.

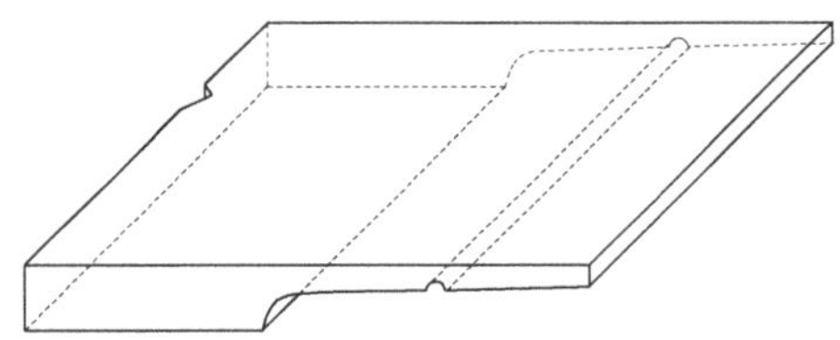

The **grooved** version of shoulder skiving is intended exclusively for folding. It requires a specialized presser foot, manufactured for a specific skiving width and equipped with an adjustable grooving pin that creates the channel by pressure.

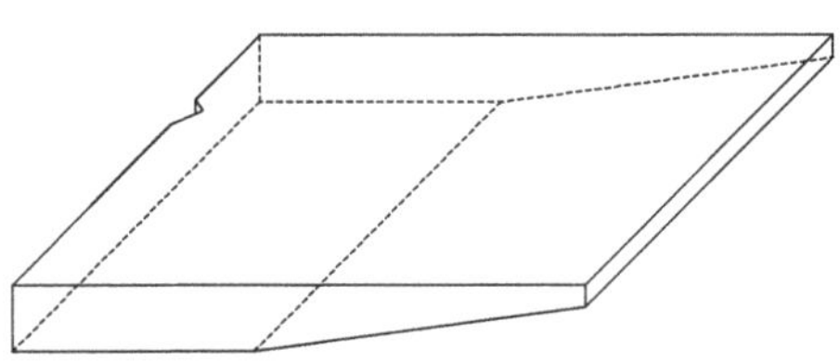

Wide skiving with some thickness left on the edge. Overlapping part of the turnseam, trimming allowance, etc

Seam Types and Edge Finishes

As mentioned before, the way two parts come together and how the edges are finished is part of the design and patterning process. Some footwear models have their dedicated seams and edge treatments that contribute to their signature appearance and altering them would undermine the traditional look associated with those models. However, for the most part, knowing the proper techniques and creatively using them to enhance the design will surely pay off in the end.

Raw cut edge

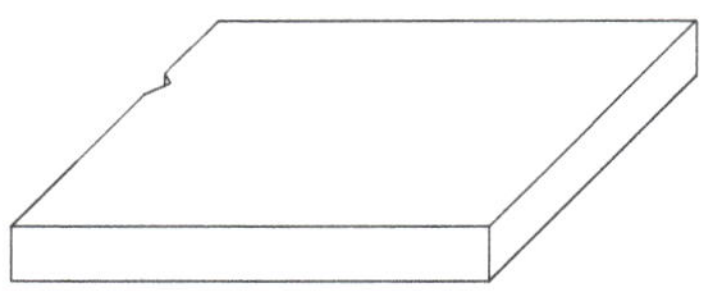

The raw-cut edge is commonly used for lap seams and, when the leather allows, even for toplines. With thicker leathers, a raw edge can be more durable than a folded one. A raw edge does not mean an unfinished edge – dyeing, burnishing, and other treatments can produce a clean, professional finish.

Brogue/gimping

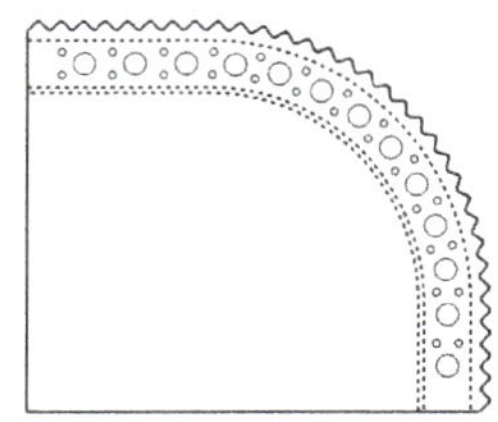

Brogueing is a decorative row of holes placed along the edge, often arranged in patterns that combine different hole sizes. Gimping is a zigzag-shaped cut applied directly to the edge, also available in various profiles. Both treatments create a specialized form of raw edge and typically require piping to reinforce the topline or seam. Traditional dress shoes commonly use these elements. Brogueing, in particular, visually suggests an edge line of its own, making it an effective and versatile design feature.

Folded edge

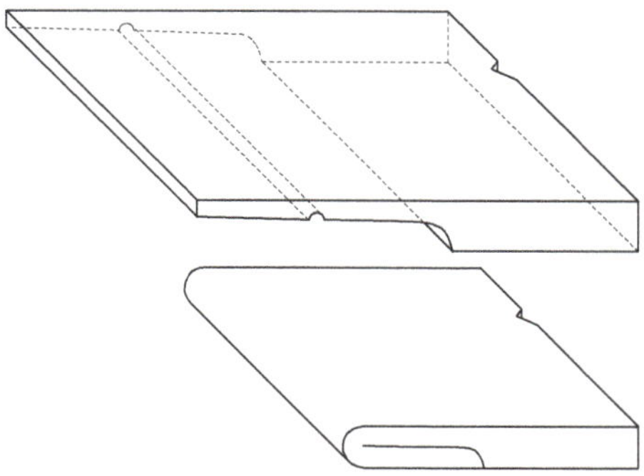

The folded-edge treatment is one of the most widely used finishing methods in the footwear industry. It creates a clean, polished, and structurally strong edge, making it suitable for toplines and many other exposed edges. The process is relatively easy to execute, whether by hand or with a dedicated folding machine.

Skiving for folded edges can follow several profiles. A straight, wide skive – typically achievable by hand – provides a simple reduction in thickness. A short-shoulder skive folds the entire reduced section inward, making it an excellent option for thicker leathers, since it preserves the original edge thickness. The most advanced method is grooved skiving, which forms a recessed channel for the fold, resulting in an exceptionally smooth, crisp, and uniform finished edge.

Piped edge

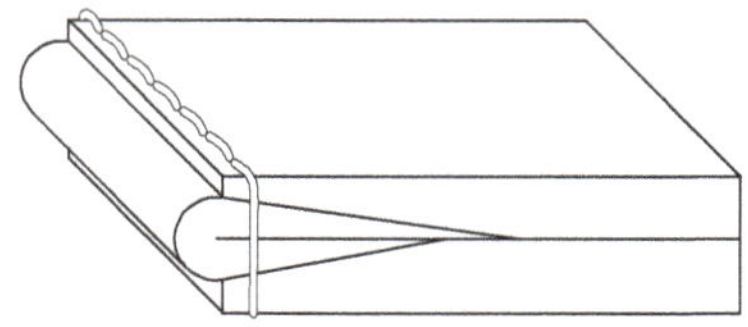

The piped edge is an excellent alternative when a folded edge presents structural or material challenges. A typical example is the wholecut upper: between the lacing rows there is no space for a folding allowance without opening the lacing line, which would compromise the style. In such cases, piping is an ideal solution. It also allows for the introduction of contrast or accent colors, making it both a functional and aesthetic edge treatment.

Lap Seam

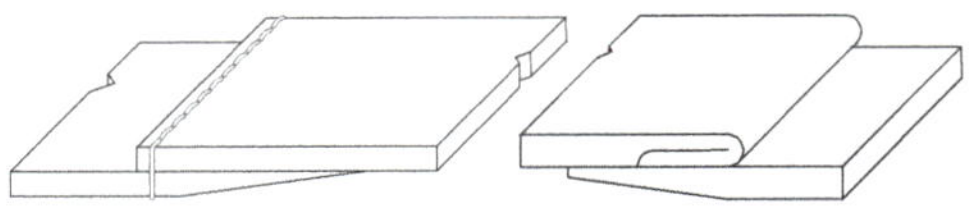

A lap seam is one of the most common methods for joining two upper components. One material edge overlaps the other and is stitched through both layers, resulting in a clean appearance while keeping the seam flat and controlled. It can be executed with raw-cut edges, folded edges, or gimped edges. This seam type is often the only viable option when decorative stitching or brogueing is applied directly along the seam line.

Lap seams usually require skiving to reduce bulk and ensure a smooth transition between layers. Proper skiving is critical, as excessive thickness can negatively affect flexibility and wear comfort. Structurally, the lap seam provides good strength by distributing stress along the overlap rather than concentrating it at a single edge. From a production standpoint, lap seams are efficient to sew and relatively forgiving, making them suitable for both handwork and industrial manufacturing.

Reverse Seam

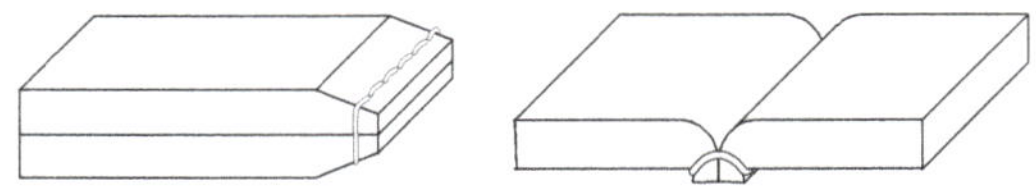

Two components are stitched together flesh side to flesh side and then turned, leaving the seam allowance hidden inside. In some special cases, the process can be executed the opposite way. This construction results in a clean exterior surface without visible overlap or edge buildup. Reverse seams require minimal seam allowance. Skiving is often used to reduce bulk and ensure a smooth transition after turning, although in certain situations better results can be achieved without skiving.

Reverse seams are most commonly used for backseams. Structurally, this seam relies primarily on stitch integrity rather than overlap strength. For this reason, stitch density and the rubbing process are critical. Reverse seams are less forgiving than lap seams, but when executed correctly, they provide superior visual cleanliness.

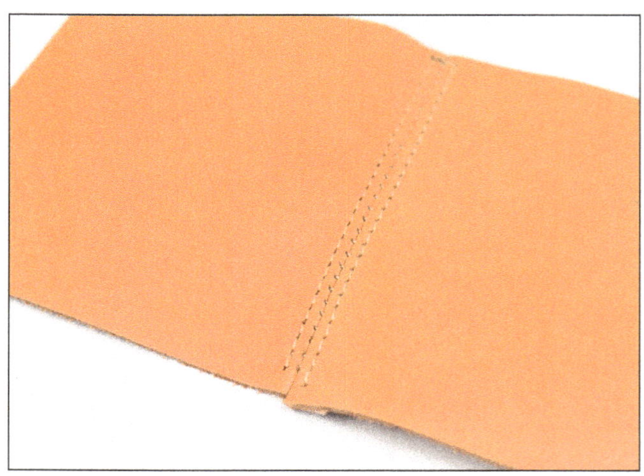

Turned Edge

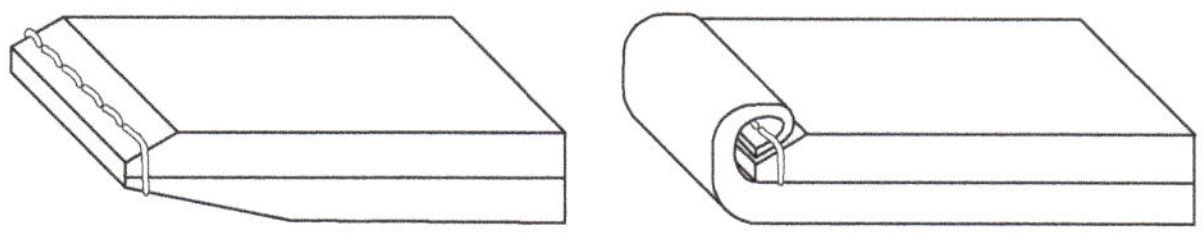

A turned seam edge is an edge-finishing method in which the raw edge of the upper is folded to the inside and secured with a seam, leaving a clean, uninterrupted exterior line. The technique is commonly used on toplines, openings, and seams where a refined appearance is required without adding separate binding material. By turning the edge inward, the construction protects the raw cut from abrasion and wear. Proper skiving is essential, as excessive thickness will prevent the edge from laying flat and may distort the pattern geometry. The turned seam edge slightly shortens the visible opening, which must be accounted for during pattern development. Stitch placement and tension are critical, since the seam controls both edge stability and final shape. This method provides moderate structural reinforcement while maintaining flexibility. It is well suited for fine dress shoes and lightweight footwear.

Turn Seam

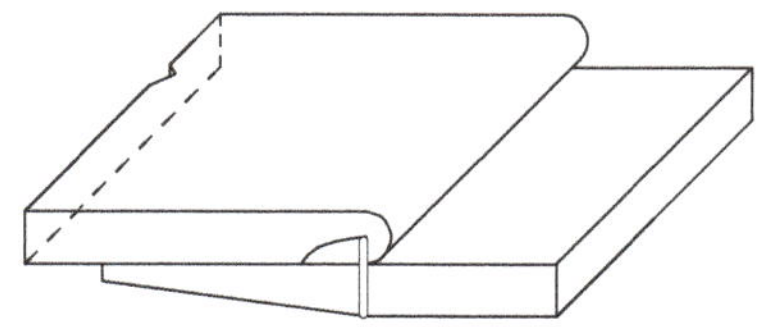

The turn seam is somewhat similar to a reverse seam, but it serves a different purpose. Both technically and visually, it is closer to a lap seam with a folded edge on the top piece. The key difference is that the folding is created by the seam itself, which strongly limits the shapes that can be used. Tight curves, sharp corners, or complex shapes are generally not feasible; the seam works best on straight or nearly straight lines.

In return, the turn seam does not require over-stitching, allowing for a very clean appearance. From a technical standpoint, it is possible to construct an entire upper without any visible stitches by combining reverse seams, turn seams, and turned edges.

Butt Seam

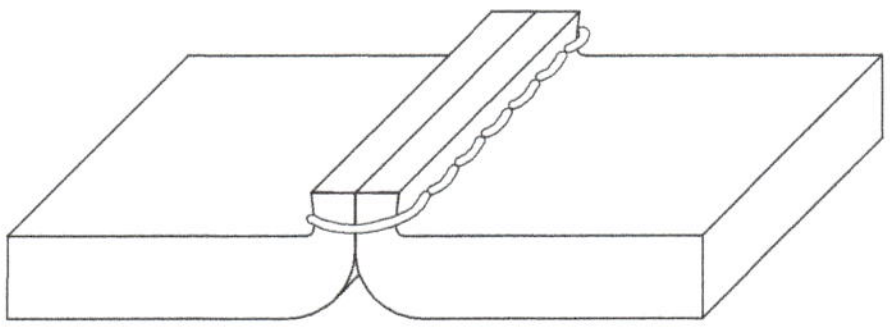

The butt seam is a variation of the reverse seam and is technically very similar, but there are important differences. The leather pieces meet flesh side to flesh side, the stitching usually remains visible, and skiving may not be necessary. This seam can also function as a literal reverse seam for the backseams of unlined boots, where it is left as is and covered with a backstrap.

Moccasin seam

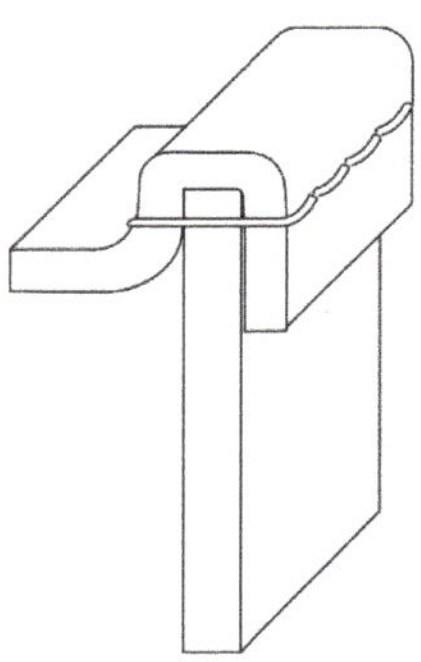

There are several variations of this seam. The basic idea is to stitch two leather pieces together–usually at an angle–so that the seam contributes to the shaping in some way. A very simple version of this is essentially a butt seam.

This seam can be executed by machine or by hand, including hand lasting from the bottom. It can also be made off the last using pre-punched holes, which requires very precise and demanding pattern work.

U-binding

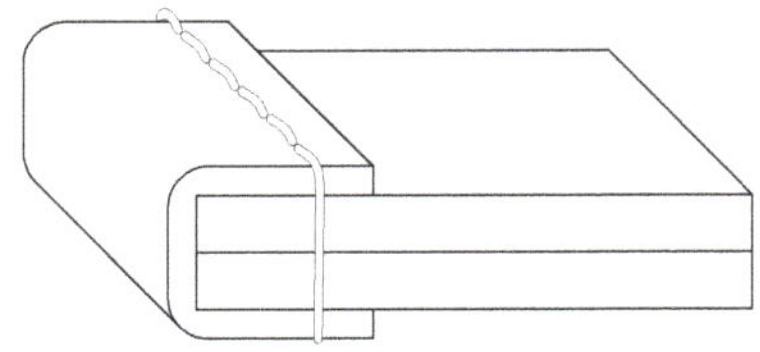

U-binding is an edge-finishing technique in which a narrow binding encloses the edge of the upper in a characteristic U-shaped profile. The binding can be applied in several configurations, most commonly hugging both the upper and lining layers simultaneously. This symmetrical version is well suited to machine application and can be efficiently sewn using a dedicated U-binder attachment. Alternatively, the binding may hug only the upper layer, or be applied asymmetrically by hand, depending on design intent and material behavior. In all variations, the binding stabilizes the edge while protecting the raw cut from wear and deformation. U-binding introduces minimal bulk and preserves flexibility, which is critical in lightweight footwear.

French Binding

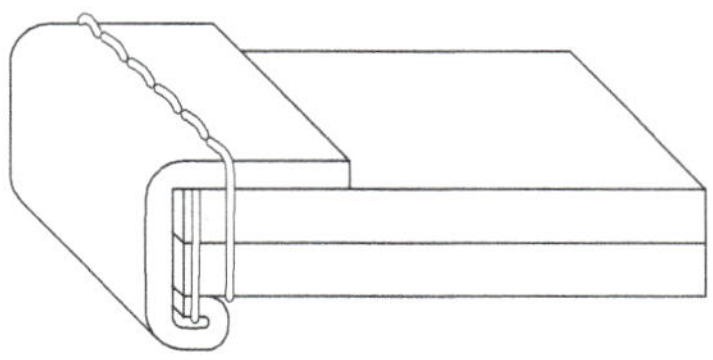

French binding is an edge treatment in which a narrow strip of material – typically leather or textile – is folded over the raw edge of an upper component and stitched in place, fully enclosing the edge. Unlike a simple folded edge made from the same upper material, French binding uses a separate binding strip, allowing greater control over thickness, color contrast, and edge behavior.

The technique results in a thin, uniform edge with high resistance to fraying, stretching, and wear. When properly executed, the binding becomes visually unobtrusive while significantly improving edge durability.

Italian Binding

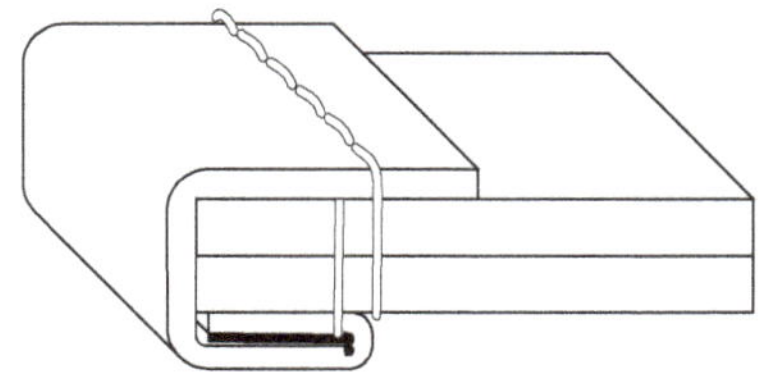

Italian binding is essentially a wider variation of French binding. The main difference is the exposed narrow flat fold, which makes it a strong alternative to U-binding. Compared to French binding, it is slightly more difficult to apply on tight curves.

When a contrasting color is used for the binding material, it creates a distinct decorative effect. While French binding is generally perceived as more elegant, Italian binding has a more casual character, although this ultimately depends on the designer's intent. The greater width of the binding strip also presents a technical challenge, as trimming the lining becomes more demanding.

Surface Decorations

The next few techniques are primarily used for surface decoration. They are not commonly applied in dress footwear, but in fashion they offer strong options to visually enrich shoes and boots. In footwear construction, overlay, inlay, and trapunto describe three fundamentally different ways of introducing additional material into an upper. While these terms are sometimes used loosely in practice, understanding their precise meaning helps clarify design intent and avoids confusion during pattern making and production.

Appliqué (overlay)

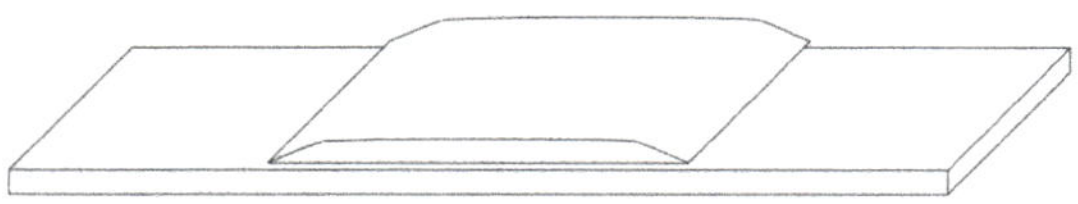

An overlay is a pattern piece applied on top of a base upper component and attached by stitching, bonding, or both. Overlays are clearly visible from the exterior and may serve decorative, structural, or protective functions. Typical examples additional logos, decorative elements, or reinforcement patches in high-wear areas. From a pattern-making perspective, overlays require careful consideration of edge treatment, stitch allowance, and material thickness, as they directly influence flexibility, bulk, and lasting behavior.

Inlay

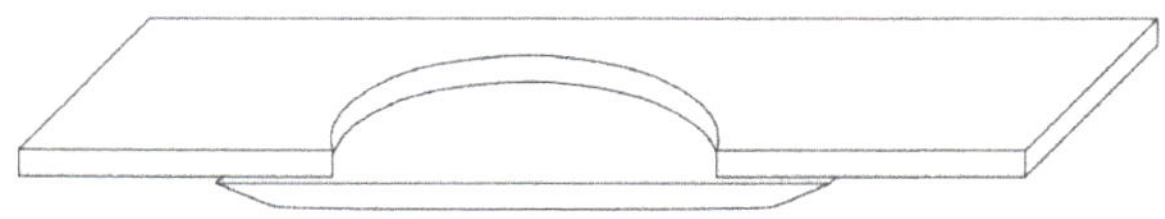

Inlay is most commonly associated with a decorative, graphic function, particularly in traditions such as Western boots, riding boots, and heritage work boots. The inlay is meant to introduce contrast and visual interest while remaining largely flush with the surrounding material.

Decorative inlays are typically executed by cutting a shaped opening into the base upper piece and inserting a second material from behind or between layers. The inlay may contrast in color, texture, grain, or finish – smooth against rough-out, light against dark, or exotic against plain leather. In Western boot shafts in particular, inlays often follow sweeping, organic lines or figurative, colorful design elements that have no effect on structural integrity but is much more about the artistry of the maker and the visual storytelling.

From a pattern-making standpoint, inlays demand a high level of precision.

Trapunto

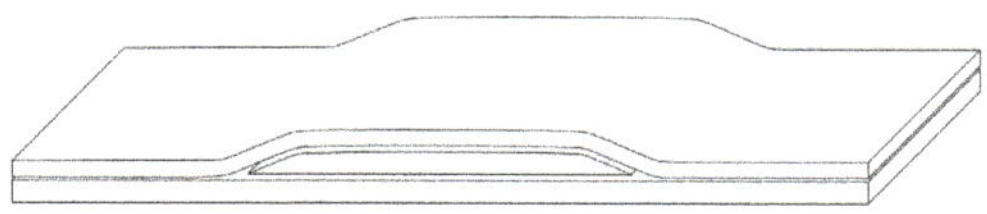

Trapunto is a term borrowed from the world of textiles and upholstery, and its use in footwear reflects a shared construction. The word comes from Italian – derived from trapungere, meaning literally to prick through.

In its original context, trapunto developed in Italian decorative arts and later became widespread in European upholstery, leather goods, and luxury interior work. The defining characteristic of trapunto is relief: the design is not merely stitched or layered, but physically elevated by an internal filler such as wool, cotton, felt, or foam. Unlike appliqué or overlay, where thickness is introduced externally, trapunto achieves volume from within the structure.

When applied to footwear, trapunto describes any construction where a (padded or hard) insert is enclosed between upper layers to create a raised pattern or panel. Common locations include quarters and boot shafts. The effect can be subtle or deliberately pronounced, forming a bold, three-dimensional graphic element.

In contemporary footwear, trapunto is most often associated with luxury shoes, sneakers, and fashion-driven designs

Trapunto-Inlay Combination

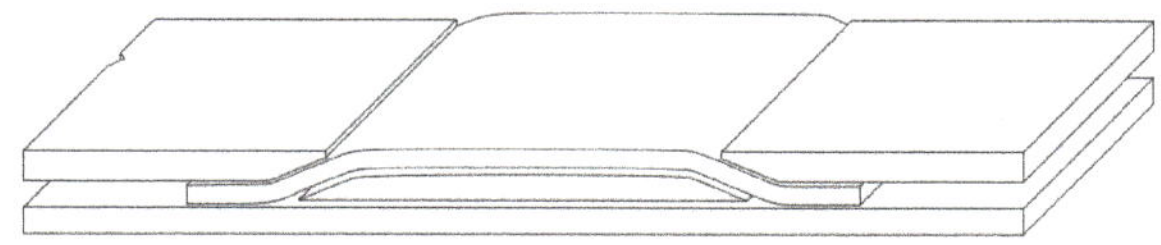

Trapunto and inlay are often combined to create a unique, three-dimensional graphic element. This provides the designer with a powerful tool to elevate a design not only through color, but also by introducing depth and texture. The inlay portion does not need to remain flat; it can be shaped, gathered, or wrinkled–resembling rose petals or other organic forms–to achieve a sculptural effect.

Using these elements thoughtfully allows decorative features to become structural and tactile components of the upper, rather than purely graphic additions.

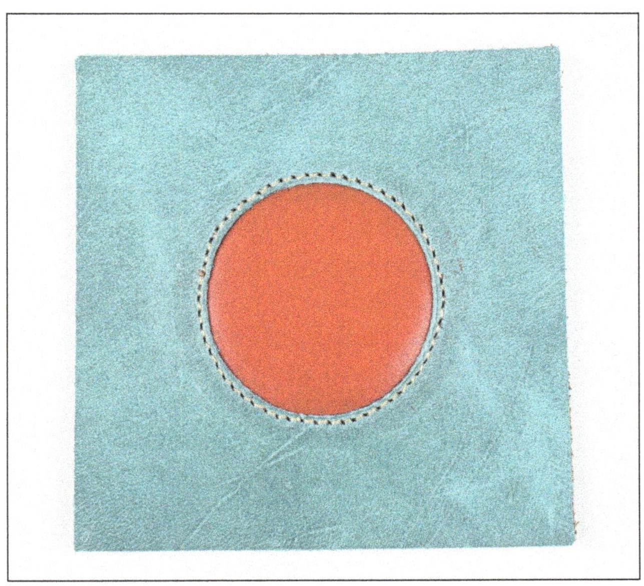

Machines in Upper Making

While the smallest workshops usually rely on a few essential machines – such as a post-bed sewing machine and a skiving machine –factories operate a wide range of highly specialized industrial equipment.

Machines designed for leather differ significantly from those used in textile production, particularly in penetration power, needle standards, material feeding systems, and overall construction. Within the scope of this chapter, the focus is limited to the most important machines relevant to footwear making.

Sewing Machines

Apart from skiving machines, the majority of the upper work is conducted on sewing machines. Since the invention of the sewing machine, footwear manufacturing has benefited enormously from its use. These machines dramatically increased production speed and made it possible to reproduce hand-sewn processes much faster, more consistently, and often with superior quality.

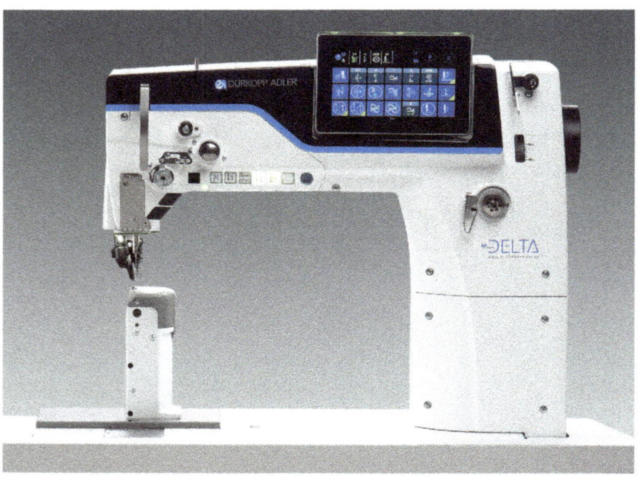

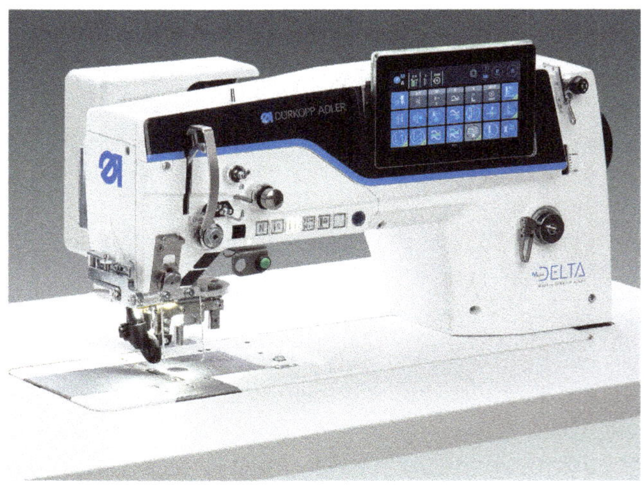

Post bed sewing machines

This is arguably the most important sewing machine in a footwear workshop. It features a narrow, vertical post that allows excellent access to three-dimensional components. The roller-foot mechanism provides a clear view of the seam, enabling the operator to sew stitch lines very close to the edge with a high degree of control. Post-bed machines are available in a wide range of configurations for different applications, and they are indispensable in situations where flat-bed machines would be impractical or ineffective.

Flat bed sewing machine

Flat-bed machines usually play a complementary role to post-bed machines in the footwear industry. Their stable working surface makes them ideal for two-dimensional components and sub-assembly operations, where they are often more effective than post-bed machines. Flat-bed machines are available with a wide range of functions required in footwear production, including double-needle, zigzag, and computerized configurations.

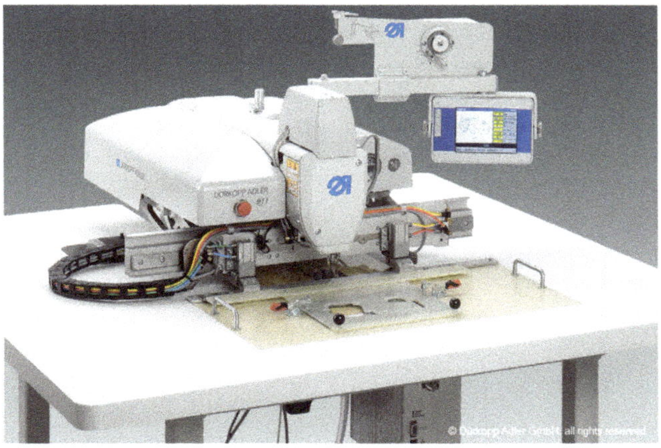

Programmable Sewing Machines

Programmable sewing machines are fully or semi-automated automats designed to execute pre-programmed stitch patterns with high precision and repeatability. In footwear manufacturing, they are primarily used for operations where consistency, accuracy, and speed are critical – especially in high-volume or brand-sensitive production.

These machines operate on flat-bed or clamping systems that hold the material in a fixed position while the sewing head follows a programmed path. This eliminates operator-dependent variation and ensures identical results across large production runs. Typical applications include decorative stitching, logo stitching, reinforcement patterns, complex seam outlines, and repetitive construction details that would be difficult or time-consuming to reproduce manually.

Skiving machines

This machine thins the edge of materials used in the footwear industry, such as leather, vinyl, and rubber. The tapered edge reduces bulk, making it ideal for edge finishes, folded edges, and seam constructions. Skiving is typically performed by a fast-spinning bell knife working beneath a presser foot, which controls the remaining thickness, skived width, and cutting angle. Skiving machines range from simple mechanical models to high-end computerized versions that store skiving parameters, allowing results to be replicated with consistency.

ATOM die clicking machine

The Die Clicking Machine

Die-clicking machines are widely used across all leather trades. They typically operate using hydraulic force, pressing a pre-made steel-rule cutting die to produce high-volume, consistent components from upper leather, sole leather, linings, textiles, and various man-made materials. Compared to hand cutting, die clicking is significantly faster, more precise, and far more consistent, making it highly efficient for production environments. The main limitation of this technology is its lack of flexibility, as each design requires a dedicated – and costly – die-cutting knife.

Comelz COM52 folding machine

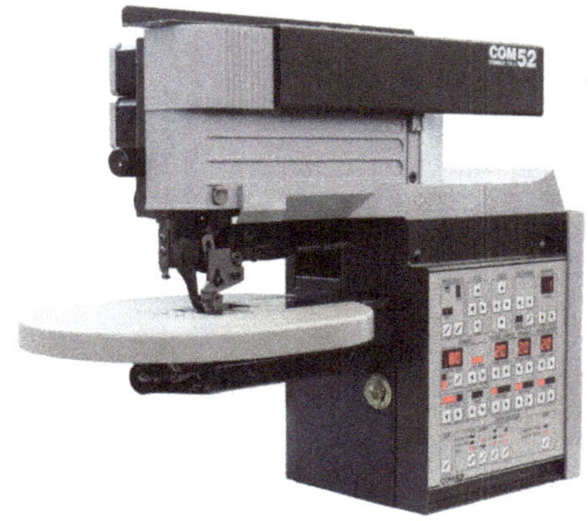

Folding Machine

This high productivity folding machine represents the next generation folding machines in the upper making industry. The numeric control adjusts the folding styles, the glue flow, the guide, the turning device, etc.

CAMOGA splitting machine

Leather Splitting Machine

Leather splitting machines are widely used in the footwear and leather goods industries to reduce material thickness uniformly across an entire hide or a cut component. They operate by feeding the leather between a rotating blade and a pressure roller, removing material from the flesh side with high accuracy and consistency.

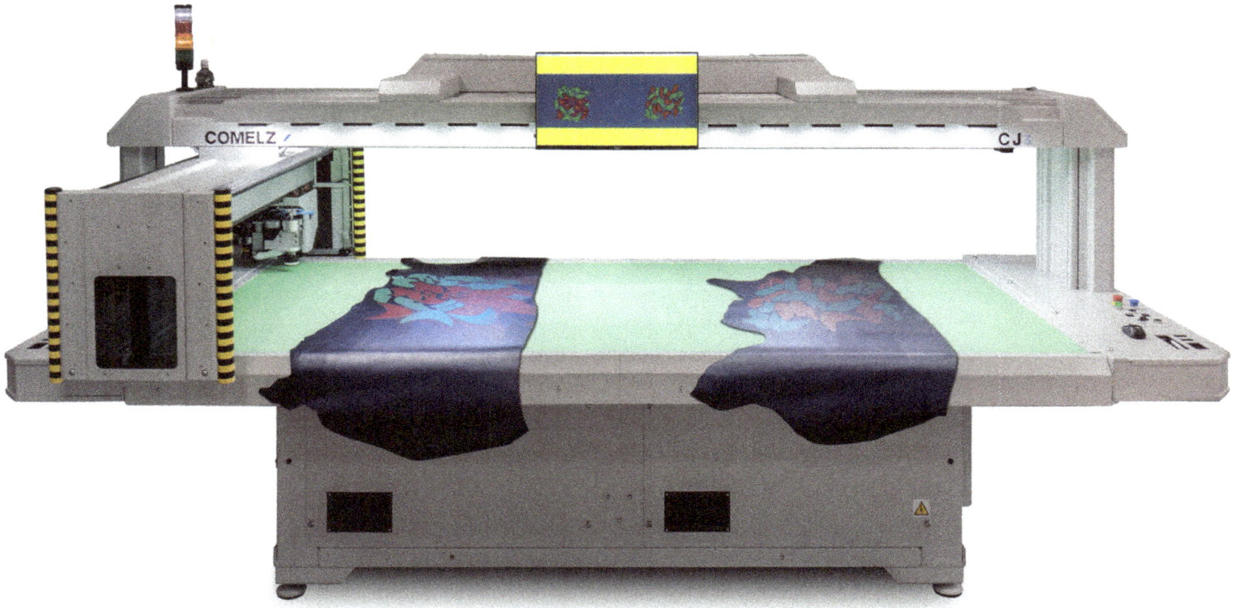

Comelz CJ3 machine

CNC Cutting Machines

Digital cutting machines are increasingly used in the modern garment and footwear industry. Their main advantage is versatility: they can be used not only for upper materials–leathers and non-leathers alike–but also for soles and other high-density materials. Using high-precision oscillating knives or rotary blades, these machines operate directly from digital files and do not rely on physical cutting dies. In addition to cutting, they can be equipped with tools such as hole punches, marking devices, and pens, and some systems use dual cutting heads to further increase productivity.

This technology allows rapid design changes, efficient prototyping, and improved material utilization through automated nesting. In high-volume operations, leathers can be graded and pattern pieces linked to specific hides, which are then identified and tracked using barcodes. In such setups, a single digital cutting system can replace multiple hydraulic clicking machines.

The illustration shows a state-of-the-art Comelz fixed-table CJ3 system with additional features.

Extra Features of the CJ3

7 tools cutting head

projector system

Vision system

Leather

Many source material is available on the topic of leather, so I will touch on the topic only briefly. Leather remains the choice of material for the footwear industry for uppers because it combines durability, flexibility, breathability, and the ability to adapt to the foot over time. In uppers, leather functions as both a visual surface and a structural element – full-grain leathers preserve the natural fiber structure and offer the best aging characteristics, while suedes and nubucks provide softness and visual depth at the cost of somewhat reduced structural strength. There are many ways to categorize leather, but the one I want to highlight is the tanning method because it directly influences the behavior of leather during cutting, skiving, stitching, and lasting.

Chrome-tanned leather is soft, elastic, and highly resistant to moisture making it ideal for uppers that require comfort and flexibility. It lasts easily, tolerates repeated flexing, and accepts dyes uniformly. The center of the leather may remain bluish-white, as dye cannot fully penetrate the cross-section, and this is entirely normal and technically acceptable.

Vegetable-tanned leather, by contrast, is a bit more firm, offering excellent shape retention and long-term stability. It develops a natural patina with wear and is less forgiving during lasting. Veg-tan is commonly used for insoles, welts, and structural components. Chrome-tanned and vegetable-tanned leathers are often combined, each used where its material properties are most effective.

True **oil tanning** predates chrome tanning. The earliest oil-based approach was most likely brain tanning, one of the oldest known tanning methods. It relies on natural emulsified fats – traditionally derived from the animal's own fat-rich organ – to lubricate and stabilize the collagen fibers of the hide. **Brain-tanned** leather, however, is not suitable for modern footwear construction. It cannot maintain lasting tension, has poor abrasion resistance, and offers limited durability compared to contemporary tanning technologies.

Later, around the 17th century, oil tanning evolved into a more controlled process by repeatedly working animal oils, most commonly fish oil, into the hide and allowing oxidation to stabilize the fibers. This method produces what is commonly known as chamois leather, a distinct category characterized by high oil content, exceptional softness, flexibility, and water resistance.

In modern footwear terminology, "oil-tanned" almost always refers to chrome-tanned leather that has been heavily fat liquored or oil-stuffed after tanning. The chrome tannage provides fiber stability and strength, while the oils add flexibility, water resistance, and pull-up characteristics.

The Sustainability of Leather and Leather Alternatives

The leather industry is a global manufacturing sector that transforms animal hides – by-products of the meat and dairy industries as well as dedicated animal stock – into the durable and versatile material widely used in the footwear industry and countless application in other fields, such as apparel, upholstery, automotive interiors, etc. Nowadays, it's far from being a single craft or trade, rather a complex system that integrates agriculture, engineering, and design, operating at scales ranging from traditional workshops that still use processes that take years to finish, to highly industrialized tanneries with quick turnaround.

Each industry's need regarding the leather is unique – for footwear leather have to withstand constant flexing, abrasion, moisture, and long-term mechanical stress, while being pliable and workable, and aesthetically pleasing all at the same time.

Environmental Considerations

As with all human activity, the leather industry has been under intense scrutiny to gauge the extent of its impact on the environment. In response, to comply with modern environmental laws, tanneries have to operate under strict frameworks, employing controlled water usage, advanced wastewater treatment, and rigorous chemical management systems. In many regions, leather production is among the most heavily regulated manufacturing activities, reflecting both its complexity and its environmental responsibility when properly managed.

Synthetic leather

Synthetic leather refers to a broad category of man-made materials designed to imitate the appearance and limited functional aspects of natural leather. These materials are typically produced from polymer systems–most commonly polyurethane (PU) or polyvinyl chloride (PVC) – applied to textile backings. Synthetic alternatives to leather have an established place in the footwear industry, not only because of their lower cost, but also because, for certain types of footwear, they are simply more suitable. While they offer advantages such as uniformity, predictable thickness, and lower initial cost, their performance, aging behavior, and repairability differ fundamentally from natural leather.

"Vegan leather"

As the words "synthetic" and "faux" have become ever less appealing to be associated with leathers, the past few decades have seen the rise of the "vegan" leather. While these products often have a plant-based component, these materials are predominantly petroleum-based plastics. Their sustainability claims often focus on small amounts of bio-based content while ignoring issues that stems from manufacturing plastic material. From a materials perspective, their environmental footprint is not remarkably better than that of other plastic-based consumer products.

Conclusion

Both leather and man-made leather alternatives' production leave a footprint on the environment. It's almost impossible to have a clear-eyed view of the exact statistics, as an objective, global evaluation is non-existent. On a personal level, when making choices, it's always safer to err on the side of longevity, repairability and natural fibers. Loud "green-washing" claims can lull the consumers into a faster fashion cycle, overestimating the offset of their consumption. In a perfect world, the products are ethically sourced, used for as long as possible, and in the end, can go back to the earth.

Other supplies

Threads

Threads play a critical role in upper making. In addition to their visual impact, they are essential for durability, holding components together over years of wear while resisting flexing and abrasion. Unlike most apparel or accessories, footwear is subjected to constant bending and concentrated stress points. For this reason, thread stability - especially in high-stress areas - is of fundamental importance.

Historically, linen thread was the first choice for makers for obvious reasons, but nowadays the industry is using synthetic ones, thanks to their strength and resistance. **Polyester** is the most common, offering a balanced combination of strength, elasticity, abrasion resistance, and dimensional stability under repeated flexing. **Polyamide** (nylon) provides very high tensile strength and elasticity, but its greater stretch requires careful control in upper construction.

Core-spun threads, typically a synthetic filament core wrapped in staple fibers, combine high strength with a traditional surface appearance and are often used for visible seams. **Bonded threads**, usually polyester or polyamide with a resin coating, improve abrasion resistance and sewing stability, especially in high-stress areas.

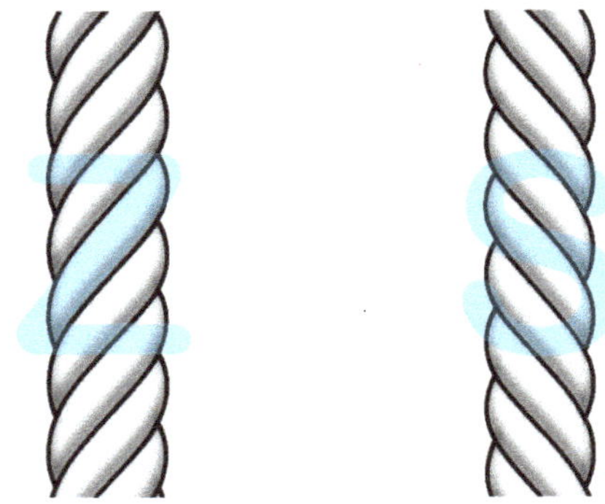

"Z" twist "S" twist

The two standard spin directions are ***Z-twist*** and ***S-twist***, defined by the slope of the fibers when the thread is held vertically. Z-twist threads are generally preferred for industrial machine sewing, as their twist aligns with the rotation of most sewing machine hooks and prevents untwisting during operation. Using an incompatible spin direction can lead to irregular stitch formation, reduced strength, and premature seam failure.

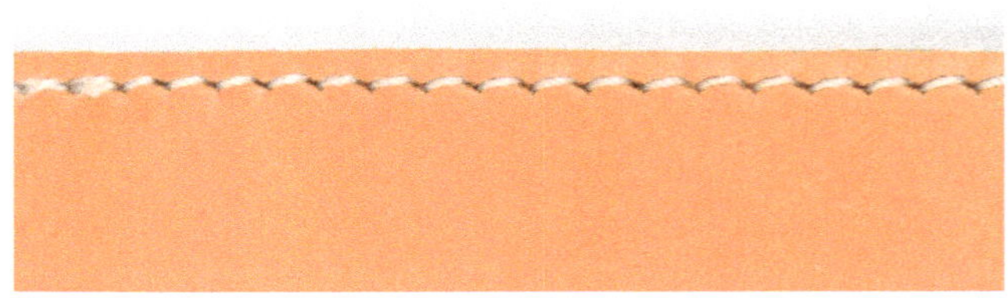

Elongation and elasticity define how a thread reacts to load and repeated flexing in an upper. A thread with too little elongation may break under stress, while excessive elasticity can allow seams to deform and permanently change the shape of the upper. Footwear uppers require controlled, predictable stretch so the seam can move with the leather without creeping or loosening over time.

Tapes

Footwear tapes are often overlooked components of the footwear uppers, yet they play a critical role in both the manufacturing process and the long-term performance of shoes. Tapes contribute to structural stability, dimensional control and production efficiency. While usually not visible in the finished product, their influence is crucial.

The **reinforcement tape** is applied to stabilize areas of the upper that are subject to stress, stretching, or deformation, such as seams and other crucial parts. They help control leather stretch, maintain clean lines during stitching, and reduce long-term distortion during wear. Reinforcement tapes can stabilize skived areas which are exposed to high stress during lasting and/or wearing.

The **topline tape** is a special narrow reinforcement tape for, as the name suggests, toplines. Thanks to their narrow shape and special weaving they can follow curvy toplines and provide stress relief for folded edges.

The **backline tape** is specifically used under reverse seams. While their tear resistance is not as impressive as the reinforcement tape, they can hold the rubber seam in center and provide structural support.

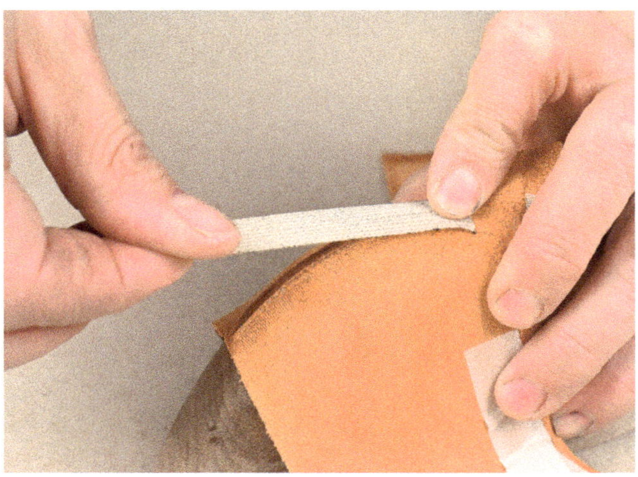

The backline tape is an essential component to create quality footwear, providing support for reverse seams on the backlines

The **lacing reinforcement tape** is a non-woven, wider tape under lacing. This is an extremely strong, tear resistant material for withstanding the constant stress of the shoe laces and provide extra strength to this area.

Double-sided adhesive tapes of various widths are widely used during upper assembly to temporarily hold components in place before stitching. They are particularly useful for positioning facings, straps, piping, and dog tails. This method improves accuracy, reduces handling time, and keeps the work and surfaces clean.

Seam-sealing tapes are applied over seams on waterproof shoes to guarantee water resistance.

SECTION TWO

4 THE MULE

Now that the basics are established, the next step is to apply them to the construction of actual uppers. This section begins with a very simple model: the mule. The classic mule has a long history spanning centuries and continents. It's been the emblematic footwear across classes, from ordinary men and women of the Middle Eastern cultures, to Roman patricians to the French nobility in the 16th century. Maybe it is because of the simplicity of the style that makes it so versatile which can be applied for a range of heel heights and for a variety of occasion. Consisting of only a single piece of vamp, the mule's pattern provides the perfect entry point for our dive into pattern and upper making. The higher the heel, however, the more challenging the vamp axis becomes. In some instances, crimping (pre-shaping) the vamp becomes inevitable. In this chapter, we will discuss the relevance of the main folding axis, allowances and an alternative solution to crimping.

This is the very first pattern in the book – and the first upper we construct – so we will keep the process intentionally simple. The geometric method wouldn't offer any real advantage, so we'll reserve that for later chapters. For this design, a pattern composed of a single topline and a well-chosen folding axis will be just as effective.

The Axis Placement

The mule is the first pattern discussed in this book, presented from initial layout to finished pattern. The model is relatively simple, using a single axis, and in some versions the same structure can be repeated for the lining patterns. This makes it well suited for introducing axis placement and pivoting. Generally speaking, the axis should be positioned to provide the highest possible accuracy for the pattern, preserving all details and proportions, including lengths and girths.

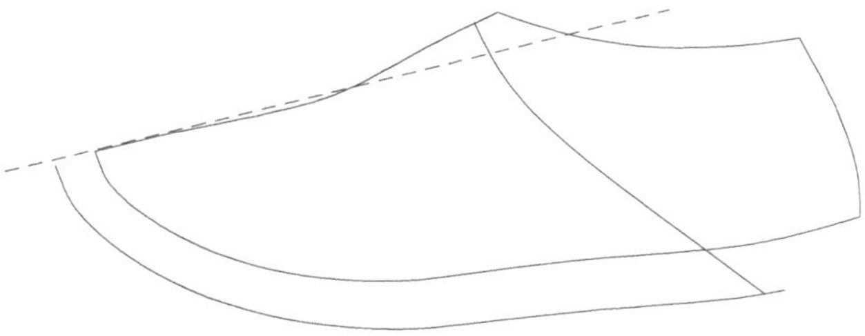

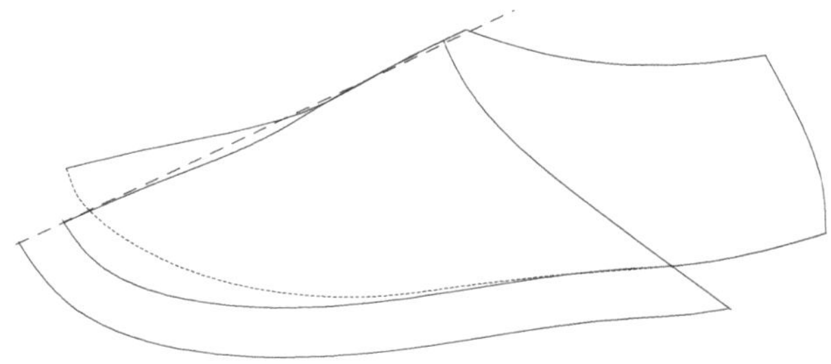

Using the front section of the last as the basis for the main folding axis is a perfectly acceptable and functional solution for a low-cut shoe. However, a mule has a higher topline, which must be taken into account.

Using the top section over the vamp point is also incorrect, as this axis would remove a significant portion of the front when the pattern is unfolded. Although this can be corrected by pivoting, doing so in this case would introduce additional complications.

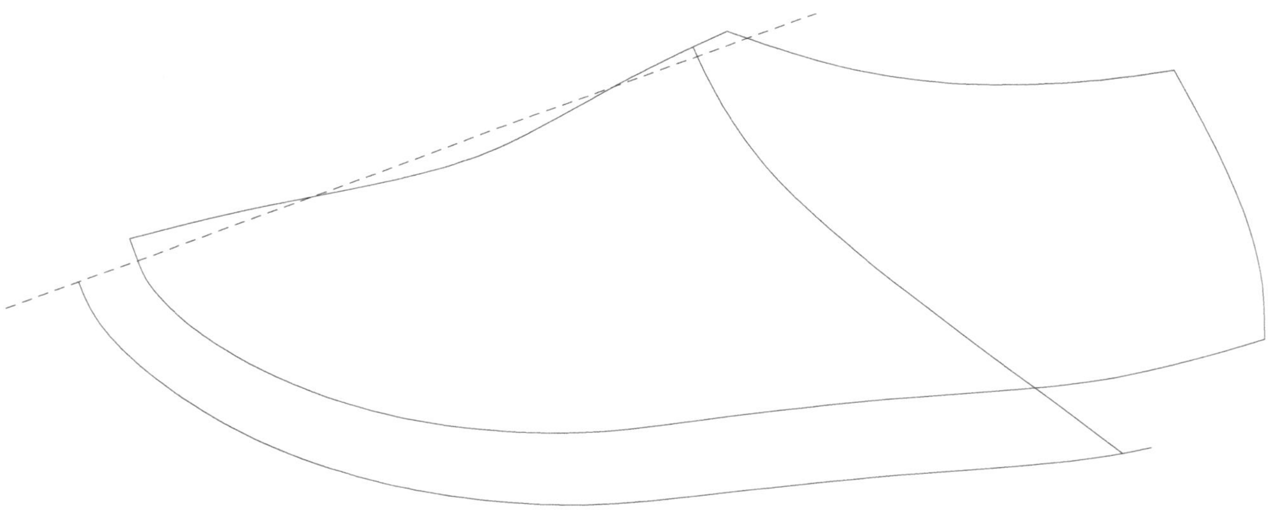

Drawing the axis along the entire curve, as close to it as possible, is the most promising method, as it avoids additional shaping or the introduction of extra pattern pieces or seams. The exact approach to determining the optimal line also depends on the leather. A stretchier, more pliable upper and lining leather combination is more forgiving and generally produces better results.

Vamp Pieces

The vamp, unfolded by the main folding axis. This very pattern can be repeated just like this for the lining pattern.

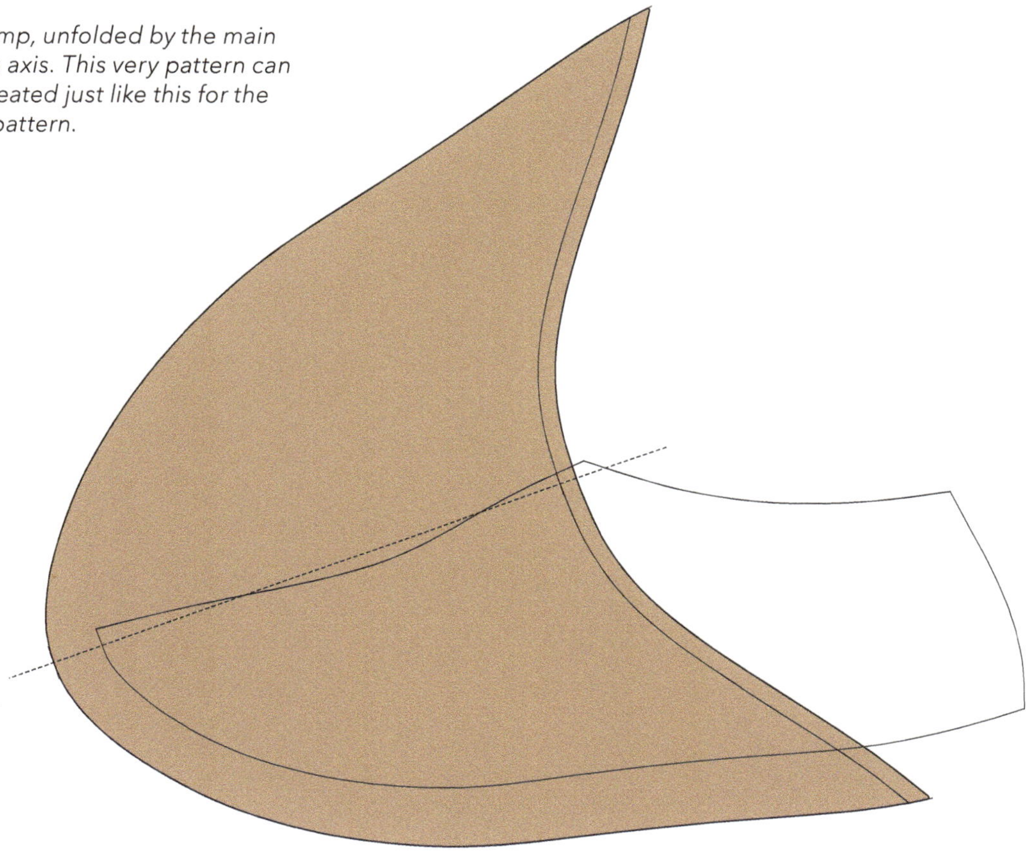

The mule pattern is relatively easy to create using the previously established axis. As this construction is forgiving, asymmetry is not discussed in this case. Greater accuracy is required only if slip-lasting technology is used, which serves as a clear example of why the pattern maker must consider the soling method when developing an appropriate pattern.

For this pattern, the heel portion of the insole is not covered, so a sockliner pattern is required. As this is an entry-level example, these details are not discussed further.

The lasting allowance always depends on the technology and the material. In this case, a uniform 20 mm (3/4 in) allowance is used all the way around.

Mule With Higher Heel

As heel elevation increases, the curvature of the front line becomes progressively more pronounced, which may require more advanced techniques. It is difficult to define a clear limit – if one exists at all – since feasibility largely depends on the properties of the leather. There are several ways to address this issue, allowing a functional upper to be developed while preserving pattern details such as a smooth topline. These solutions are widely used in the industry and are selected based on the equipment available in a factory and the technical approaches it can implement or is willing to adopt. Material choice is also a key factor when determining the most appropriate solution.

An important note: even though this footwear has a higher heel, the pattern is not affected; therefore, the heel elevation is not relevant in this case.

This portion of the pattern lies outside the mirrored area, above the axis, and therefore will not be part of the pattern piece, as the axis cuts it off. The piece is not simply missing from the vamp; instead, the mirrored lines meet at an acute angle. An axis configured in this way is not very practical.

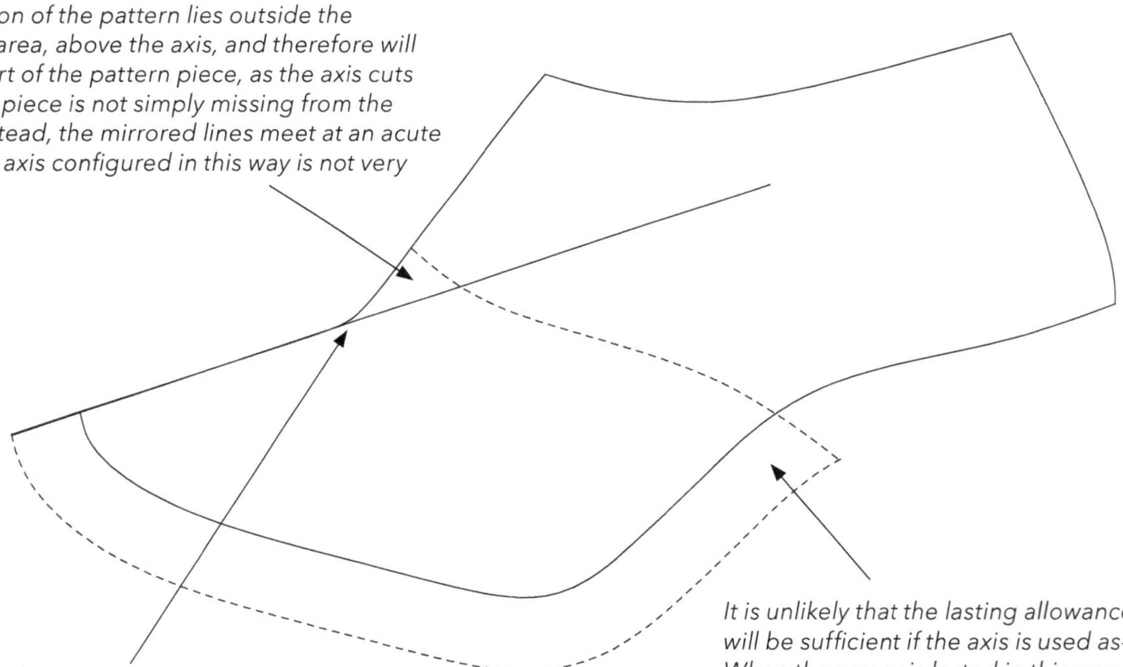

This design might remain a wishful thinking as it is very possible that the curvature will not lie on the last as intended. This will result instead in unwanted pleats.

It is unlikely that the lasting allowance will be sufficient if the axis is used as-is. When the upper is lasted in this way, the missing middle section – meaning the upper is smaller than required – will be pulled upward from this area.

In conclusion, this axis placement introduces additional challenges and leads to an unpredictable outcome. While it may be feasible with very stretchy leather, it is clearly not an ideal approach for sound technical design.

Solution no. 1: Using the Average

Using the average means placing the axis so that the overall difference between the axis and the front curve is minimized. Unsurprisingly, there is a highly developed mathematical theory behind this concept, but applying it formally would be excessive for the purpose of defining an axis. In practical industrial pattern making, a trial-and-error approach is the most effective solution.

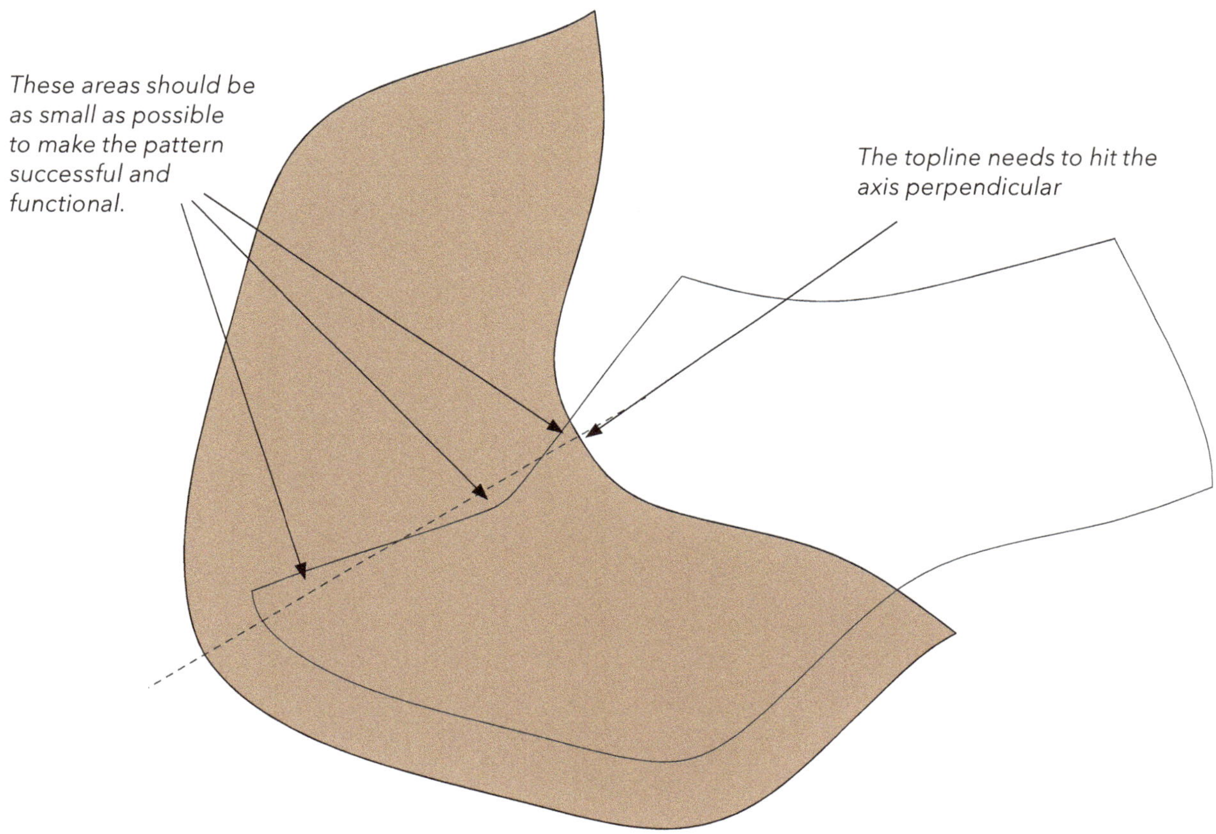

This method inherently involves compromises along the entire curve, resulting in minor deviations at the bottom, middle, and top. The only area that typically requires adjustment is the topline. Even if the line slightly cuts in, the pattern can still function correctly. The small amount of "missing" girth is compensated for by material stretch and, at the bottom, by the lasting allowance.

By definition, the topline is perpendicular to the front curve, but it may no longer be perpendicular to the newly placed axis. This must be corrected; otherwise, the topline will not appear smooth and continuous. This is a very common adjustment in pattern making. In practice, whenever an axis is drawn, the last-copy reference lines usually need to be modified to align correctly with it.

Solution no. 2: Two Axes

If the design allows the use of two pieces, dividing the vamp and using two different folding axes is a very effective and accurate way to address the problem. Assembly may be slightly more challenging, as one of the pieces must be forced to match the other. As a rule of thumb, the smaller the allowance, the easier the assembly. With this in mind, the allowance can be reduced to a minimum of 5 mm if necessary, particularly with firmer, harder leather.

The lining can be constructed in the same way or by using almost any of the solutions discussed here.

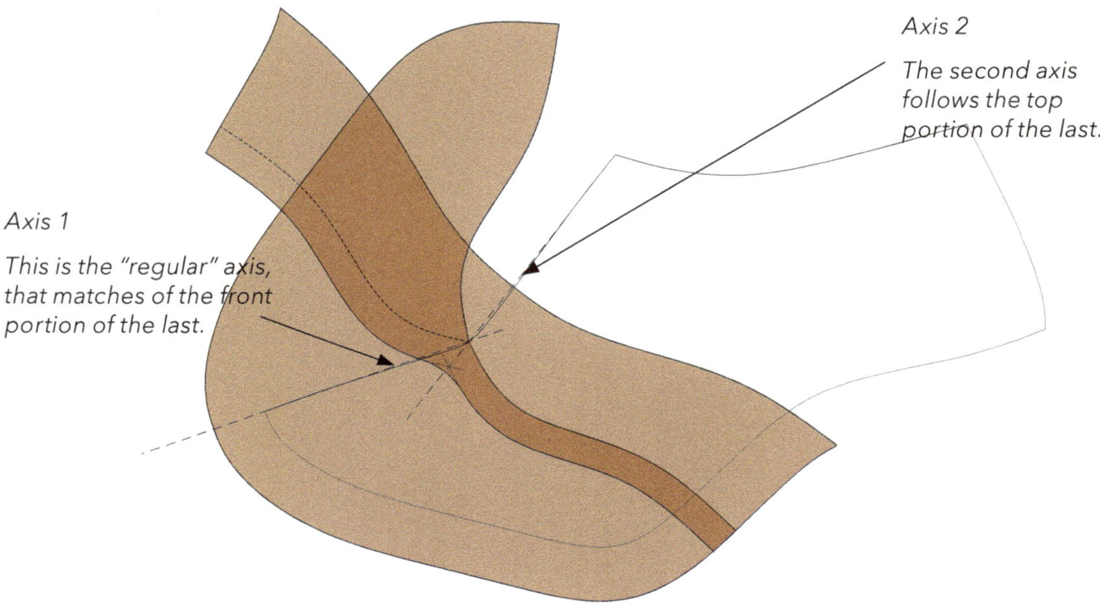

This is how the he bottom portion of the vamp looks like based on Axis 1.

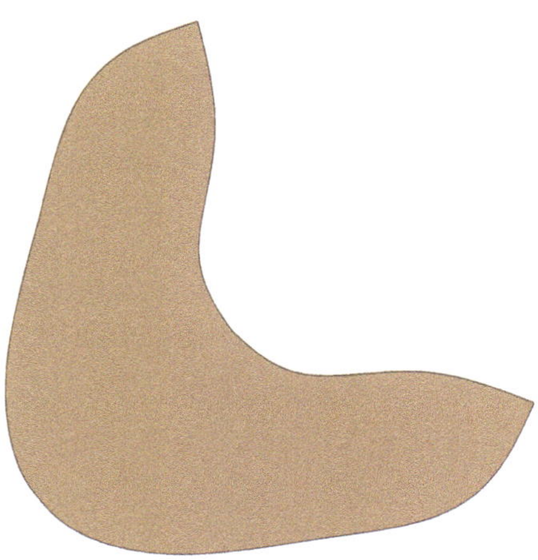

This is how the top portion of the vamp looks like based on Axis 2.

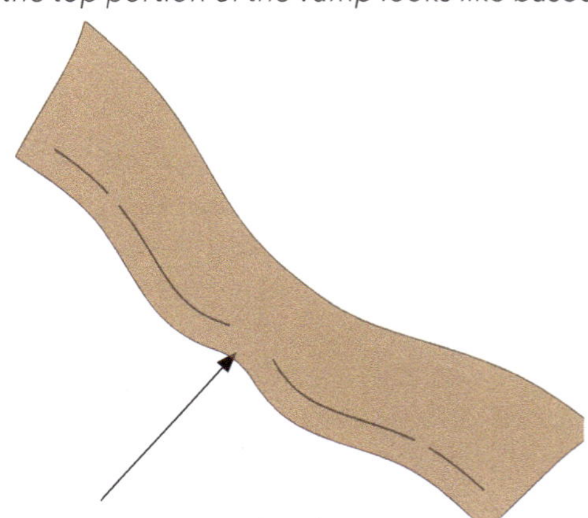

Solution no. 3: Altering the Pattern

A small cutout can eliminate the overlapping problem very easily. When this pattern is lasted, the cutout will open as much as necessary, becoming much wider than the flat pattern indicates. For clarity, a simple axis dictated by the front of the last is used here, but it is not fixed and can be modified slightly.

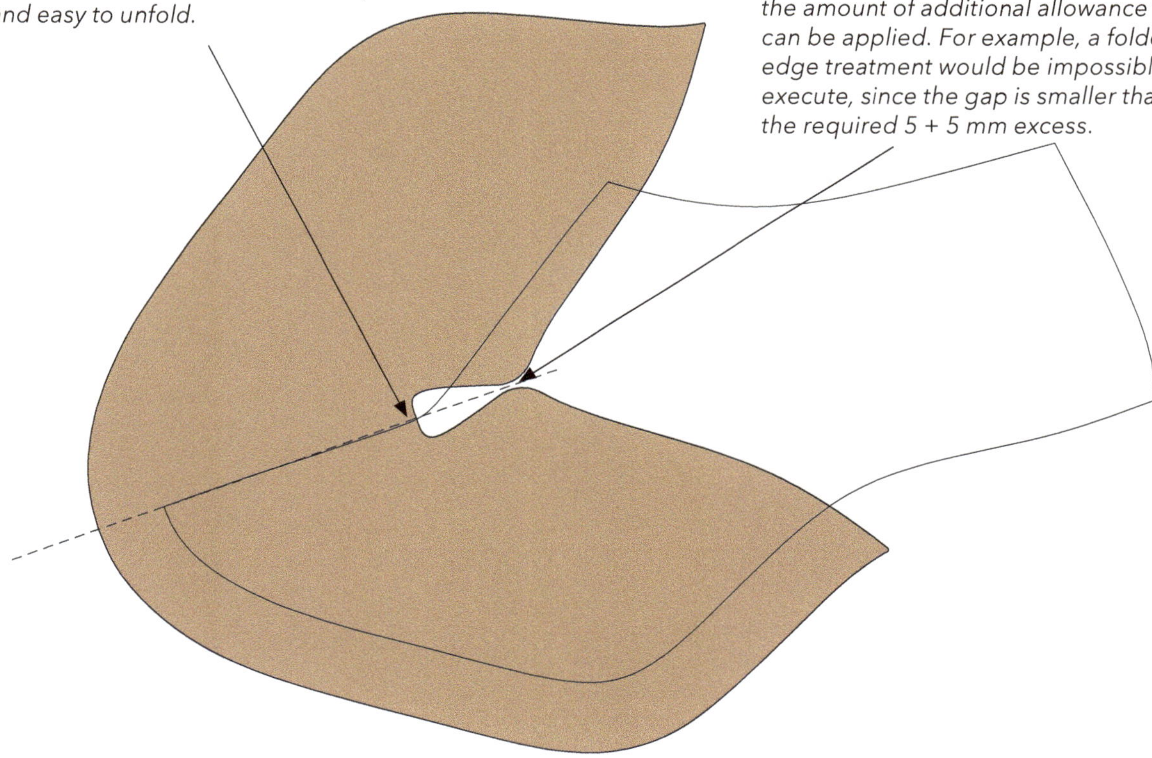

Why would this be done? One obvious reason is edge treatment. If, for example, the technical specification requires a folded edge, at least 4 mm is needed. In this case, there is not enough space on the pattern between the design line and the axis (or 8 mm between the design lines). Tilting the axis upward slightly would resolve this issue. In practice, with such tight curves, a folded edge may not be the most suitable choice.

Solution no 4: Medial Insert

This solution is very common in boot patterns when crimping is not an option for technical reasons or when the material itself does not allow it. From the lateral side, the footwear appears to be made from a single piece, while the insert enables the required curvature. This method is widely used for Chelsea boots and similar models. Although it's not considered a sophisticated solution, it's effective and saves time while reducing the risk of errors.

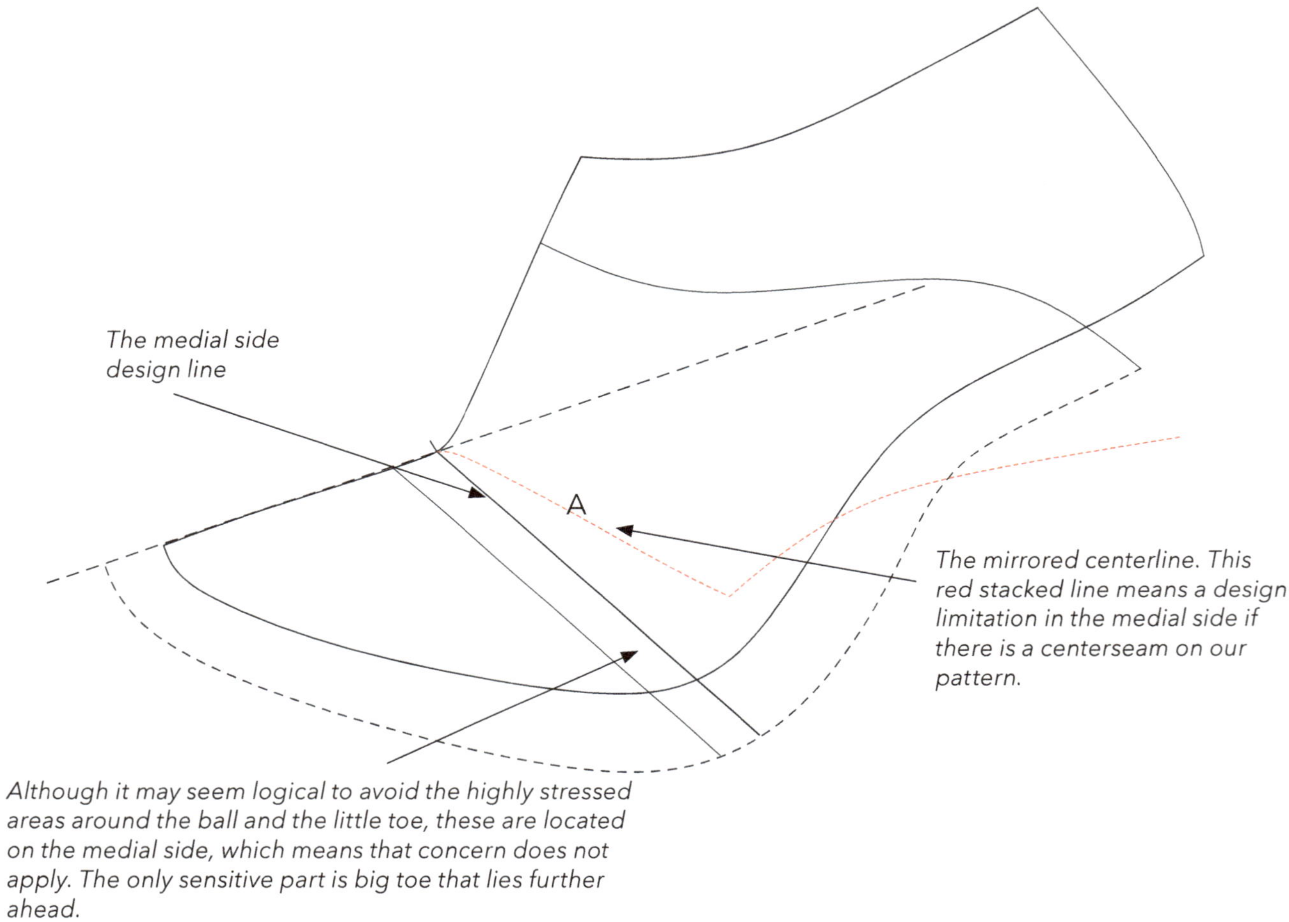

Above is the master pattern with the intended appearance. The stacked line (A) indicates the centerline of the pattern; this line defines the limit beyond which the insert cannot extend (otherwise it would go to the other side). Crossing it would cause the pattern piece to overlap itself. A line is drawn anywhere between this centerline and the toe – ideally not too far back, in order to preserve a clean front appearance – and then mirrored to the opposite side. As long as the line does not cross the centerline, the construction remains feasible.

Unfolding the piece will result in a pattern piece like this. Opening the slit (or just turning down the medial side, where it should be) gives a chance to sew in the medial side insert, so it will fit the last perfectly, has the desired curvature and gives the illusion from the lateral side, that it is one solid piece.

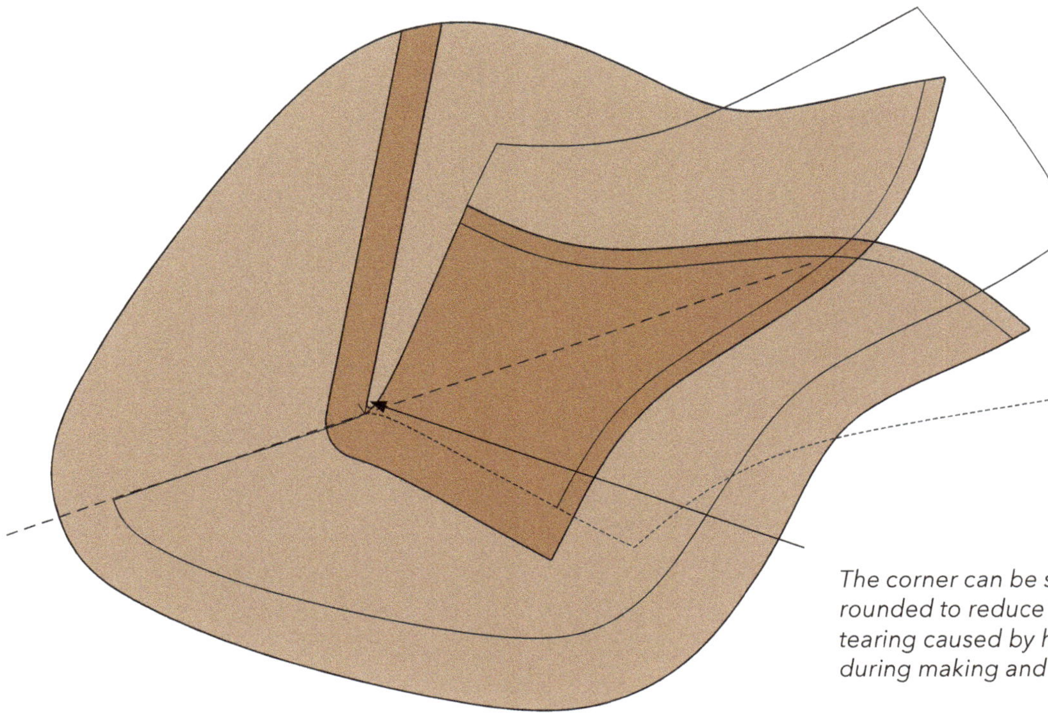

The corner can be slightly rounded to reduce the risk of tearing caused by high stress during making and lasting.

This piece is mirrored to the opposite side and sewn together along the center.

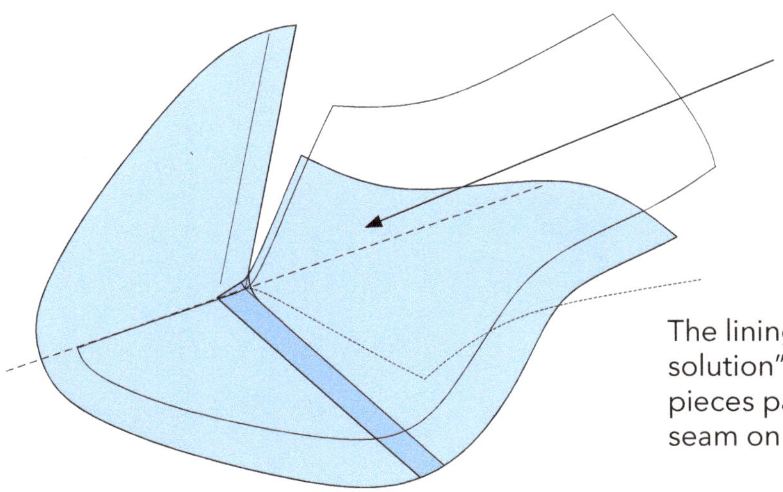

The lining can be made similar with a "2 axes solution", discusses beforehand or with a 3 pieces pattern like this, using a reverse or silk seam on the front.

Solution no 5: Centerseam

This solution is simple and very accurate, but not all materials allow it. It uses a precise center seam along the front line. The most common choice is a reverse seam, but a butt seam or any other method that securely joins the two sides is also possible. Edge finishing and lasting requirements may impose some limitations.

Technically, this is the simplest solution, which is why it is often used in lower-priced footwear and for fit testing, where time-consuming crimping can be avoided. The lining can use any of the available construction methods.

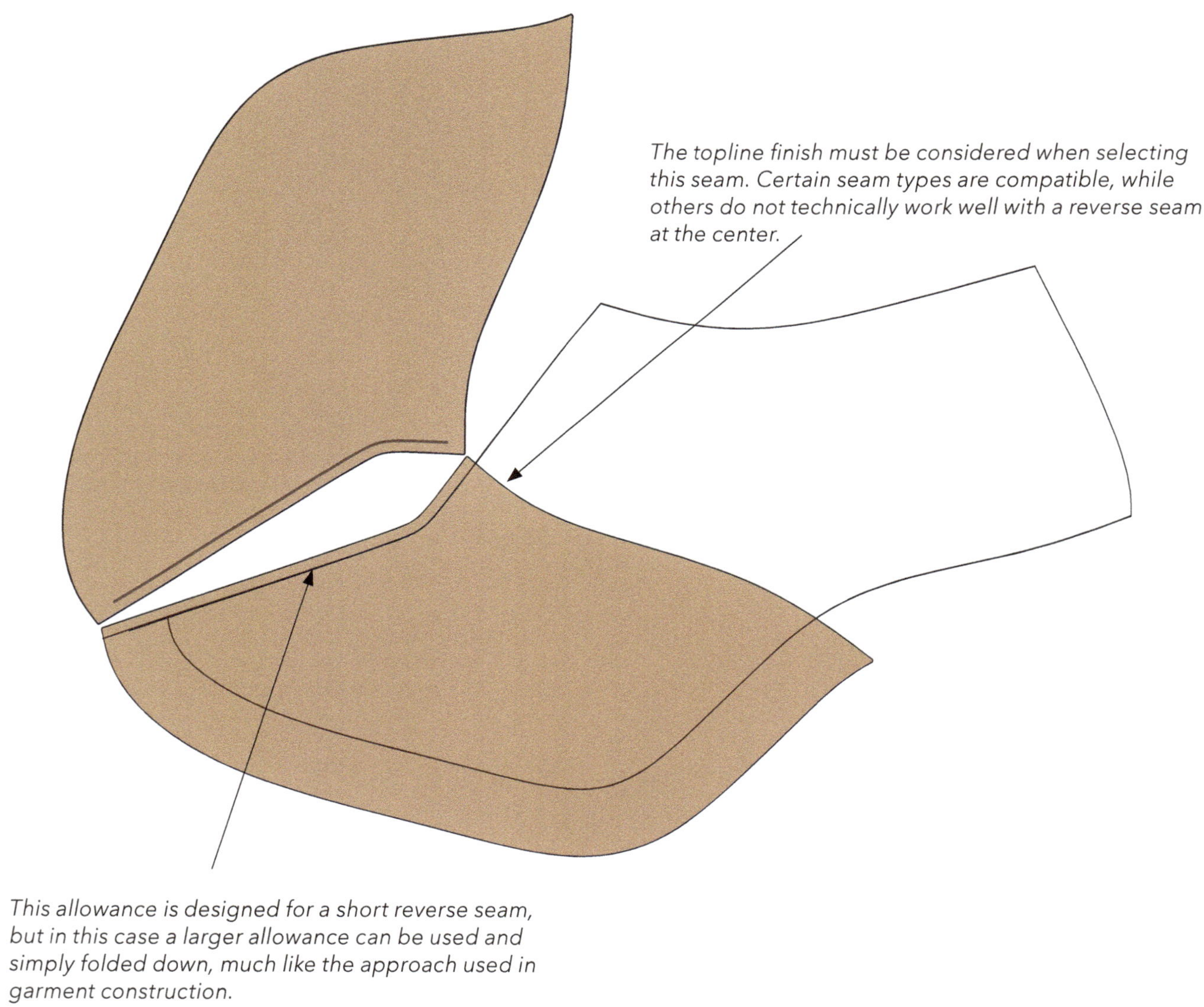

Solution no 6: Pleating Solution - The Shelby Mule

Pleating is a creative way to deal with the two inch heel problem. We create pivoting points on the bottom of the lasting allowance, keeping it the same distance and "open the topline. This method obviously will elongate the topline and this is what make the pleats possible.

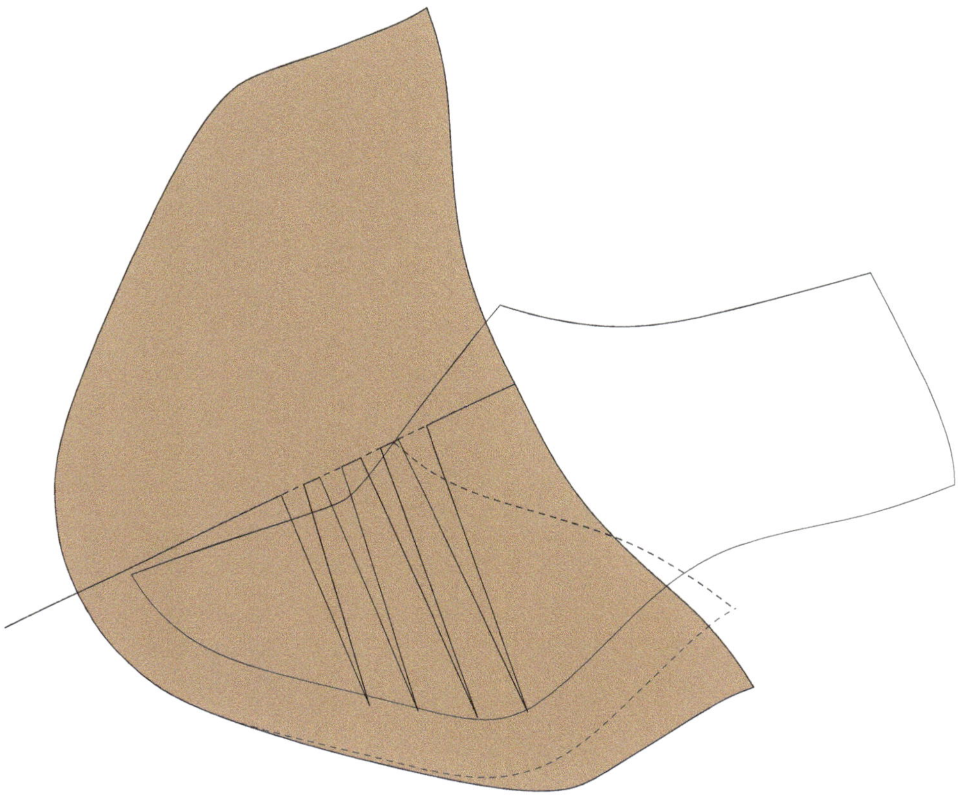

This is a good example of resolving a problem without creating technical difficulties, instead approaching it elegantly and producing a result that is both remarkable and aesthetically pleasing. Adding these pleats may not be possible with every type of leather; the material must be pliable and sufficiently soft. During the process, the maker has control over the formation of the pleats. Importantly, the lining pattern does not follow this structure but instead follows the previous one–the center seam, for example, aligns very well.

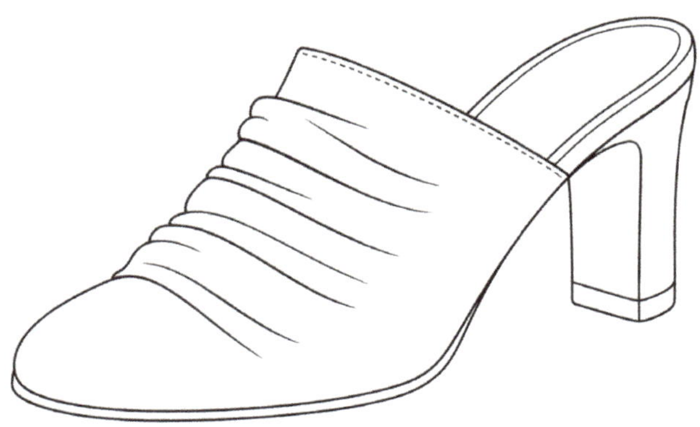

Assembly order

1. Clicking, skiving
2. Turn gran side to grain side and sew the topline
3. Cement the topline and turn

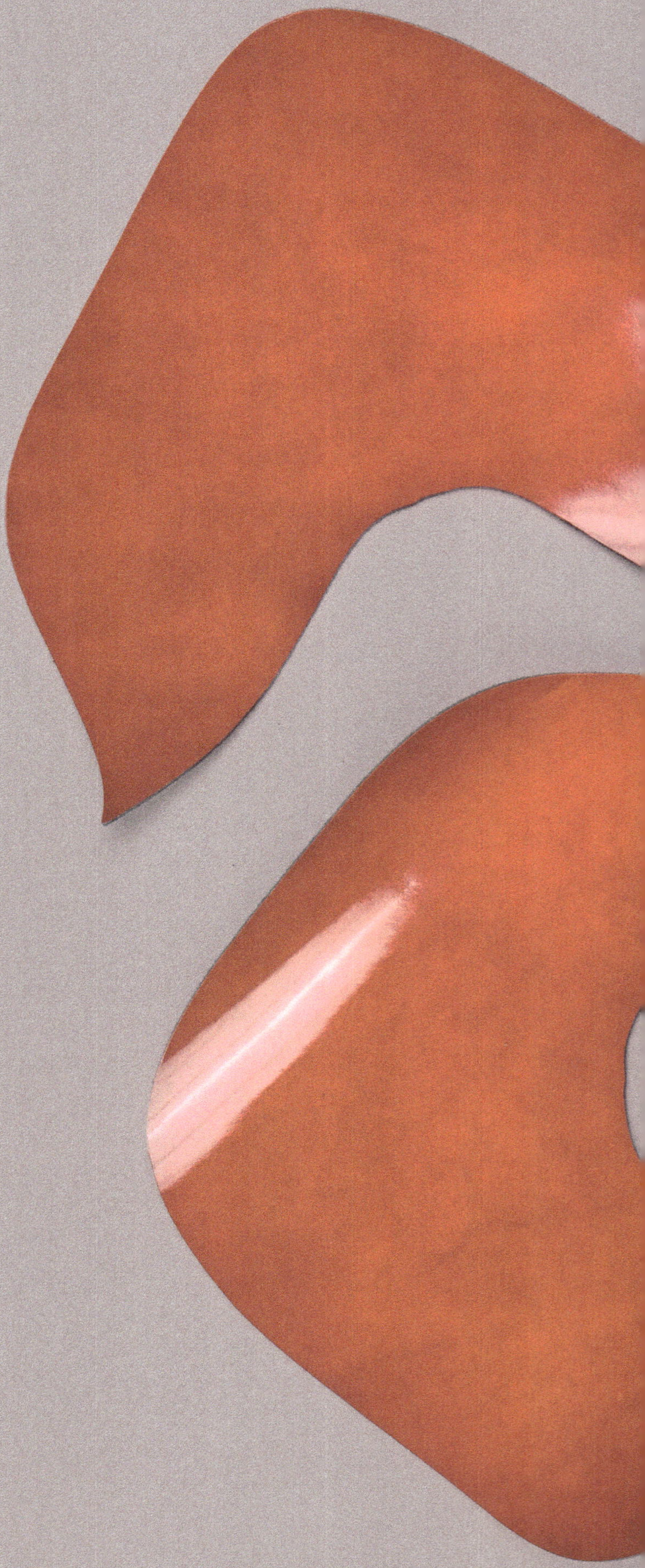

Skiving

Skiving for a turnseam has very specific parameters, depending on the kind of look we want to achieve. In most cases, the weaker material tend to do the turning. But in this case, as mentioned, the thicker upper material is expected to do the turning thus protecting the sewing, the skiving needs to happen accordingly.

An experienced upper maker can execute this skiving without much test and adjustment, but learning the trade, working with new materials, it's prudent to do small material swatches, as discussed in the ***Seams and edge treatments*** chapter. Following some trial-and-error, make sure to make a note of those skiving parameters. The upper needs a little longer skiving since the lining can stay intact or only very lightly touched.

Skiving lining can be tricky. Make sure the skiver is very sharp in order to deal with the soft, thin leather.

Numeric control skivers, such as the Comelz SS20 is able to repeat the same accurate result by saving the skiving parameters. If a shoe design moves into production, achieving consistent results every time is critical. While this level of consistency is technically possible with a manual skiver, computerized machines make the process significantly easier and more reliable.

Assembly

In order to hide the stitching and expose the grain sides of the two layers after turning, the sewing needs to be done with the flesh sides facing out. This type of turnseam is a relatively simple compared to some other, more challenging ones in the industry. The pieces are lined up exactly for the sewing process.

The sewing is relatively easy, which makes this a perfect practice model. Keeping an equal distance is not only important because a smooth edge is desirable, but because varying thickness when turning will amplify any unevenness.

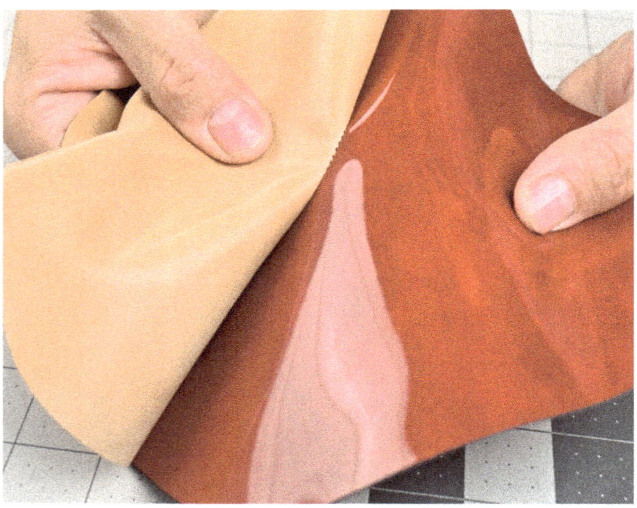

Once the stitching is done, it's worth to stop for a quick second and acknowledge that this seam is vulnerable in this phase, therefore it must be handled with great care. Check if the stitches caught both layers before continuing the work.

Not all materials need trimming. For instance, those that are prone to fray need more excess to stay structurally sound, as well as symmetrical turnseams and padded edges. In this case, however, trimming will yield a more even and tighter turn.

Following trimming, a light brushing of glue will ensure the seam stays as strong as possible and in place after turning.

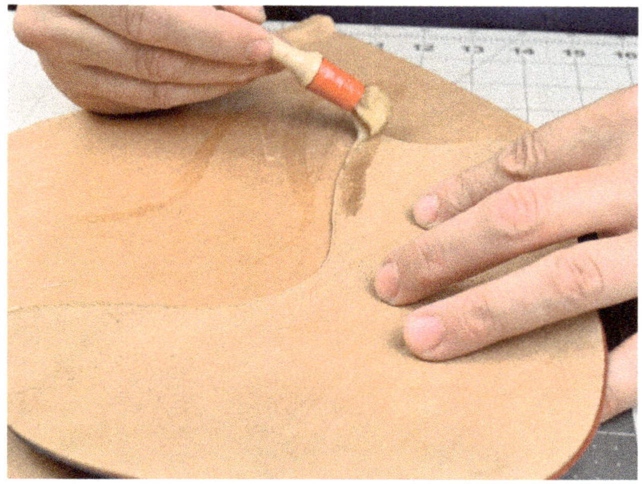

Add glue of 10 mm width on the edge, on both sides. For this very seam, I suggest not to wait the full drying time. Normally, a strong bond would be a great feature, but it might cause problems during turning.

It is very important to start the turning at the center. The upper layer is tightly folded, with the lining laying slightly under the top edge. The desired look here is very similar to a folded edge but the lining is not exposed and without a visible stitchline. For some shoe style, this turned edge is part of the look.

Where it is needed, a pointy tool, such as the folding hammer's awl tip can help push the material to its proper place. Pointing to the desired turning point, in this case the seam on the upper side, and gently turn the lining. This will help to achieve the even tension all the way through the whole length of the topline.

With the topline almost finished, the turnseam is secured. A wider glue layer helps to make sure it stays this way.

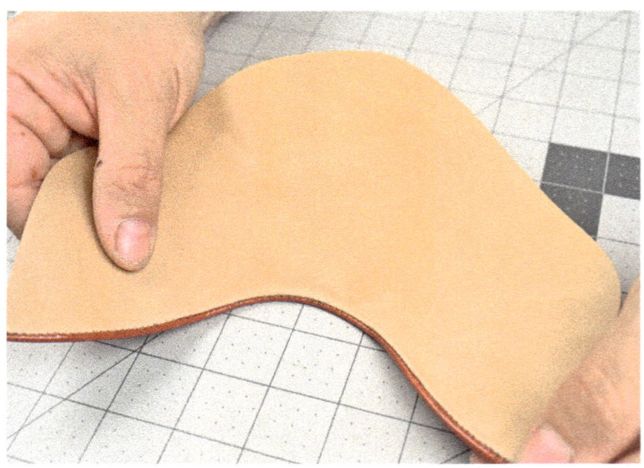

After the glue has dried, push the layers together, use a gentle hammering and our topline is ready.

With that, the uppers are done, ready to be lasted. The beauty of this simple, low-cut design is that the topline hardly passes the vamp point therefore eliminating the need to preshape the upper, so lasting will be perfectly enough.

5 THE BALLERINA SHOE

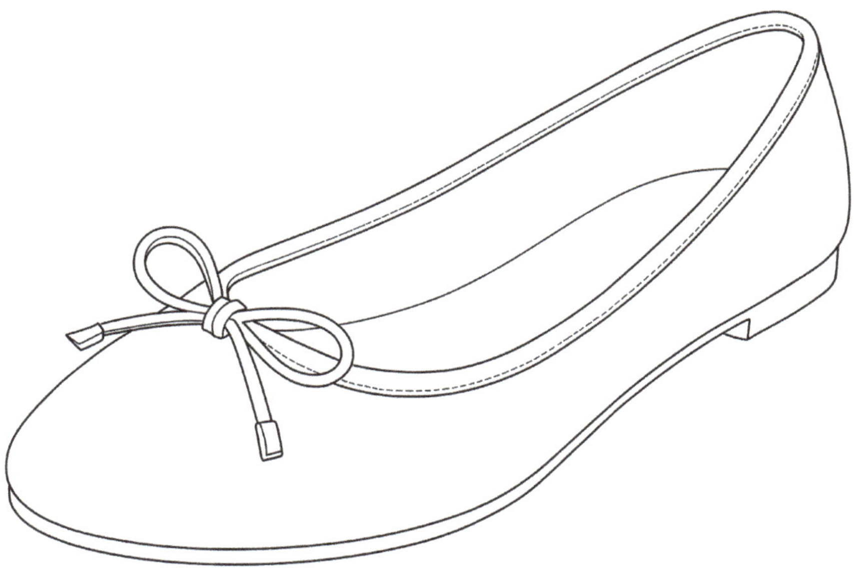

In this chapter, we continue our studies with a simple ballerina shoe, also called a ballet shoe. The ballerina shoe, as the name suggests, was used by dancers, and originally included heels. In the 18th century, the construction shifted to heelless, flat shoes, which served as inspiration for 20th century designers who came up with a street-ready, casual style. The design was popularized by Audrey Hepburn, who, being a former dancer, often wore this type of footwear in her movies in the 1950s. Since then, ballerina shoe has become a staple item for the consumers. The low heel, rounded toe shape and minimal decorative detail make it versatile for both formal and casual occasions.

The pattern for the ballerina is pared down enough to allow us to broach a number of new topics such as reverse seam, folding allowance, as well as how to deal with asymmetry with pivoting, without much distraction. And unlike the mule, the lining pattern starts to deviate from upper, most notably due to the orange peel-effect. These principles will be recurring considerations going forward as we tackle ever more complex constructions.

The Geometric Approach

The geometric method is a great tool for designing a ballerina pattern, although the work can be done entirely without it by relying on a few key measuring points. The first – and arguably the most important – is the counter point. The reason is simple: footwear is only functional if it supports the basic mechanics of the human foot. One of these is walking. A ballerina shoe is essentially a half-coverage model, so slipping off during gait is a real concern.

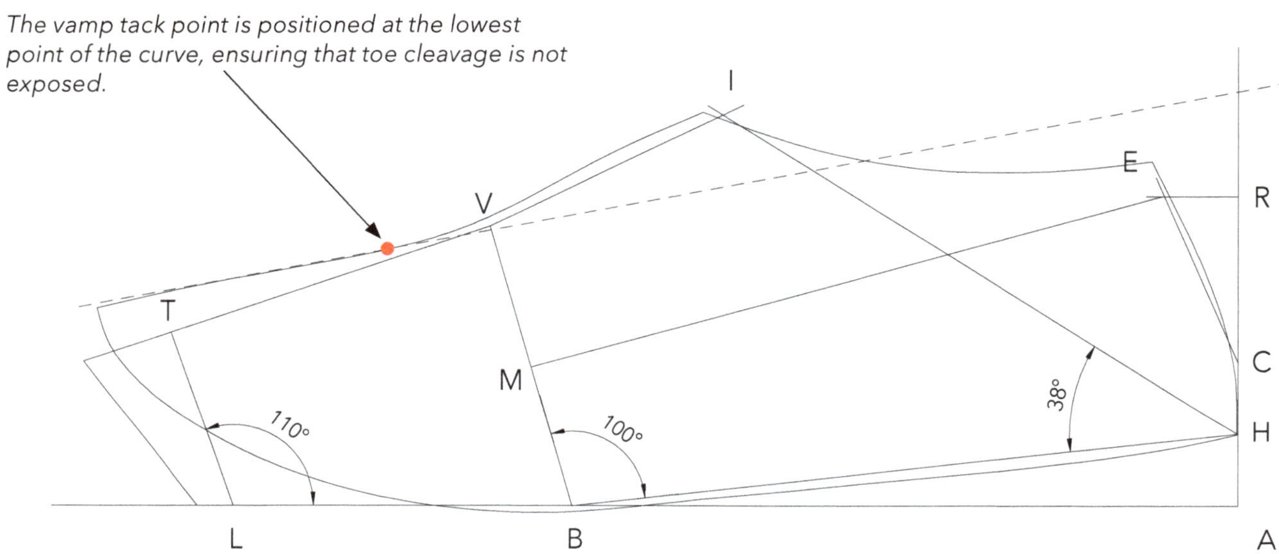

Pattern engineering has several ways to address the counter point. Makers resort to elastic toplines; specially shaped lasts with very narrow top sections so the counter can grip the foot; post-production shaping to tighten the counter further; suede-side linings, or other high-friction synthetic counter materials, etc. But none of these solutions will perform well if the counter is in the wrong position. Even shoemaking traditions that shorten the counter slightly compared to Knöfel's original formulas never reduce it by more than 4-5 mm, precisely because going lower compromises hold.

The other crucial decision is the front cut. Historical pattern books often bring this cut all the way to the vamp point which is a structurally sound choice, but fashion dictates otherwise today. The real question becomes: Do we want to expose toe cleavage or not?

This introduces an important point on the last: the vamp tack. This is where toe cleavage begins. Bringing the topline below this point will expose it; keeping the topline above will hide it. Technically, both choices are equally valid. The decision is purely aesthetic, which means a useful tool for the designer and pattern maker.

The Main Folding Axis

When it comes to choosing the main folding axis – one of the first and most important decisions in the entire pattern construction process – the ballerina offers two equally valid solutions.

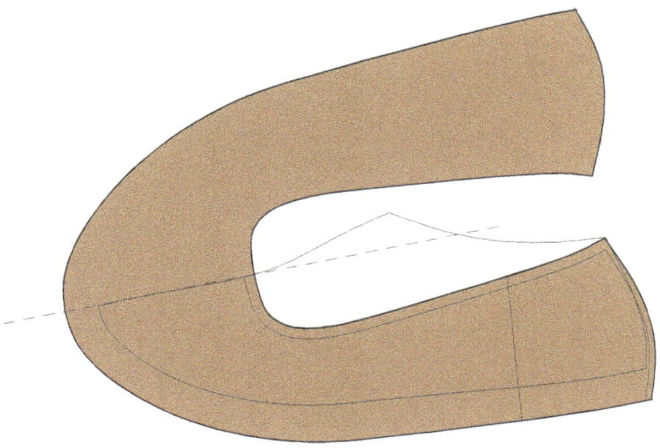

Using the last copy as a base for the axis gives an accurate, perfectly fitting pattern.

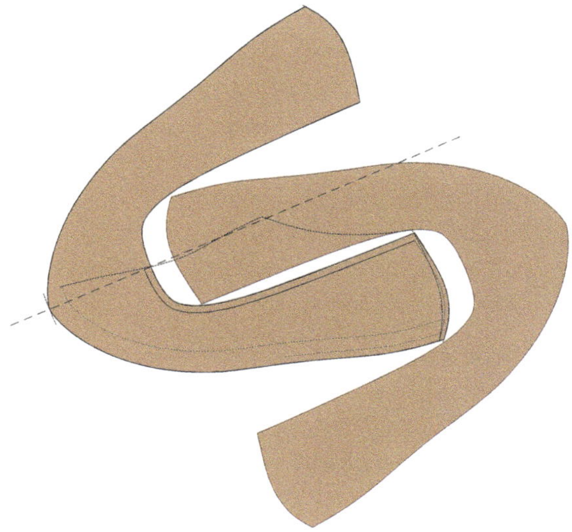

The pivoted axis results in a better, more efficient yielding.

The first option is the last copy line itself. For handmade, custom, one-of-a-kind footwear, this is an excellent choice.

The second option is the pivoted axis. Pivoting the axis allows the vamp pieces to be nested far more economically on the leather, significantly improving material yield. For factory work or any situation where leather optimization matters, this method can offer measurable savings.

Both solutions are technically sound – the decision depends on workflow, production scale, and priorities.

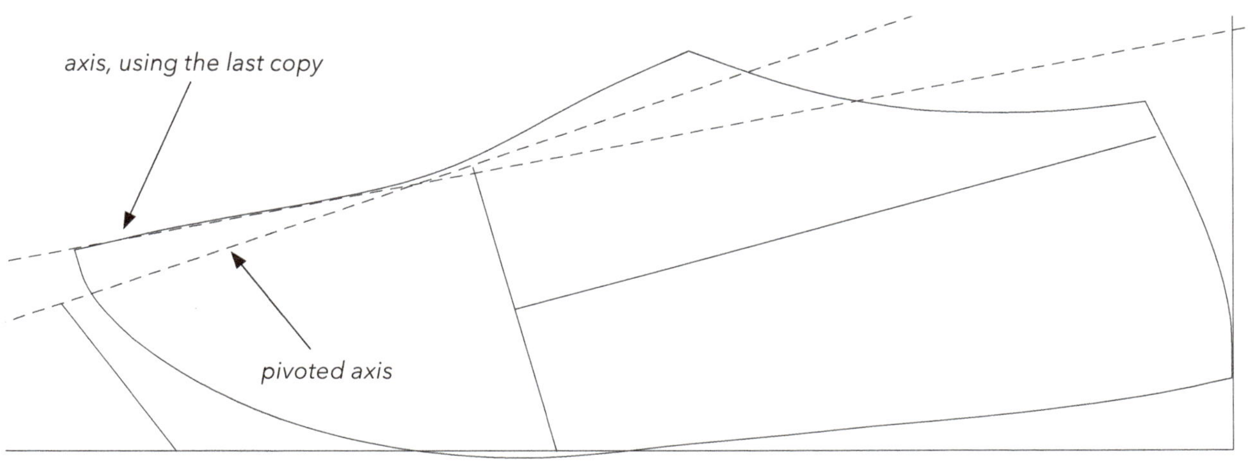

The Topline Position

The topline position– as stated earlier –can vary widely. In most footwear, the topline is one of the most defining elements of the design (along with material, construction method, sole, and embellishments). In some shoes, such as this ballerina model, there are very few additional design features, which means the topline becomes the primary opportunity for the designer to leave a personal imprint on the product. For this reason, the topline is often the most distinctive characteristic of this style.

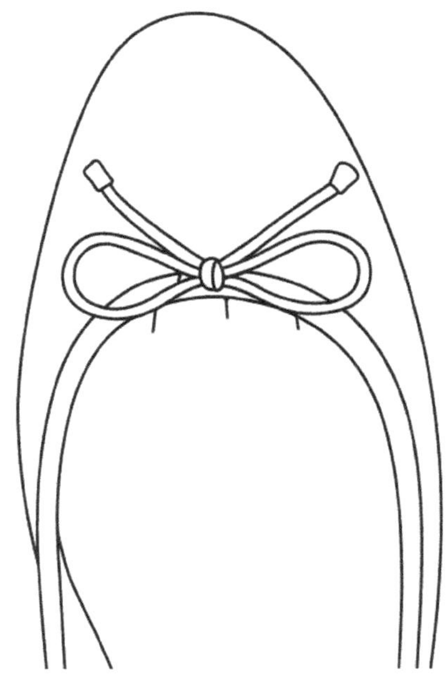

The diagram below illustrates the upper limit where the topline can still function properly and remain within the ballerina category, as well as the point where we will place it. There is no strict formula for determining this exact position; it depends largely on functionality. As long as the topline securely holds the toes at the front and maintains a pleasing appearance, it is acceptable.

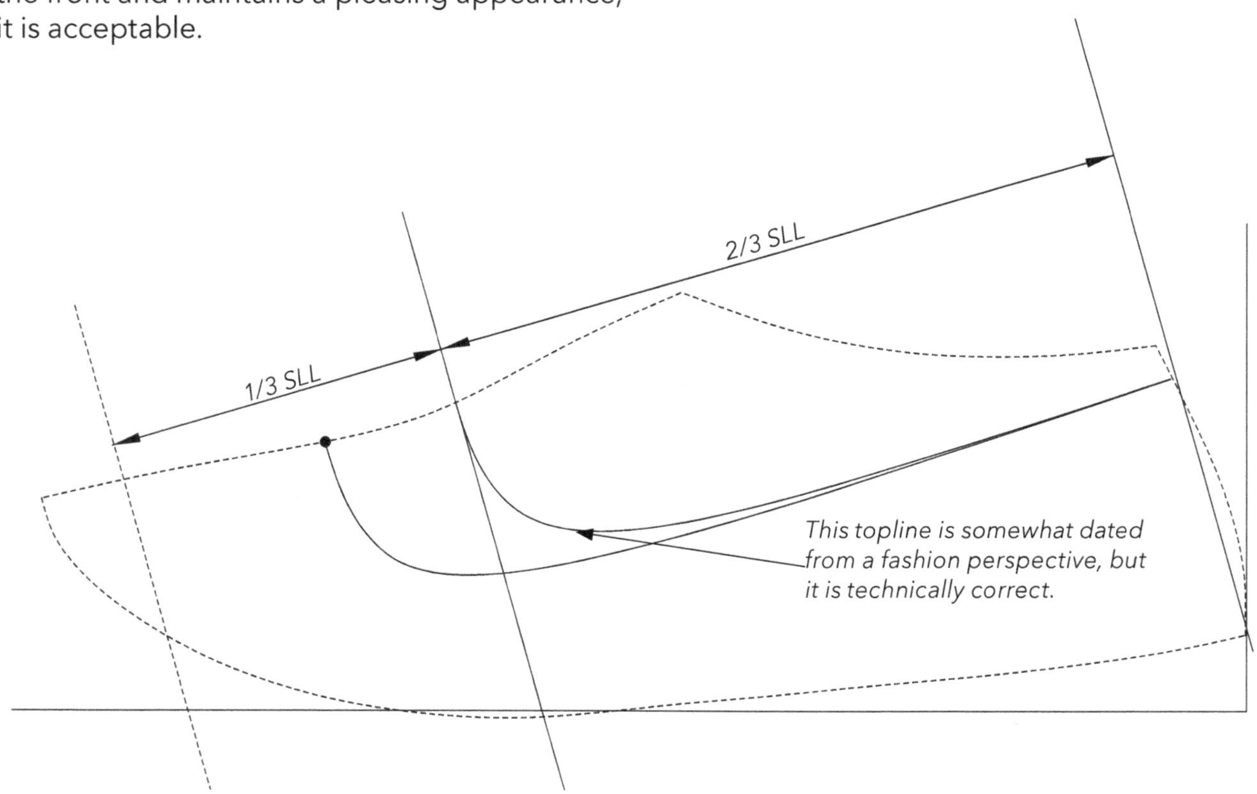

The Lasting Allowance

The lasting allowance is always determined by the chosen construction method. The same amount of excess will not work for every technology, so it is essential to use the appropriate allowance for the specific process. The values shown here are suitable for hand-lasting with a relatively thin insole, a cemented outsole, and a medium-stretch leather.

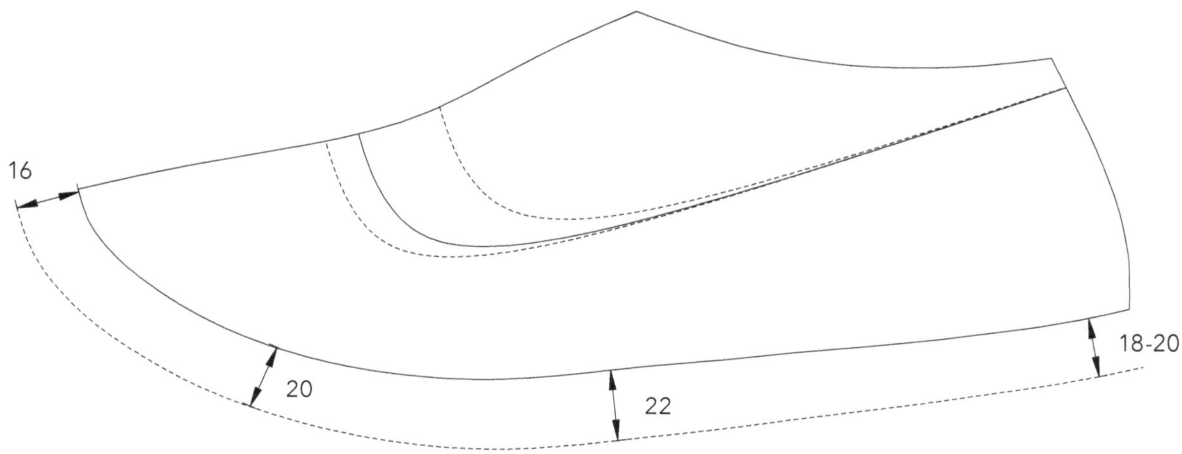

The Backline Adjustments

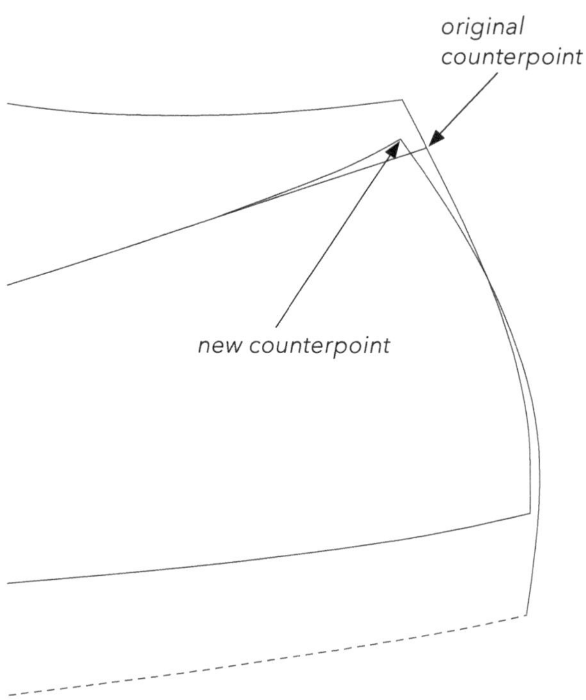

Adjusting the backline is an overlooked and underestimated part of the process, mainly because many pattern makers don't make the shoes themselves and miss out on feedback regarding minor misalignments. While a decent backline can still be lasted, it requires extra effort from the maker. To fix this, my solution is to move the original counterpoint inward and upward by 3 mm, and the topline should be adjusted to meet the backline at a right angle.

This small change results in a taut topline without any gapping. Finally, the bottom line should be shifted outward by 2-3 mm to accommodate the heel counter, ensuring the backline fits perfectly.

Upper Piece

Allowances depend on the chosen seam types. Since a reverse seam is used on the backline, a 3 mm allowance should be added, while the topline requires 5 mm for folding. This same 5 mm margin serves as the trimming allowance, allowing the additional line to cover both purposes.

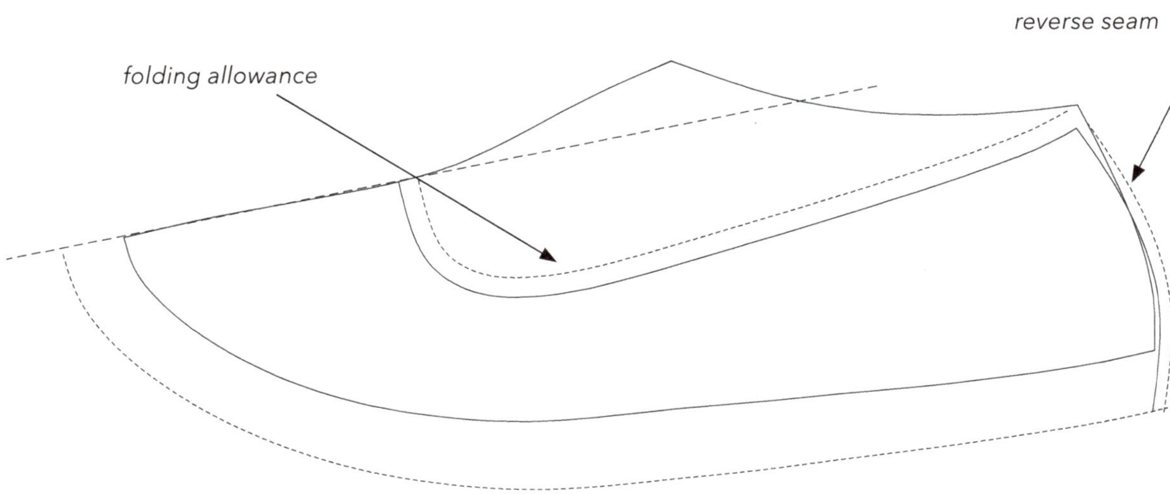

It is debatable whether the pattern piece should include these allowances at all; for one-of-a-kind shoes, the clicker can add the excess manually, letting the original pattern function as the folding template. In small operations focused on samples or custom orders, time is saved by not creating separate pattern variations for marking, clicking, and folding. Instead, the same pattern is reused with a "+5" notation at the edge, indicating that the clicker must add the extra amount during the process.

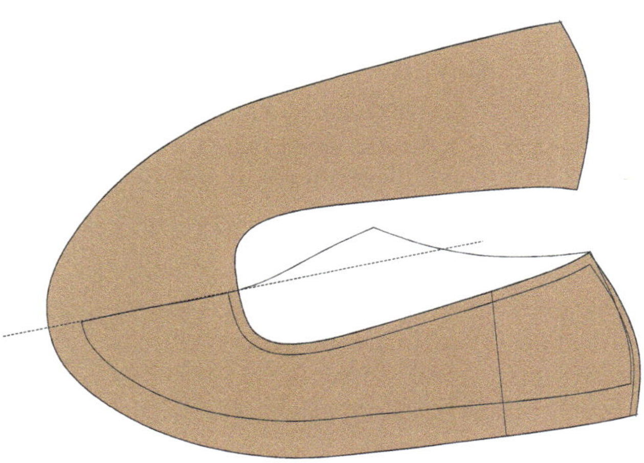

The unfolded pattern piece should include all necessary allowances. Every edge must be double-checked after cutting to ensure the required allowances are present and to prevent avoidable errors.

Lining Pieces

The lining pattern uses a lap seam for practical considerations, positioned roughly 50 mm from the top of the backline. The lap seam allowance should be added to only one of the pieces, as a double allowance can ruin the upper. The lining vamp requires a 5 mm allowance on the topline and an 8–10 mm allowance on the lap seam toward the counter.

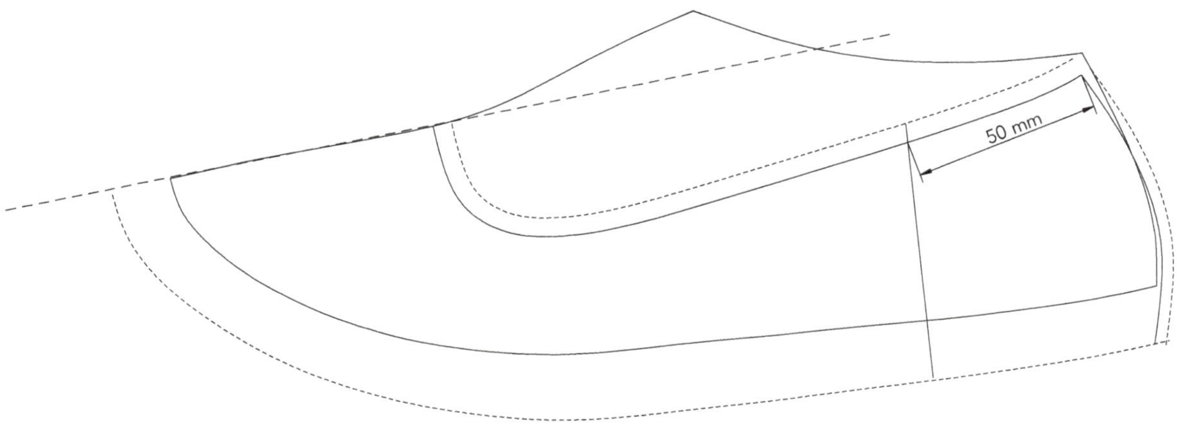

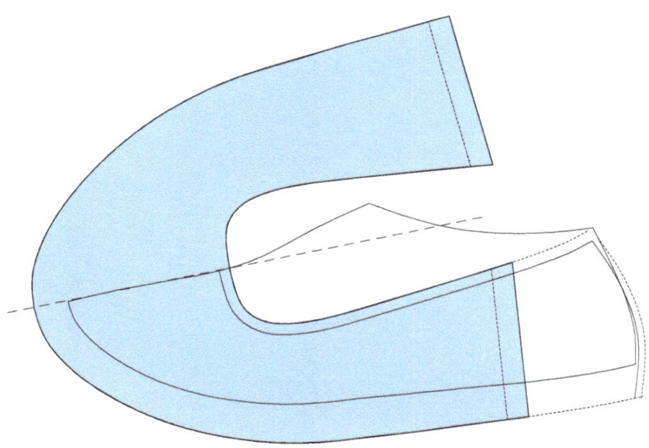

The lining vamp matches the shape of the upper vamp because the pattern is intended for hand lasting. For machine lasting, the lasting allowance should be reduced so both layers are caught and bonded securely with cement. This adjustment eliminates the need for trimming the lining allowance after lasting, a step common in hand-making.

The trimming allowance serves its purpose until the topline stitch is finished and should be removed immediately after stitching. Leaving the allowance on until lasting makes the process unnecessarily difficult and is simply not appropriate for that stage of construction.

The Orange Peel Effect

One of the most common errors beginners make is duplicating the upper pattern for the lining. This is incorrect primarily due to the "orange peel effect". This is where the inner layer (in this case, the lining) must be slightly shorter to fit inside outer layer (the upper) without bunching. If the lining is cut to the exact same length as the upper, it will cause pleats to form inside the shoe. Therefore to achieve a smooth finish, the lining layer should be made slightly smaller. While this rule can be ignored when layers are laminated flat – such as with a western boot leg panel – it is essential for components that wrap around a curve. A heel counter is simply a smaller version of this principle, and the orange peel effect must be accounted for to ensure a proper fit.

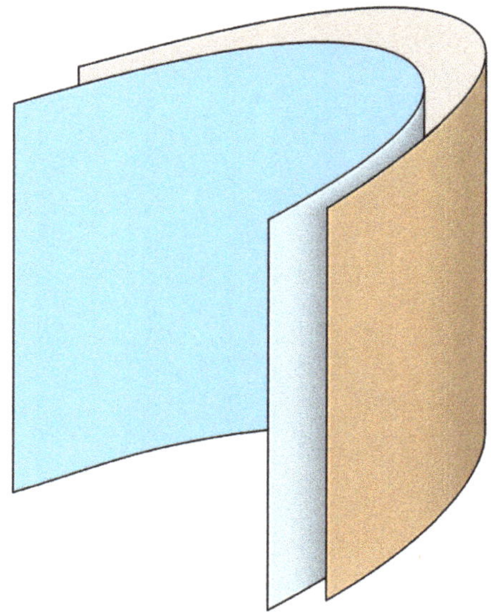

If the thickness of the two layers (upper and lining) is around 2 mm, the length difference will be around 6 mm. To express it in mathematical terms:

$$(R - r) \times \pi$$

where R represents the radius of the bigger circle the r represents the smaller one. In our case this difference is the thickness of the two layers together. For ease of calculation, instead of the exact π (3.1459...) it's perfectly sufficient to calculate using 3 – the margin of error will be insignificant.

In pattern application, the calculation must account for half of the thickness difference between the layers, as the pattern represents a folded version of the actual upper.

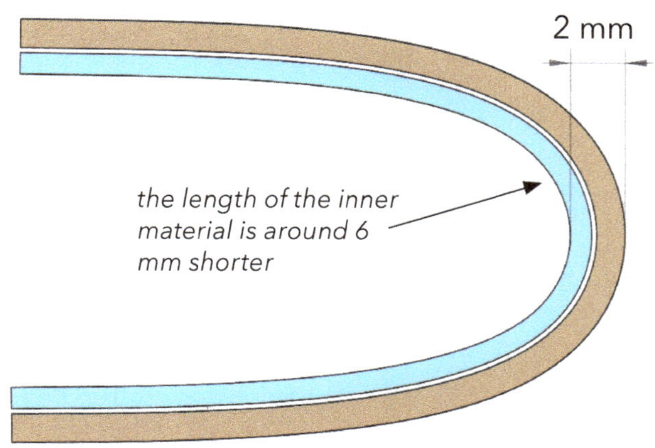

From the top view the counter looks like a half circle, which is close enough to use our formula

The Lining Counter Axis

The top of the lining backline should be shortened to account for the thickness of the materials. Based on a total calculated difference of 6 mm, the lining backline is brought inward by 3 mm. In most cases, this adjustment is sufficient to produce a well-fitting lining. However, if the upper leather is substantially thicker, a slightly larger adjustment may be required.

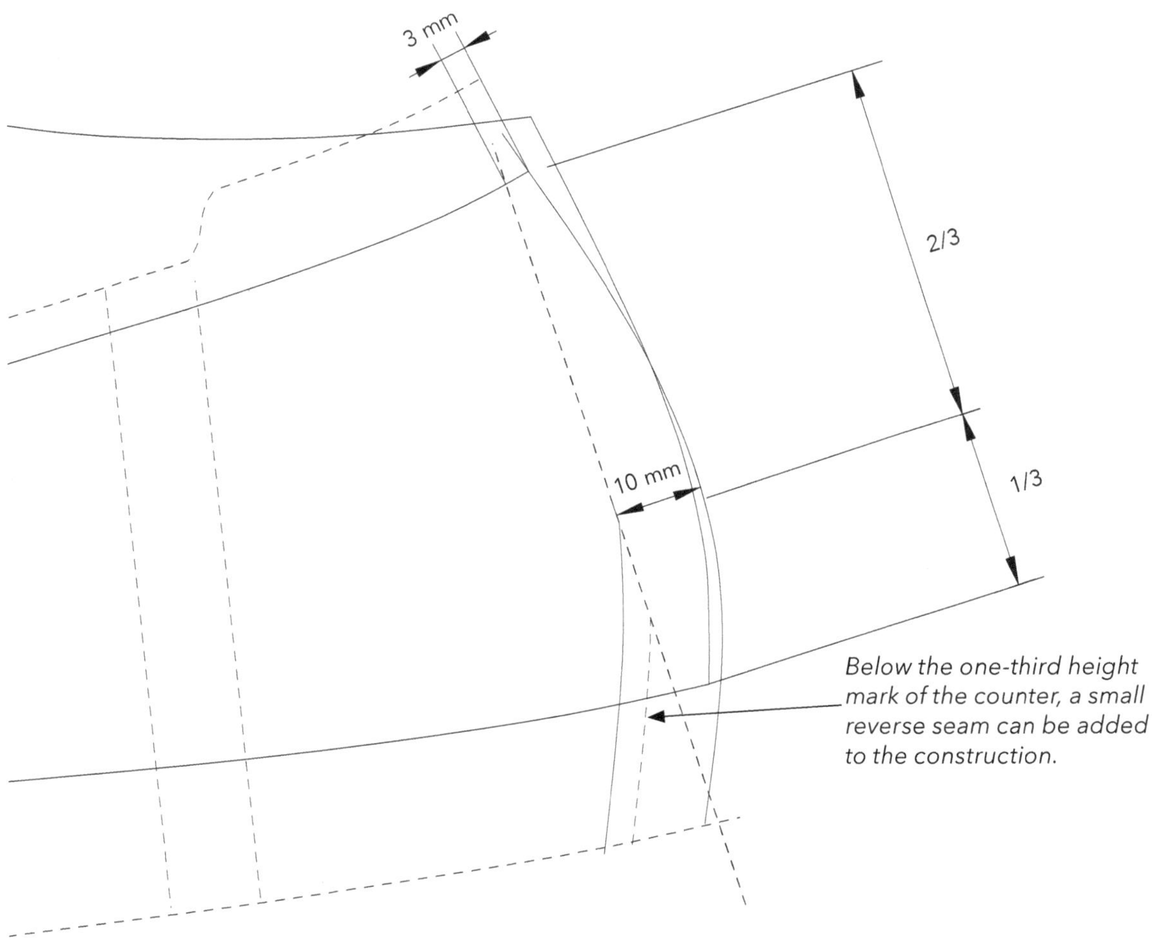

At approximately one-third of the counter height, additional allowance is required for two reasons. First, the thickness of the counter must be factored in; applying the formula to a standard counter thickness results in roughly 3.5 mm of extra space needed. Second, the inherent stretch of lining leather during the lasting process requires an additional 2-3 mm adjustment. At this height, the line can either follow the outer curve of the backline at a consistent distance or be drawn as a straight line, given that the curvature is minimal. Alternatively, the back seam on the lining can be omitted entirely if the axis is positioned far enough from the upper to prevent pleats from appearing on the visible portions of the lining.

The Counter Lining

This pattern piece receives a small extra allowance along the top edge. This is called the tacking allowance, and its purpose is to hold the upper securely in place during the lasting process. Unlike the regular trimming allowance, this section must remain on the piece during lasting which is exactly the purpose it's meant to serve. This additional material will, however, make the skiving slightly more difficult.

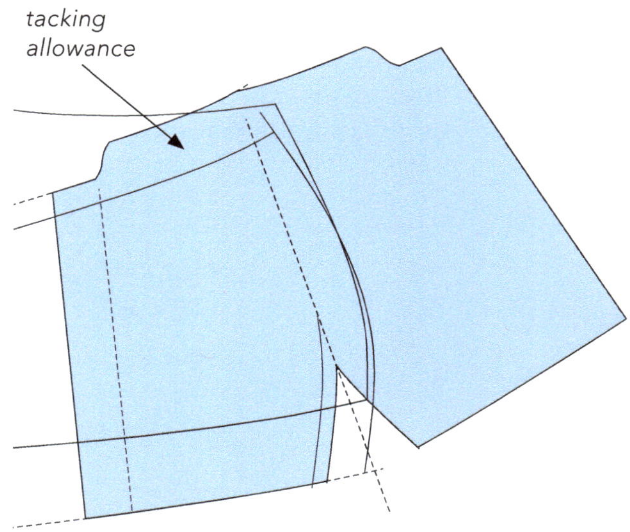

The Master Pattern

With all these considerations factored in, the full ballerina shoe pattern looks something like this.

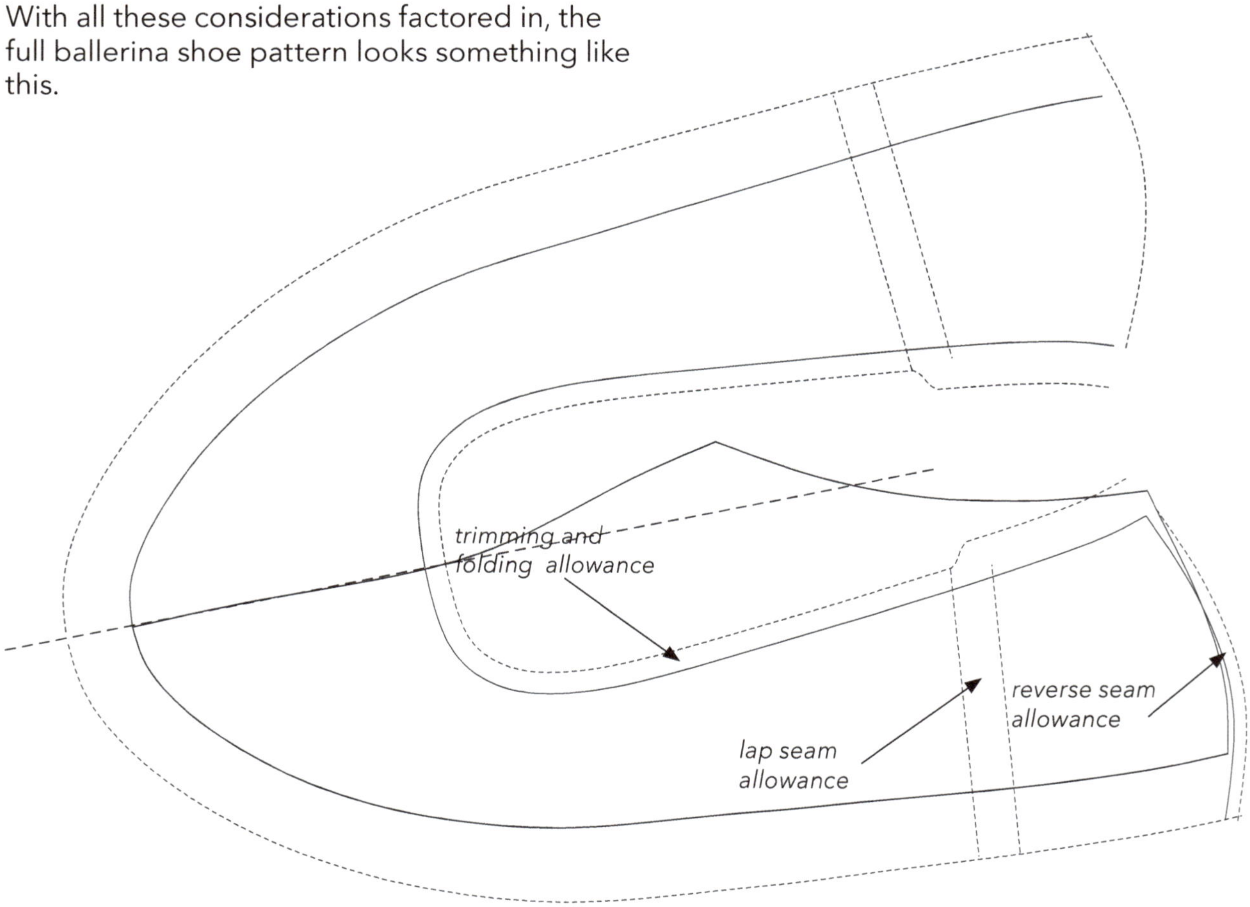

The Complete Pattern

1	NO SKIVING
2	SHORT SKIVING HALF THICKNESS
3	MEDIUM LONG SKIVING HALF THICKNESS
4	LONG SKIVING HALF THICKNESS
5	LONG SKIVING NO EDGE THICKNESS
6	SPLITTING

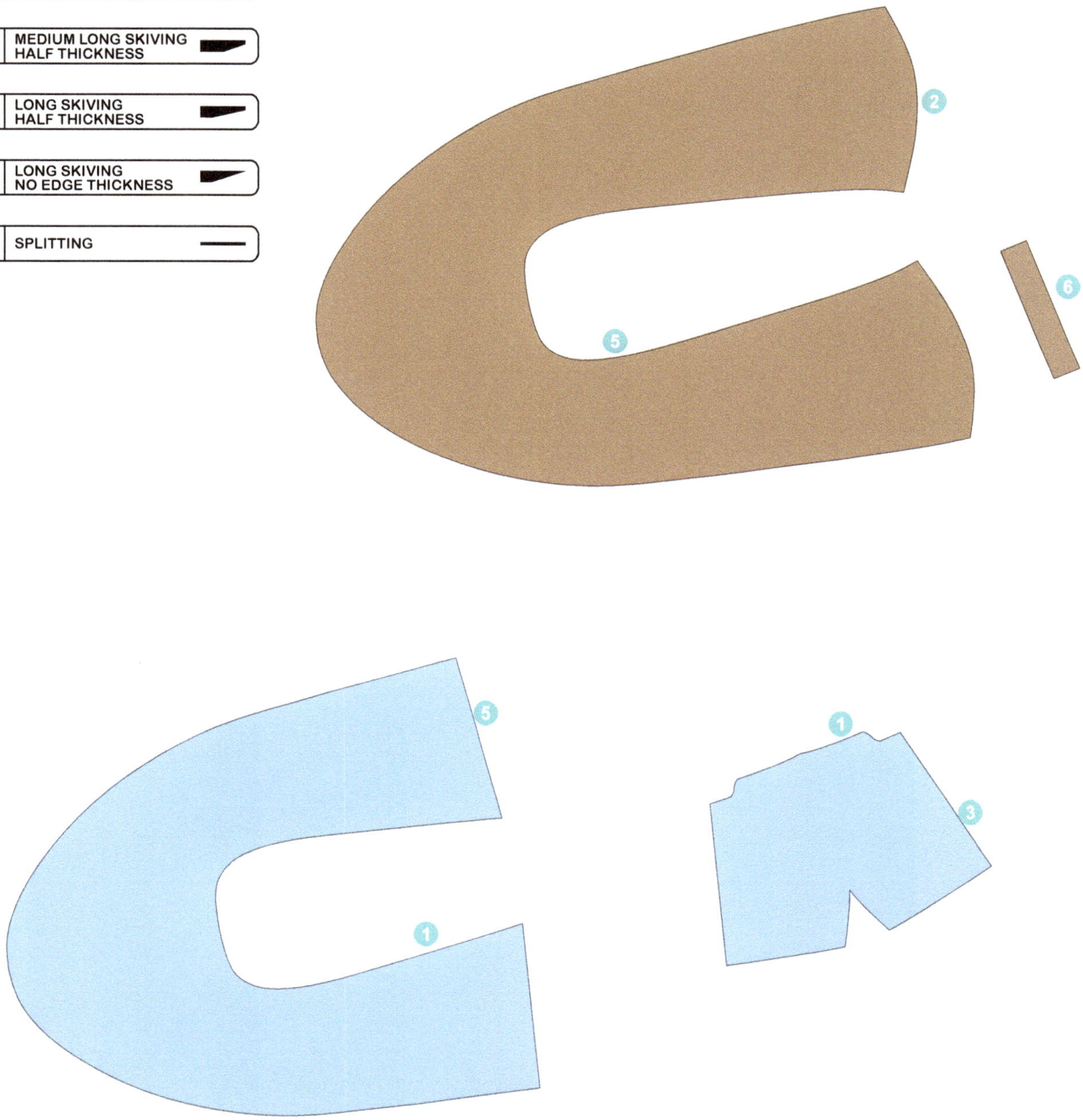

Assembly order

1. Clicking, skiving, marking if necessary
2. Reinforce the upper topline
3. Fold the upper topline
4. Reinforce the front under the topline
5. Sew the upper backline
6. Trim, rub and reinforce the backline
7. Finish folding
8. Assemble the lining pieces
9. Sew the lap seam
10. Sew the reverse seam on the counter lining
11. Add the hidden backstrap
12. Attach the lining
13. Sew the topline
14. Trim the hidden backstrap
15. Trim the lining

Folding

The topline tape helps reinforcing the leather. This is a necessary step for most folded edge, but in this construction, it's downright inevitable. The only times when it's not needed is when the leather is already very sturdy or if the stretch is calculated and built in the pattern.

Mark the desired folding line on the skived edge, suede side up. When adding the self adhesive tape, place it under that marked line so it's still visible. This way, the edge will end up on the line perfectly.

Add cement to the edge. It's possible that the solvent of the cement interacts with the adhesive of the tape, so use the minimum necessary amount. For cement options you can choose rubber cement, a sole glue or a special upper cement - whichever is available.

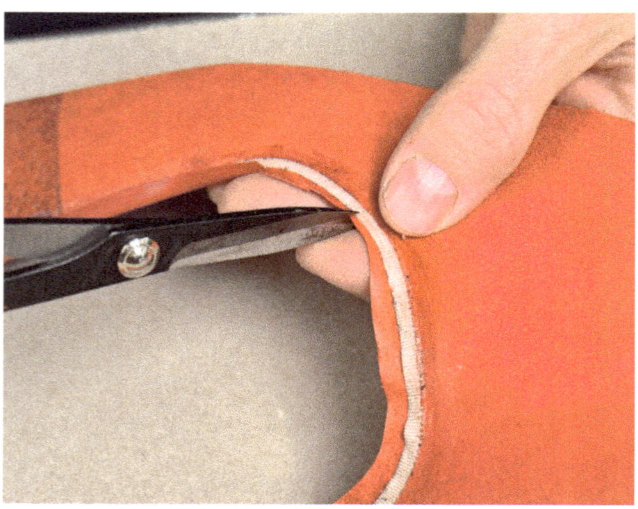

When reaching the curve, small release cuts will help spread out the material and prevents deformation. Use scissors to make tiny cuts on the curves halfway to the tape. Overcutting will result in an angular corner instead of a smooth curve, therefore it's very important to practice restraint. Another detail to keep in mind: start adding these before the turn and end them after so the curve is more even.

If the preparation was accurate and the skiving is to the correct parameters, the folding process will be easy and the result will be pleasing. Our aim is for an evenly folded topline all around. Given the order of processes, it would be logical to first finish sewing the backseam, then carry on with the folding all around. However, this is not ideal since it's more challenging working with a 3D shape. Another common mistake beginners make is folding the topline all the way to the backline, making the seam bulky.

Do this instead: fold the topline on a flat surface but leave the last little section open, then finish when the backseam sewing is done. This way, we can work precisely and comfortably.

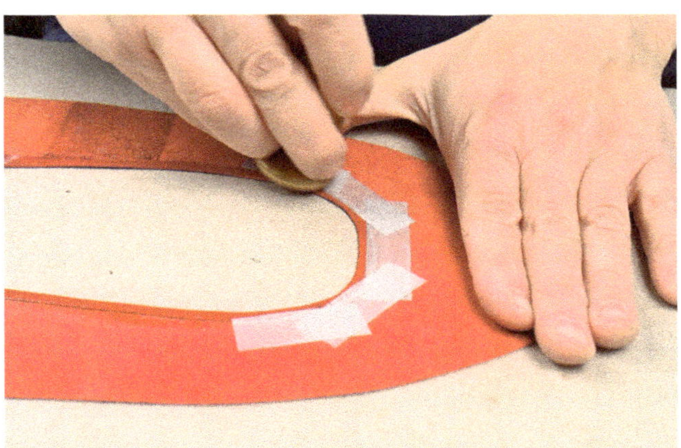

In the industrialized footwear manufacturing, a special self-adhesive reinforcement that matches the shape of the vamp is used to reinforce the topline around the front. In the bespoke/handmade trade, simply overlapping pieces of reinforcement tape (as shown in the photo) is sufficient. This helps the topline to maintain its unique design without overstretching, thus ruining aesthetics and functionality.

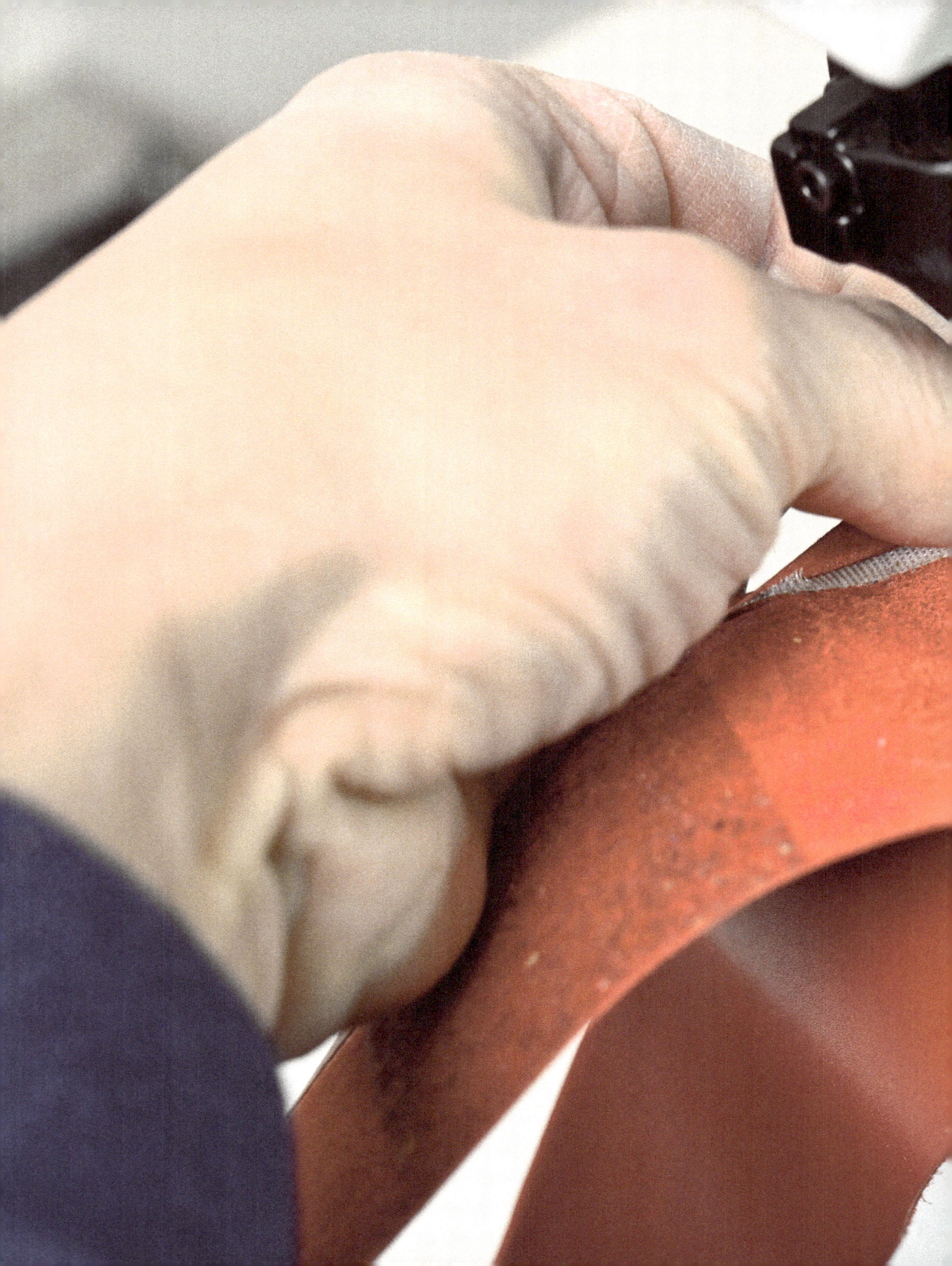

Note that the folding is not finished to the back when it's time to sew the backseam, as explained earlier. As for where to put the stitch line, there are two options. Either sew close to the edge and don't trim off the excess, or as demonstrated here, sew at a distance of 3 mm parallel from the backline then trim. In this latter case, there is very little chance that the bottom layer accidentally misaligns. The decision which option to choose has to happen during the patterning stage and the corresponding markings should make it clear for the upper maker.

A shoulder skiving is a good idea on this backseam. In which case, the stitch line runs on the skived part, but the roller foot be placed on the untouched leather. This is to ensure the foot's path will not be affected by the slanted edge.

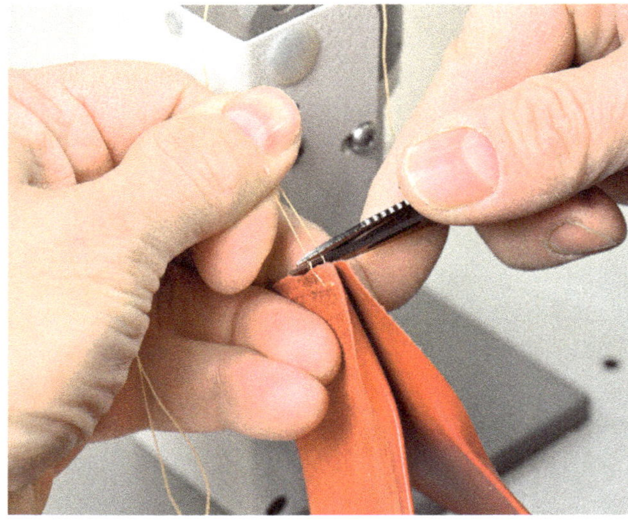

I prefer to err on the safe side and do what seems an excessive amount of backstitching – sometimes 7-8, so there is no chance the seam frays. Trim back the thread ends but leave a little extra thread for melting. As you can see, the topline is still not finished folding yet.

There is a rule of thumb for trimming off the allowance. The final width should be the thickness of the trimmed edge. Let's say the edge is skived to 0.8 mm, the two layers together will be around 1.6 mm, and that is the distance the cut should be made away from the stitch line. Leave too much and excess can flop to one side of the seam creating bumps. Trim too close to the stitch and it could affect the integrity of the backseam, causing it to pop open during the lasting process. The right amount of trimming can prevent both problems.

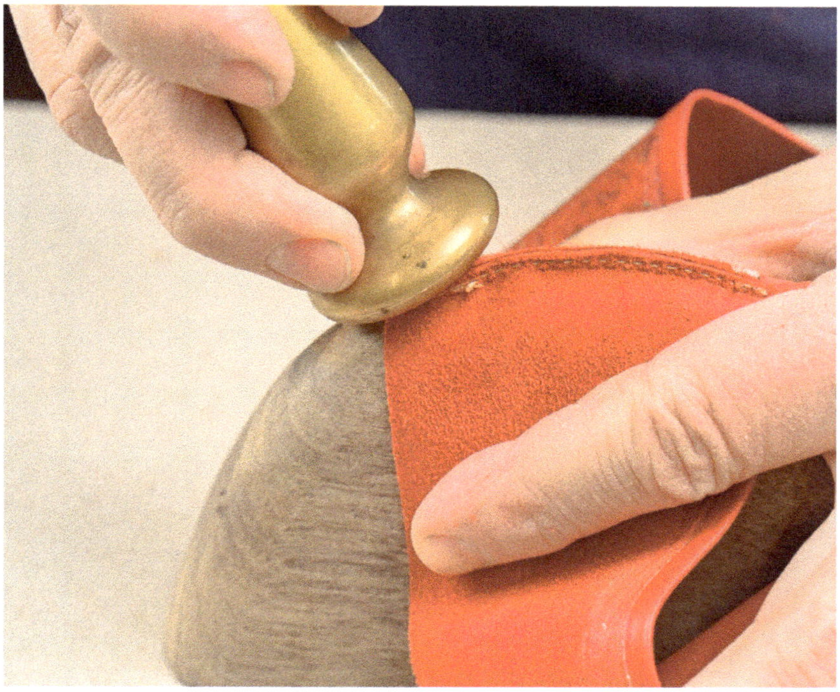

To achieve the perfect look, the seam needs to be rubbed and hammered, which is a more delicate job than it sounds. If the hammering or rubbing is done carelessly with too much force the thread, which doesn't stretch, might end up cutting the leather apart and the seam will rip open. In manufacturing, seam rubbing machines are used that can be adjusted to prevent this. By hand, however, it's up to us to know exactly how much force to use. It is also up to our knowledge of leathers to be aware that some of them are a lot more vulnerable to this process than others.

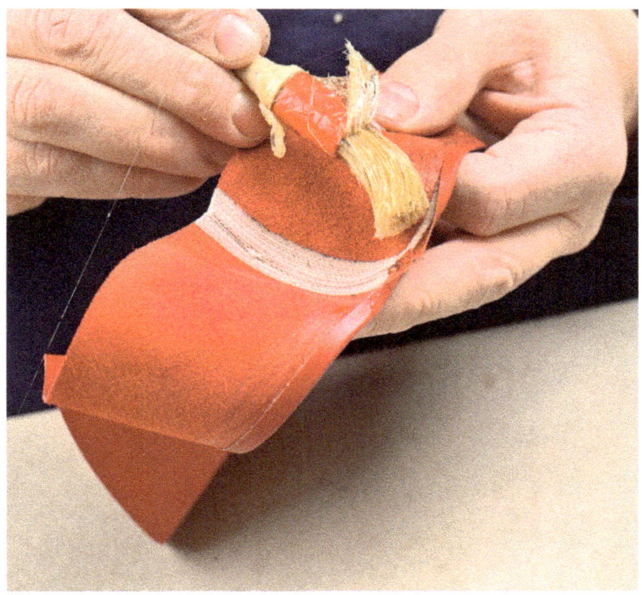

A piece of backline tape helps to protect the reverse seam and hold it in place. Now it's finally time to finish the folding process at the top of the backseam. Note that neither the topline tape, nor the backline tape go under the folding around the seam. This minute detail contributes to a perfectly even edge with no change in thickness anywhere. Add cement under the folding allowance and let it dry a little while before folding it down. Since the lining will also have a reverse seam, these two on each other means extra thickness which is tempting go and smash them flat. Needless to say, this can greatly damage the seam. Instead, consider that the backstrap will cover this, hiding minor errors and carry on with caution.

The Lining Structure

Accuracy is the main challenge for putting together the lining pieces. Any misalignment might end with a nasty pleat inside the shoe.

As previously mentioned: every assembly order varies based on the technology and the quantity. In a factory production, efficacy is king, and with that, repetitive steps in short amount of time. However, when working on one pair, there is no reason not to prepare as much as possible for sewing.

When putting the pieces in the machine for sewing, double check which piece overlaps the other. In this construction, the vamp lays over the counter, therefore when sewing the pieces together, is on the left side. Since the counter is suede side up, a typical error here would be sewing this on the other side which would end up with a stitch line extremely far from the edge of the piece.

There are two tips here that are worth mentioning. First one is fairly predictable: always do an excessive amount of double stitch on the lasting allowance, as it will likely be – depending on the lasting method – strenuously worked on during lasting. The second is less intuitive: the 5 mm trimming allowance added to the topline means all the backstitching will be cut off if they go too close to the top, rendering them perfectly dysfunctional. With that in mind, start and stop far enough from the edge.

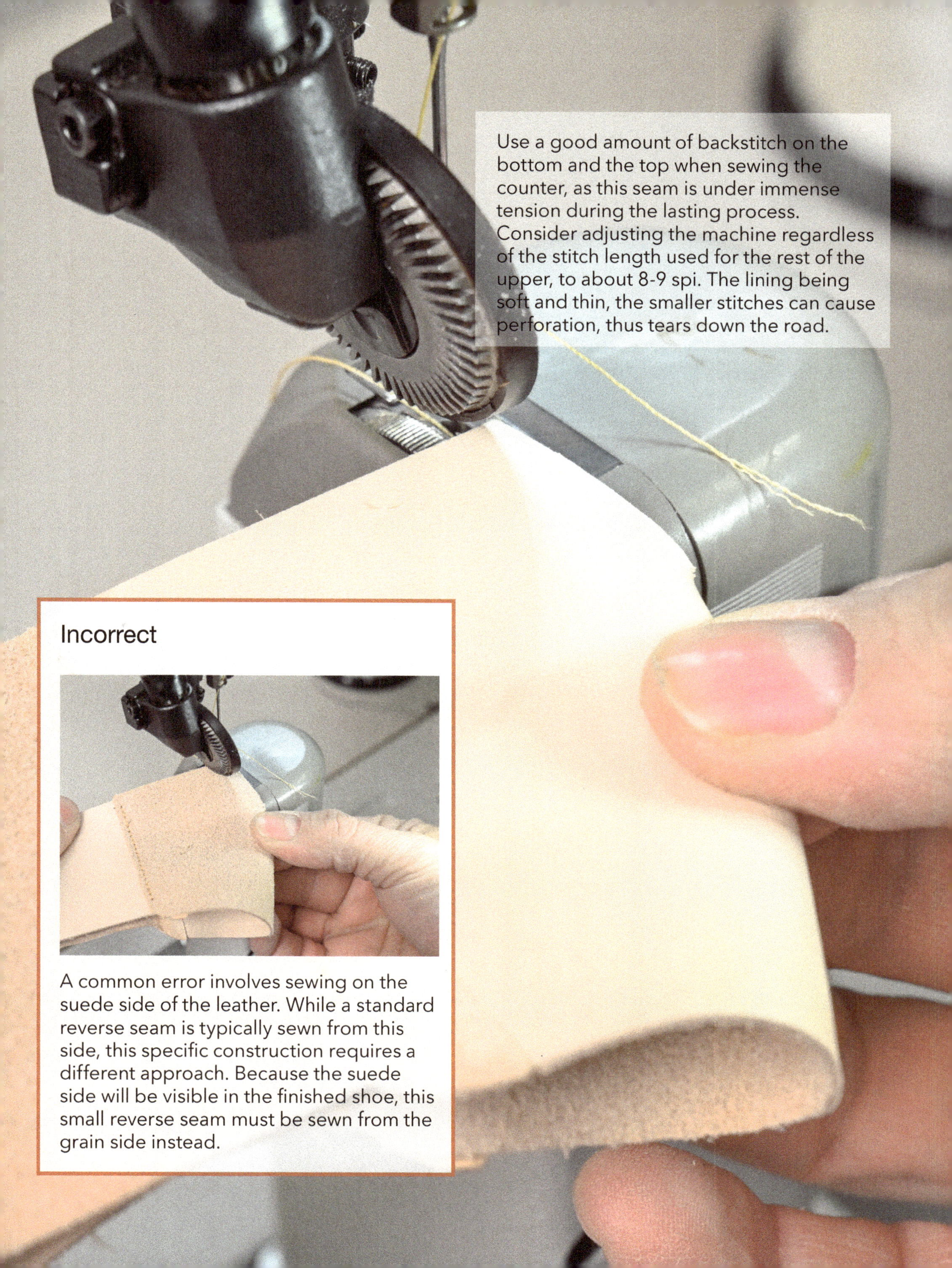

Use a good amount of backstitch on the bottom and the top when sewing the counter, as this seam is under immense tension during the lasting process. Consider adjusting the machine regardless of the stitch length used for the rest of the upper, to about 8-9 spi. The lining being soft and thin, the smaller stitches can cause perforation, thus tears down the road.

Incorrect

A common error involves sewing on the suede side of the leather. While a standard reverse seam is typically sewn from this side, this specific construction requires a different approach. Because the suede side will be visible in the finished shoe, this small reverse seam must be sewn from the grain side instead.

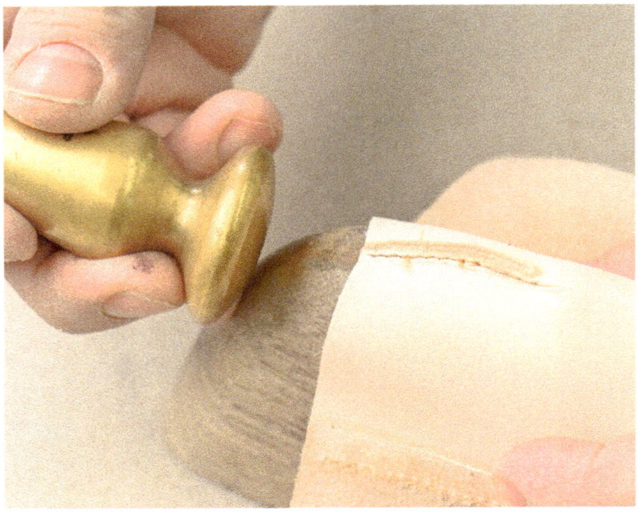

Finish the reverse seam the usual way - by flattening it using folding hammer and rubbing form. Although this step adds to the quality, it is however, optional since the heel counter is thick enough to cover this seam. The reason it is still a good idea not to skip it because this way ensures the seam stays centered.

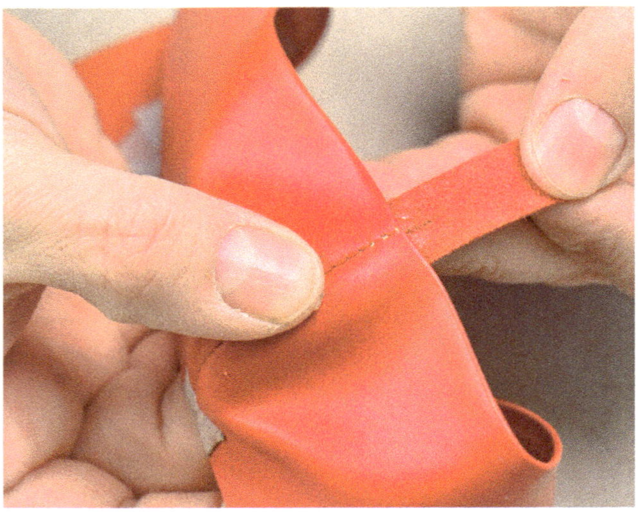

Mark the center line on the hidden backseam, and cement the piece perfectly to the center from inside, around 20 mm under the topline edge. The backline tape is already in place, so positioning this little tape is not necessary. Make sure to leave enough to be able to hold it during sewing. Leaving this too short would mean struggling keeping it tight when sewing, leading to subpar appearance. Depending on the sewing machine, a good 30-40 mm should do the trick.

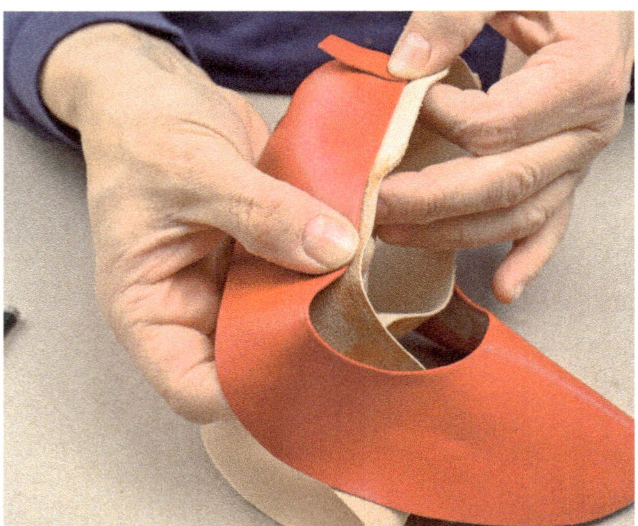

The vamp lining is ready for attachment. I prefer to start in the middle of the counter and be mindful of the allowance designated by the pattern here. As a reminder: 5 mm all around except for counter portion, which is 10 mm. Working toward the front the upper can be flattened out, while watching out for pleats piling up in the center. The trimming allowance here is quite comfortable, so there's wiggle room if the alignment is not absolutely perfect.

Start the topline stitch line somewhere on the medial side. Avoid highly visible places such as the center as the backstitches might become too obvious.

Hold down the backstrap excess firmly but not with tension, to sew through this portion. Resist adjusting the distance from the edge to offset the thickness. It is only a matter of a few millimeters at most

Trimming the backstrap is easy when you keep in mind a few basic techniques. Use a pair of sharp scissors. Hold the backstrap tight and do not change the tension during the process. Hold the scissors roughly 45° from the upper and make a relatively slow cut.

After trimming, ideally, the stitch line is equidistant from the edge and the bottom of the excess at around 1.5-2 mm. It's possible to go even shorter but it might render the piece useless. This seam sees a lot of use and abuse, which is the very reason for this type of reinforcement. This feature is especially common for ballerina and similar shoes, such as high heels. It is, however not limited to those constructions, so feel free to experiment on even chukka boots.

Trimming the vamp lining excess needs a steady hand, a sharp knife, and some practice. Hold the blade so that the upper edge can guide it but at an angle so that the knife can never cut into the upper.

I prefer to use the corner or the edge of the table so there's always a firm support for the cut without having the parts I don't need to focus on in the way.

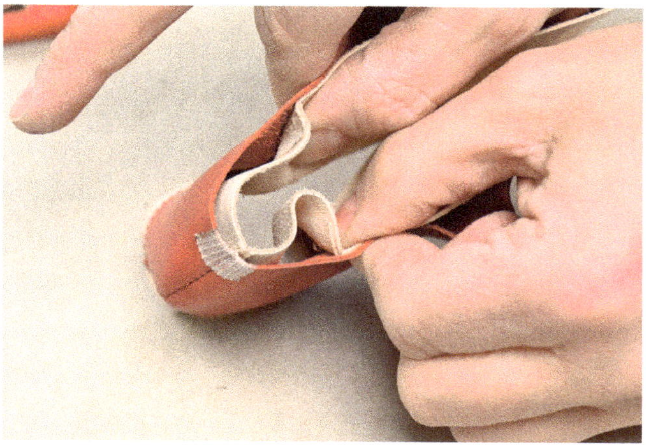

If the counter lining bunches up, it indicates that the lining axis was not properly positioned. This misalignment creates significant difficulties during the lasting process. Such errors occur frequently, often as a result of a lack of communication between the pattern designer and the shoemaker.

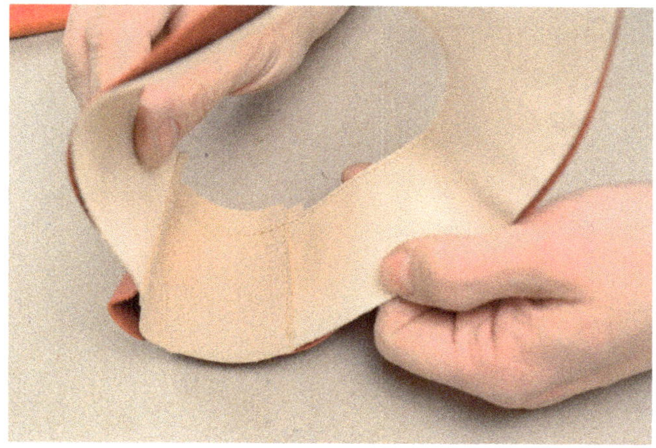

If the axis is in the right position, the lining should be laying snugly inside the upper. This means easy lasting and smooth lining. This is all thanks to the classic orange peel effect: a smaller lining fits perfectly into a bigger upper.

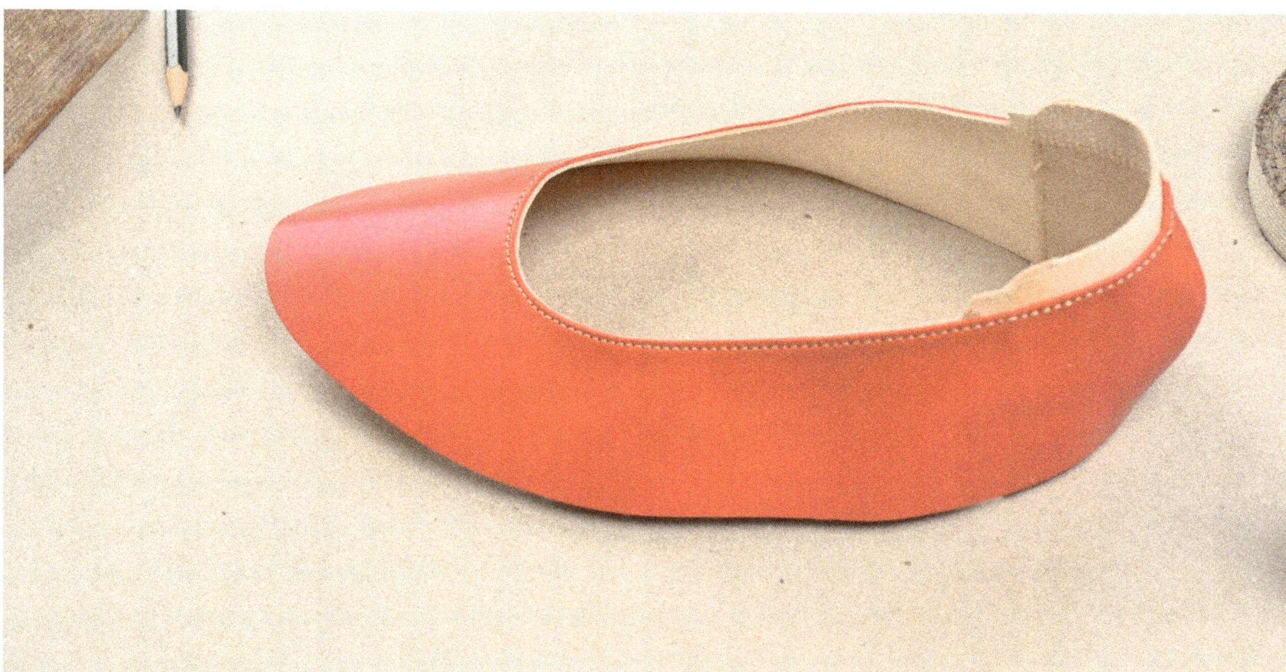

With the trimming done, the upper is ready to be lasted. Note the tacking allowance on the counter lining is intact. For this upper, a beige thread makes the stitch line more visible for teaching purposes, but stylistically a matching color would be a more desirable choice. Some further minor touches to finalize the upper: cementing certain parts together, such as the side and a narrow strip on the front around the topline; as well as adding some sizing indication somewhere – the lining on the lasting allowance is my go-to.

7 THE HIGH HEEL SHOE

High-heel shoes occupy a special place in footwear design. Unlike lower-cut styles, their elevated heel shifts the body's weight forward, demanding greater precision in both last design and pattern execution. The higher the heel, the more pronounced the changes in balance and body posture. In addition to the heel itself, these shoes place greater visual emphasis on the upper, making clean lines and precise pattern solutions even more essential.

The foundation of the upper assembly remains identical to the ballerina construction covered in the previous chapter; therefore, the step-by-step photographic illustrations are not repeated here. Instead, this chapter focuses on the specific design issues that arise as heel height increases and the topline begins to cross the main folding axis. In this chapter, two effective solutions to this problem are laid out. The first is a commercially common approach, while the second is more refined and technically advanced. By understanding both methods, the pattern maker can choose the one that best suits the intended level of craftsmanship and the design intent of the shoe.

The High Heel Pattern

The high-heel pattern can illustrate several important principles in pattern making. Technically, this pattern is not very different from the ballerina, but there is one major factor: heel height. As discussed in the mule chapter, this relates to the **two-inch heel problem**: once heel elevation exceeds two inches, the main folding axis begins to cut into the pattern, making unfolding problematic.

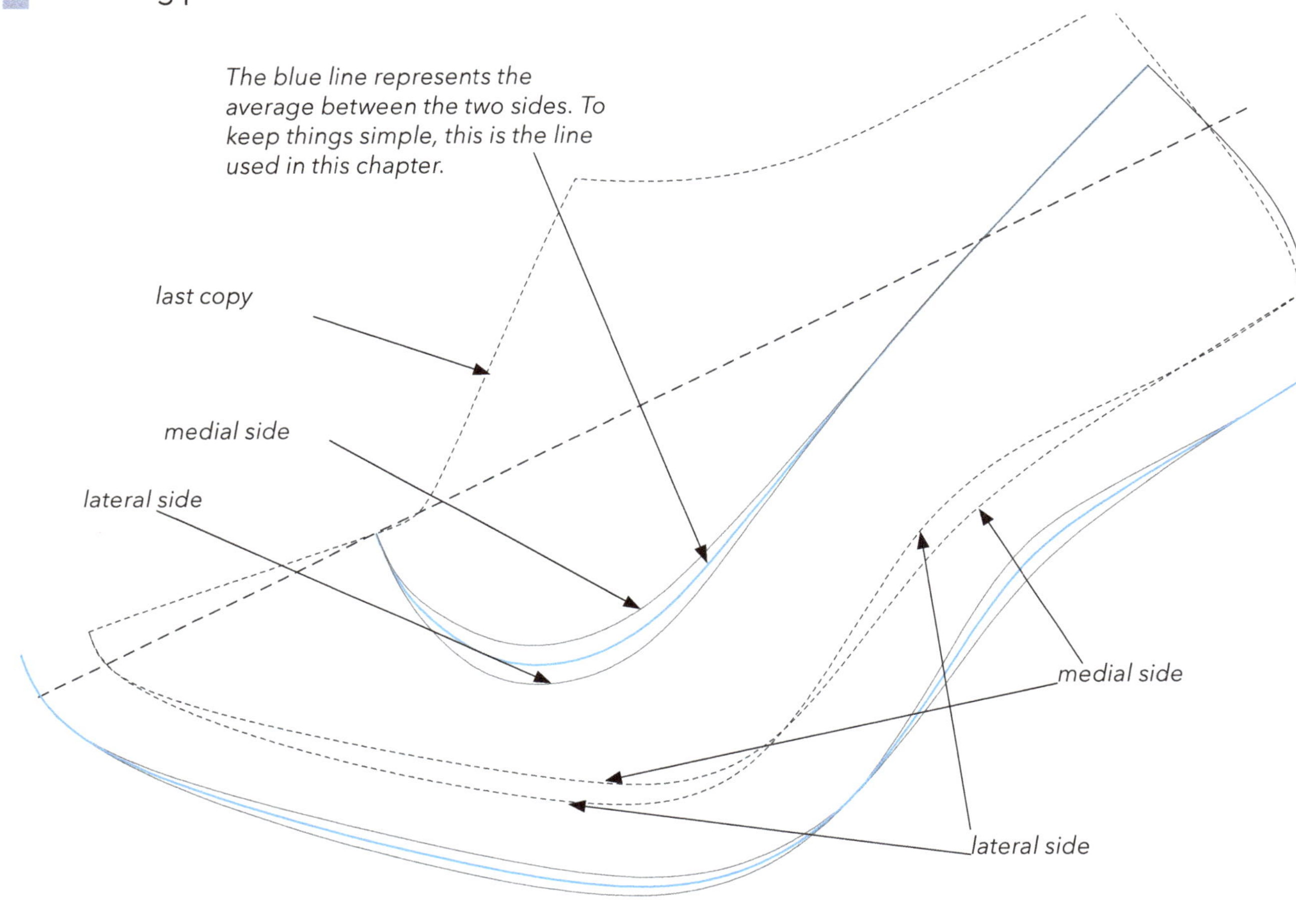

It's worth noting that this high heel pump is almost always asymmetrical. However, to keep the explanation clear, the theory is first discussed using a symmetrical example, and asymmetry is introduced later.

In this example, the work starts from by averaging the two last-copy lines on the medial and lateral sides. This is indicated with a blue line on the illustration and serves as the reference pattern line.

For asymmetrical lines, a simple rule helps identify each side, though recording this information on the master pattern remains best practice. At the topline, the higher line represents the medial side and the lower line represents the lateral side; this rule applies to all footwear and pattern pieces in the toe area, including wingtips. At the feather edge, the layout is slightly more complex as the two lines typically switch positions around the ball of the foot. The medial side generally remains straighter at the front and runs lower around the shank.

The Two-Inch Heel Problem

This illustration clearly demonstrates the issue as it appears on a high-heel pump, where the topline extends over the main folding axis. Cutting an upper like this from a single piece is not possible. There are several ways to resolve this; in this chapter, two of the most effective solutions are discussed.

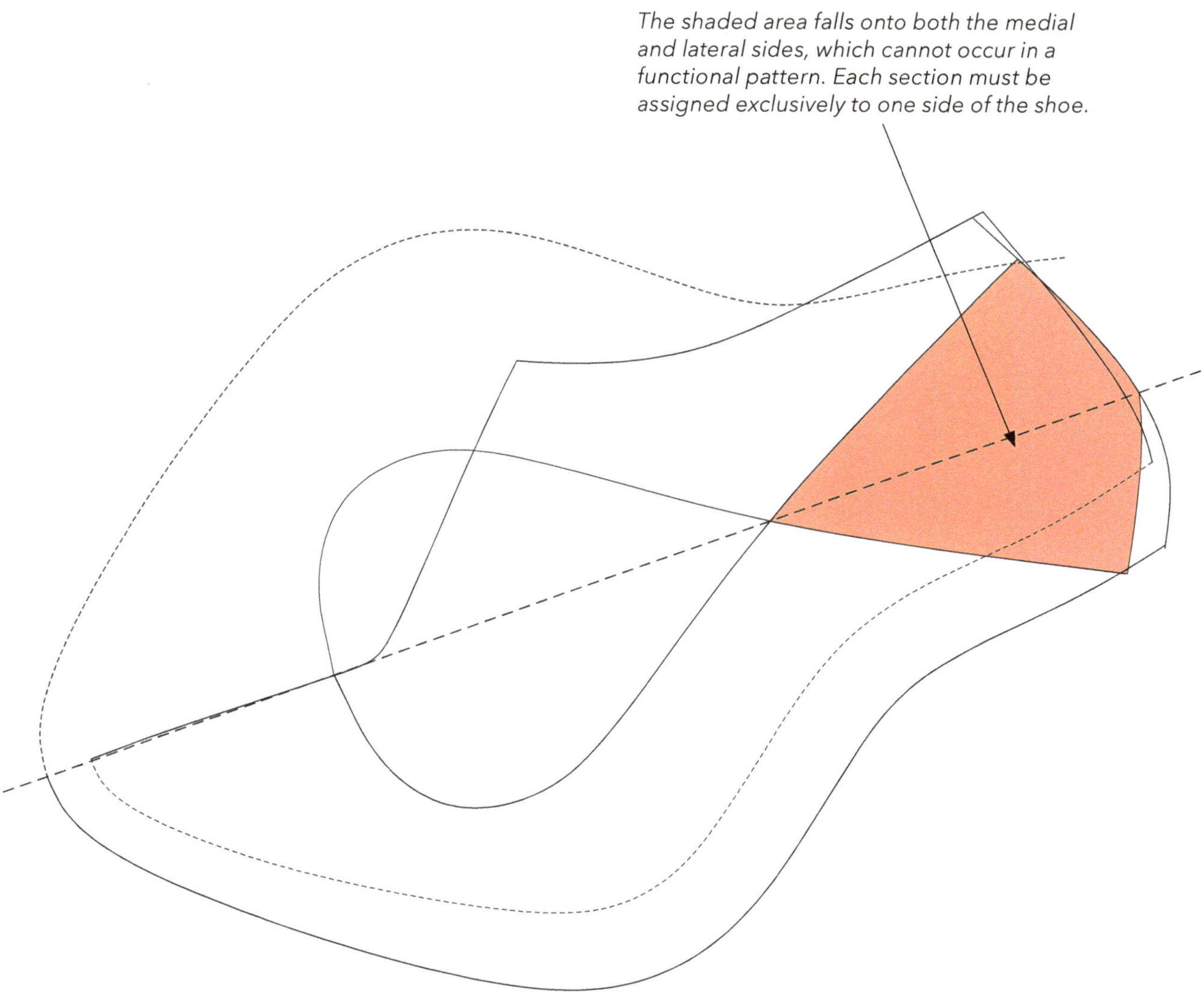

The Incorrect Solution: Rotating the Main Axis

Simply tilting the axis upward at the counter is an ineffective solution for several reasons. First, it creates a large gap at the top, adding unnecessary material. More importantly, this gap increases the length of the topline, which presents a serious technical issue. In pump construction, the topline length is critical as it defines the fit, the style, and the tension of the shoe on the last. Even a change of a few millimeters can shift the line into a position that deviates significantly from the original design.

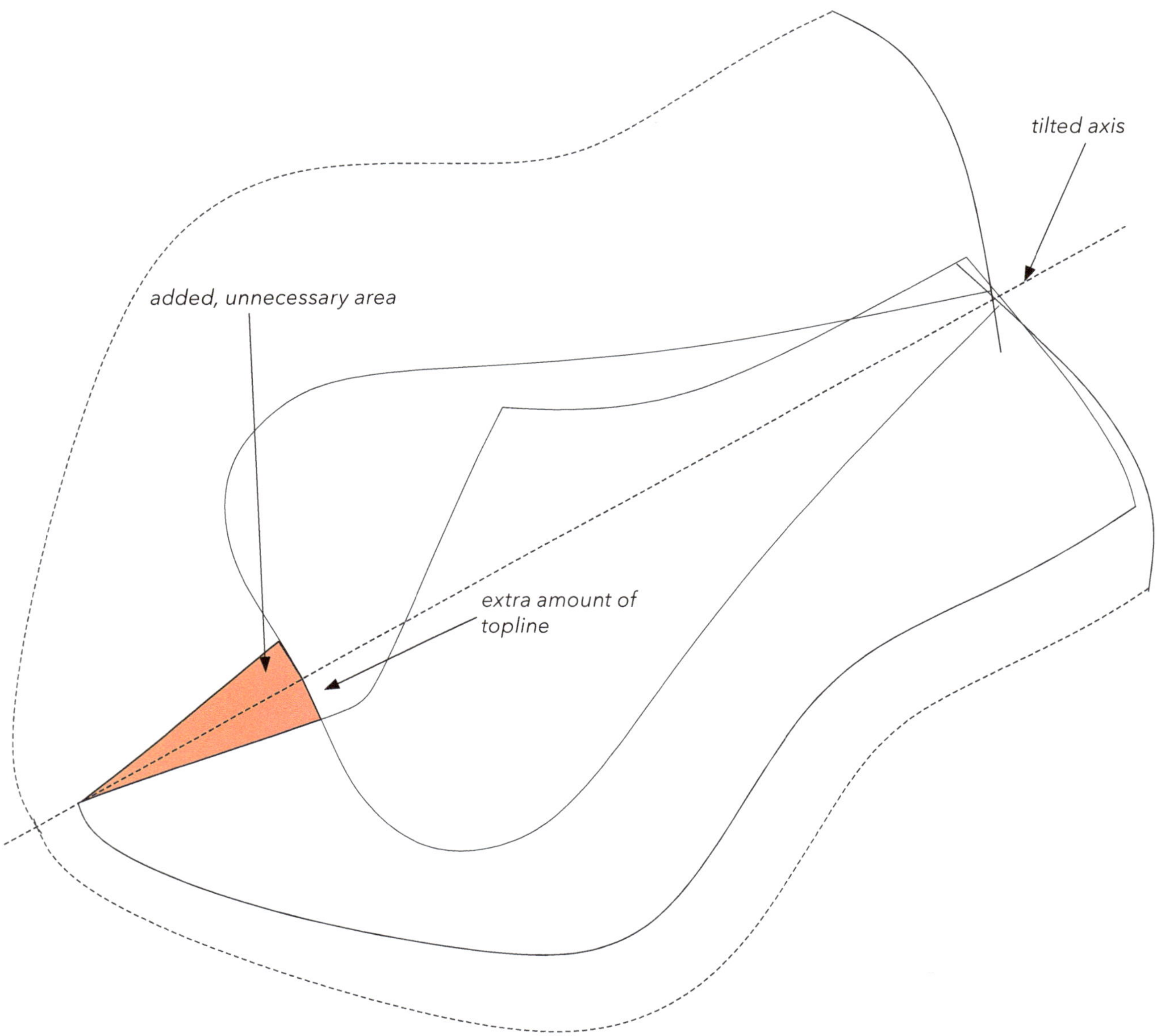

The Common Solution: Dividing the Medial Side

Dividing the vamp (and effectively the entire upper) is the most common solution to the two-inch heel problem, as the overlapping section is cut from a separate piece of leather. This approach is widely used in low- to mid-range commercial shoes because it is simple, accurate, and preserves the clean look of a high heel by keeping the lateral side intact and hiding a – usually reverse –seam on the medial side.

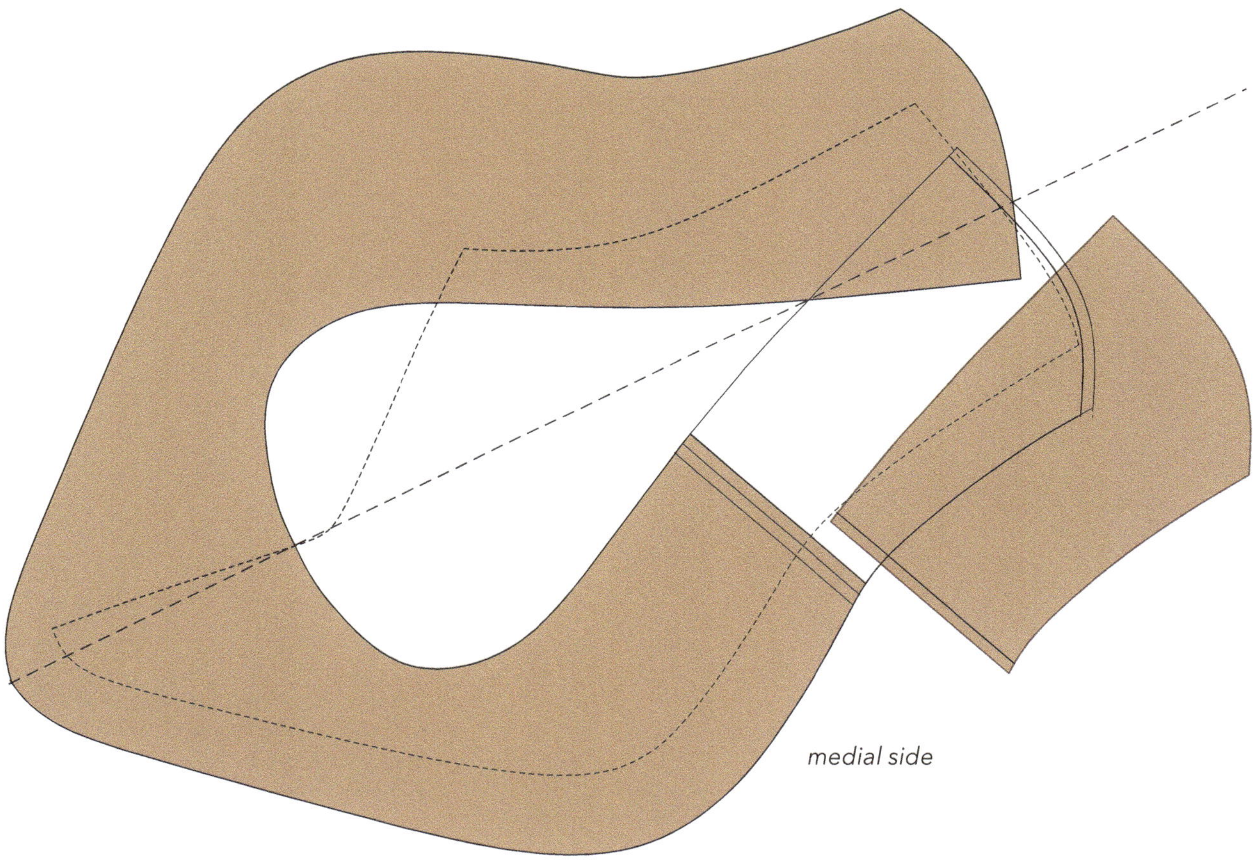

The positioning of this seam is crucial. To find the right position there are a few considerations: firstly, a reverse seam is strongest when the pieces meet the topline at a perpendicular angle; and secondly, for entirely aesthetic reasons, the seam should not visible from the top view. Since typically when worn, the shoe is most often seen from this angle, a seam on the medial side is practically invisible.

Adding the Lasting Allowance

Every shoe requires an appropriate lasting allowance that accounts for the leather type, construction method, insole thickness, and heel counter. For hand lasting, 15 mm at the toe, 18-20 mm in the midsection, and approximately 16 mm at the counter are generally sufficient, though these values vary. Typically, hand lasting requires a slightly larger allowance than other methods.

The pull direction runs through an open area, which explains why this model is very easy to over pull. High heels require special hand-lasting techniques.

The lasting allowance at the toe is crucial.

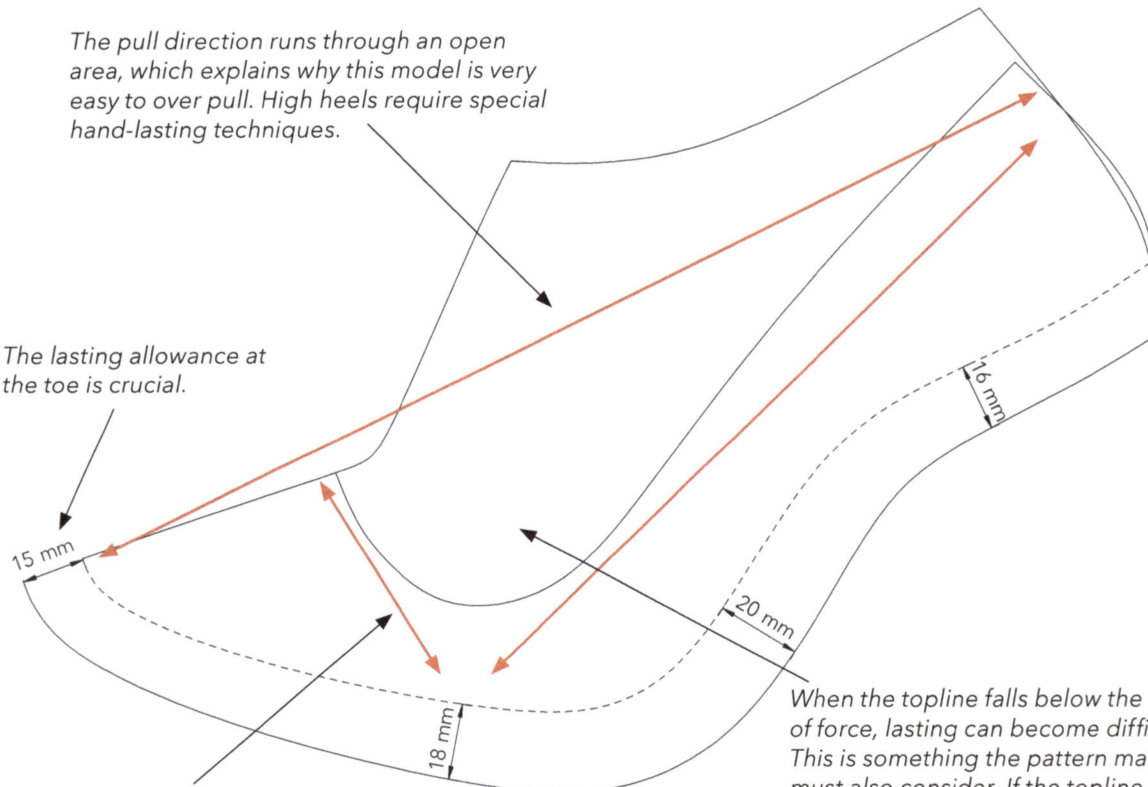

The red arrows represent the lines of force during lasting. These directions pull the upper down onto the last, allowing the stretch factor to be calculated. Unlike the initial pull direction, these lines of force act below the topline.

When the topline falls below the line of force, lasting can become difficult. This is something the pattern maker must also consider. If the topline opens too much, it becomes problematic and must be corrected. Either the lasting allowance or the topline shape needs to be adjusted.

High heels present a unique challenge due to the **toe pull**. In standard footwear where the upper covers the entire last, the structure provides good resistance, meaning only the leather's stretch - usually between 6% and 10% -must be considered. For a 240 mm upper, this stretch adds about 12 mm, making a 16-18 mm allowance more than adequate.

In high heels, the deep cut of the topline allows the opening to expand during lasting, which complicates both pattern making and assembly. If the lasting allowance is insufficient, the shoe might still be lasted, but the topline will lose its intended shape and fail to settle into the correct position.

Minimizing the Extra Leather Around the Heel

Apart from the toe pull, high heels possess another unique factor: the heel construction itself. In standard footwear, where the heel is a solid, sturdy structure, it assists in shaping the counter during lasting by pushing pleats down toward the insole. This structural integrity can often compensate for minor lasting imperfections.

High heels, however, often feature a thin plastic core covered in leather, creating a minimal wall around the feather edge.

Because this construction cannot perform the same shaping function as a solid heel, the pattern maker must compensate by reducing excess material around the heel area. The solution is to bring the backline in slightly. It should be noted that this adjustment may not be necessary for machine lasting or when using pre-shaped heel counters.

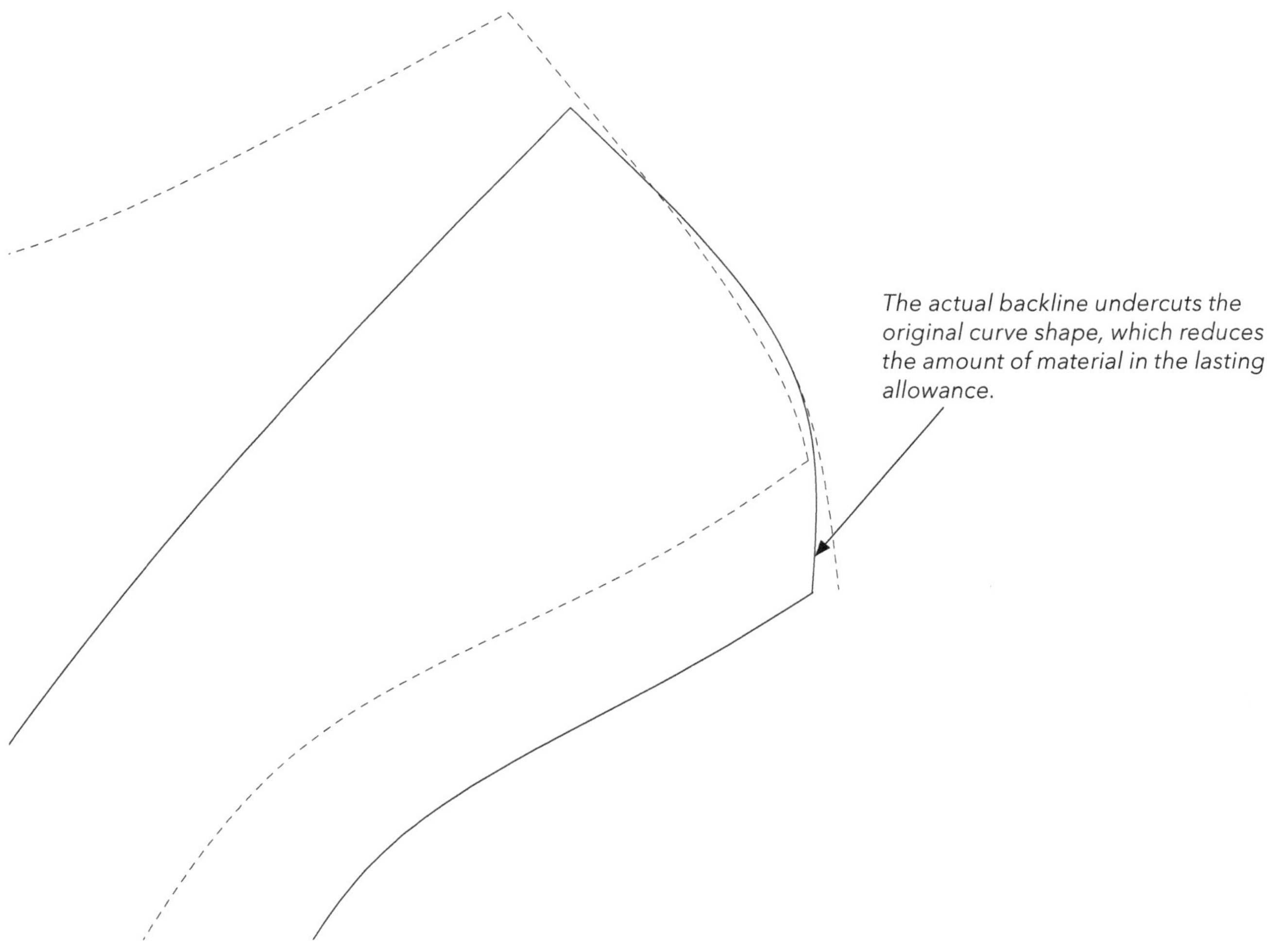

The Main Folding Axis

With the ballerina pattern, the front section of the last copy could be used as a guideline for the main folding axis. Here, a minor adjustment is required. This is not due to the upper itself, as the dividing solution eliminates that issue, but because of the lining, which will be discussed later. Rotating the axis slightly upward resolves the problem, even though it trims a small section from the toe, which is not a concern.

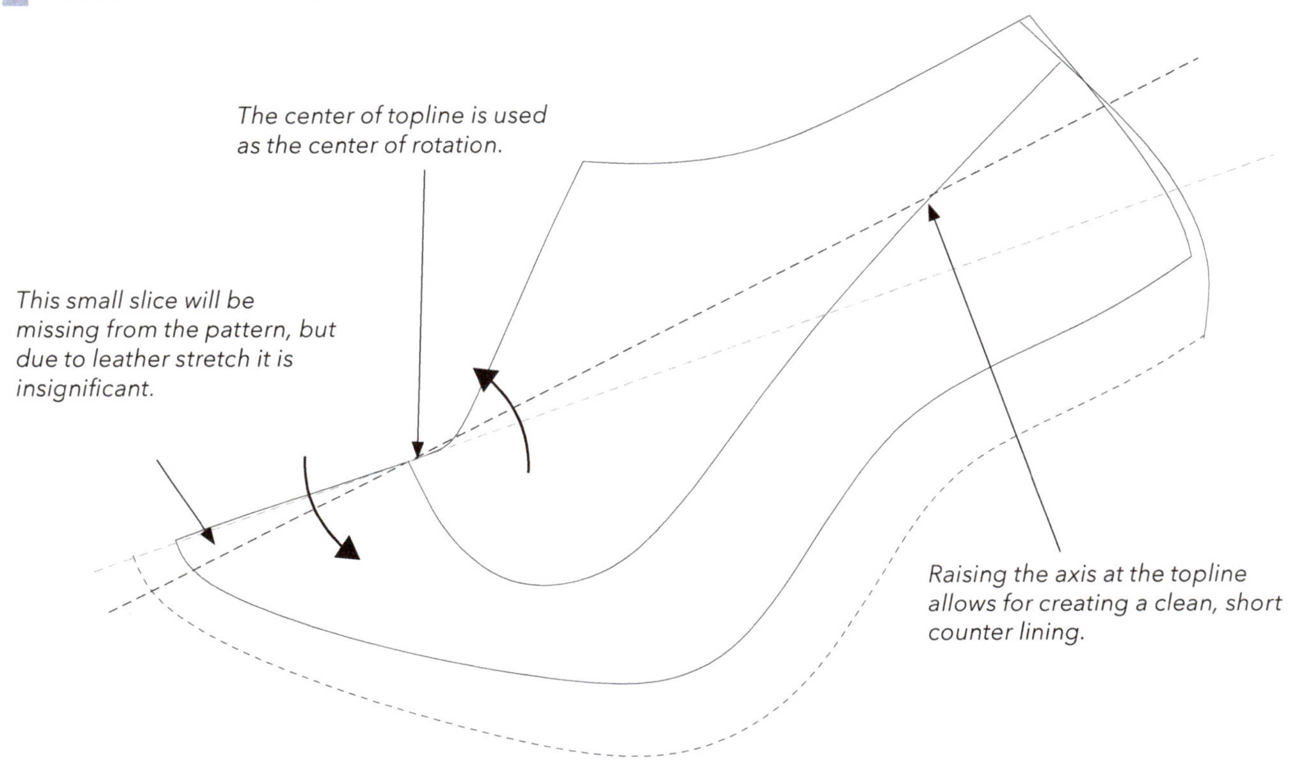

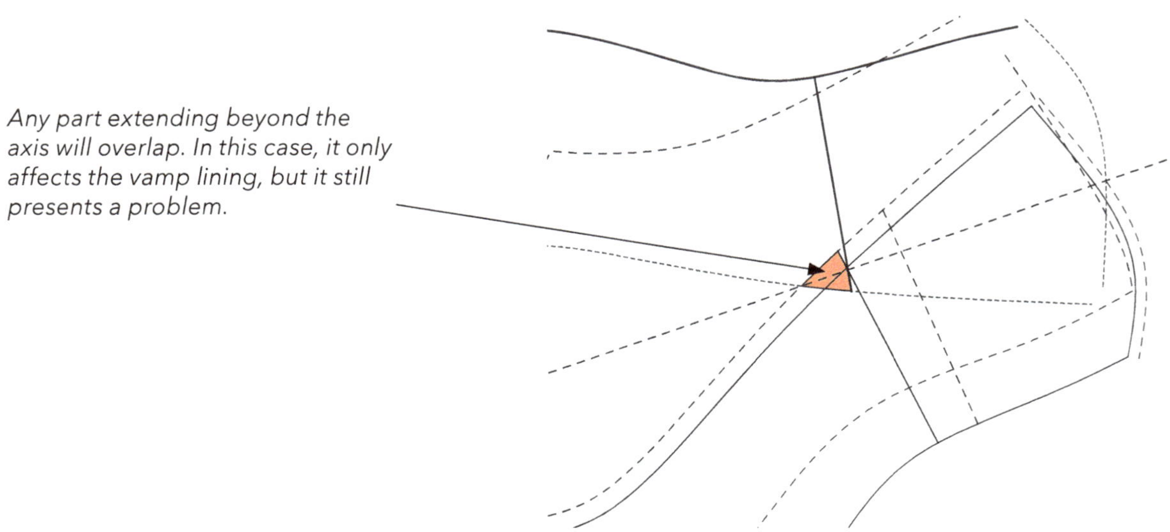

The Medial Side Seam Position

Finding the correct position of the medial-side seam is crucial. As discussed earlier, the hidden section on the last must be identified. When developing the pattern directly on a taped last, this is easy to determine visually. However, when designing the upper in a 2D CAD application or on paper, a reliable guideline is to place the seam approximately halfway along the straight portion of the topline. There is generous tolerance in this placement; as long as the seam remains concealed from the top view, it is considered correct.

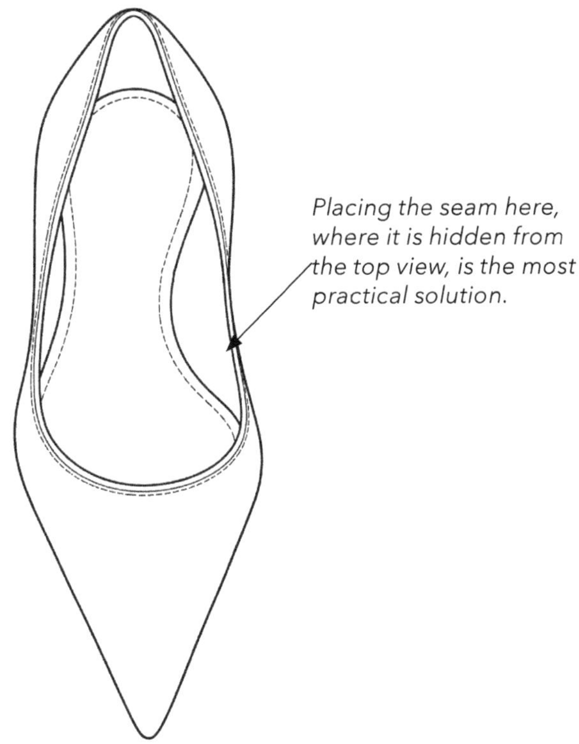

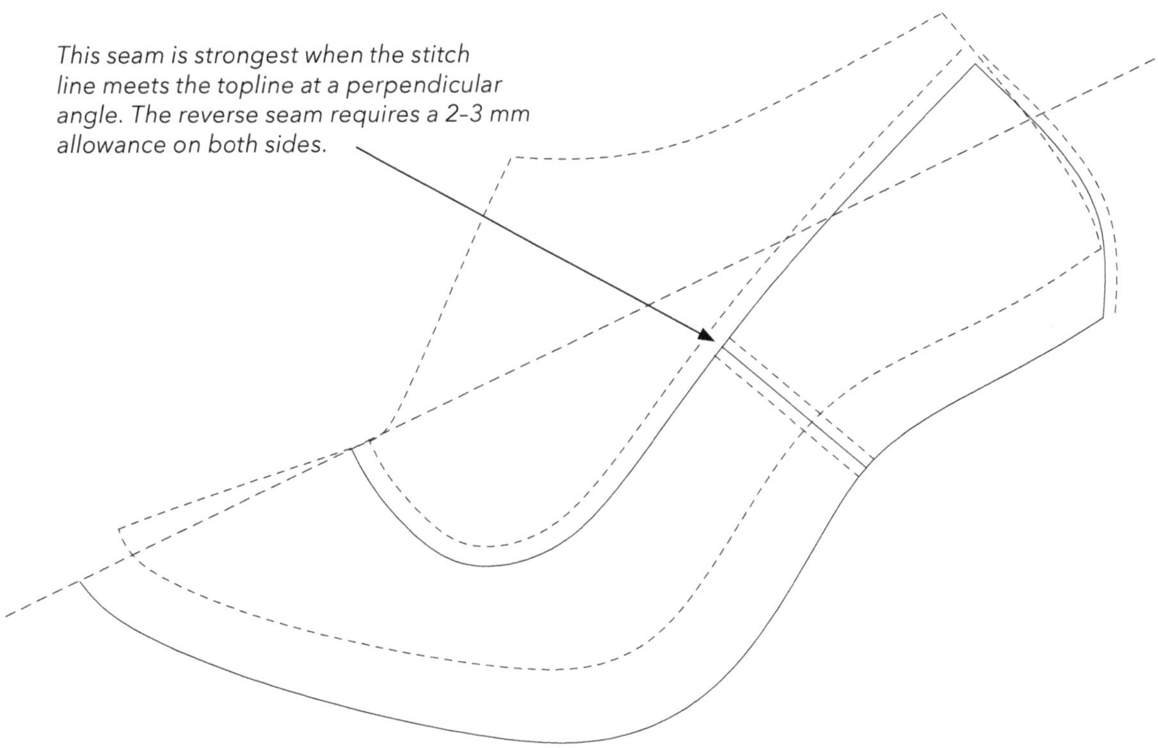

The Counter Lining Folding Axis

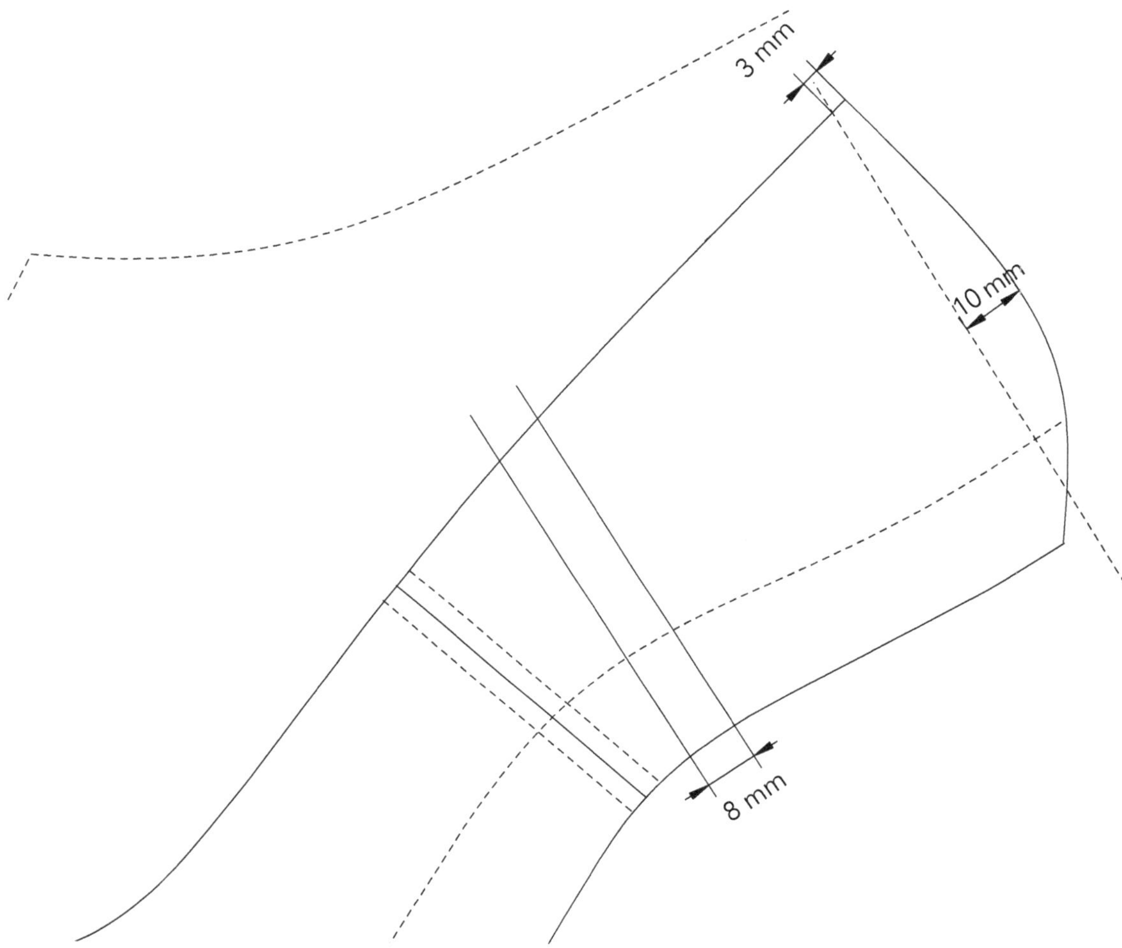

Like discussed in the ballerina shoe, the orange peel effect is very much effect in the high heel upper. The issue arises because the shoe upper follows a longer, outer curve, while the lining follows a shorter, inner curve. When these two parts are stitched together along the backline without compensation, the difference in length and curvature creates a distortion commonly referred to as the orange-peel problem. Refer back to the previous chapter for how to adjust for the orange peel effect.

In this pattern, the necessary corrections are applied here. The lining axis is shifted 3 mm inward at the counter point and 10 mm at the lower third of the backline curve, just above the feather edge. The smaller value relates to material thickness, while the larger adjustment also accounts for the lining's stretch and the presence of the counter stiffener.

Lining pieces

The Counter Lining

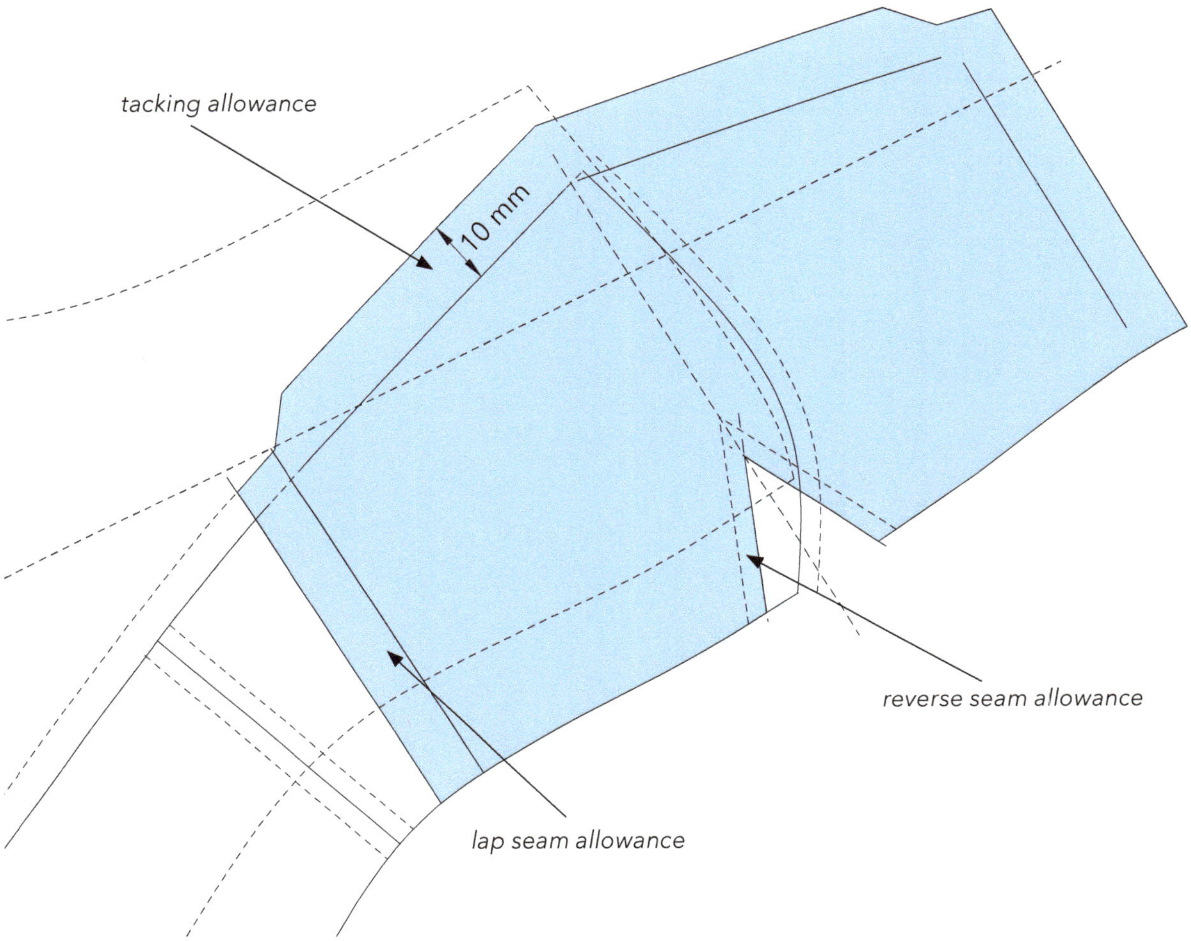

Using the corrected axis, the counter lining can now be created. Only a few allowances and one minor modification are required. At the topline, an 8–10 mm tacking allowance is added. The extended topline of a high heel makes lasting more demanding, and maintaining its exact position is critical. This additional allowance provides extra security during lasting.

At the front, a lap seam allowance is added, allowing the piece to be sewn cleanly to the vamp lining.

A small modification is made at the bottom of the backline. Although the corrected axis would technically already ensure a wrinkle-free visible lining, high heels are a special case: bulk under the heel must be minimized as much as possible. For this reason, a small reverse seam is included here, removing excess material and helping the lining sit more smoothly.

The Vamp Lining

The vamp lining is now ready. The counter's front line is placed as far back as possible. Any further and the vamp would overlap itself. It's self-evident that this is also why the lap seam allowance must be added to the counter lining rather than to the vamp lining. The top of the piece receives a 3-5 mm trimming allowance.

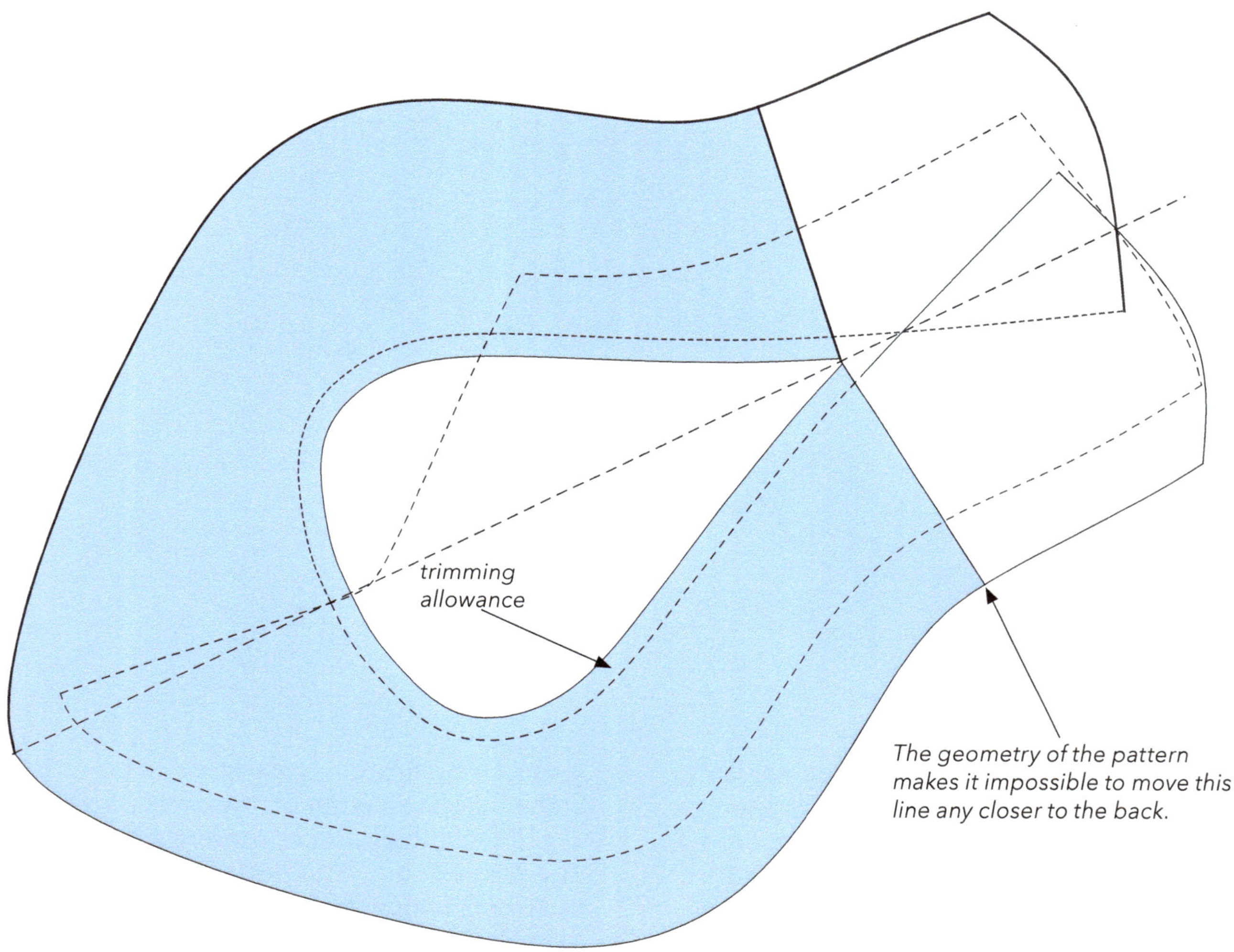

Upper Pieces

The Vamp

The vamp pieces – divided on the medial side, as previously discussed – require a 2-3 mm reverse seam allowance on both pieces. The back seam also receives a 2-3 mm reverse seam allowance. At the topline, a 5 mm folding allowance is added.

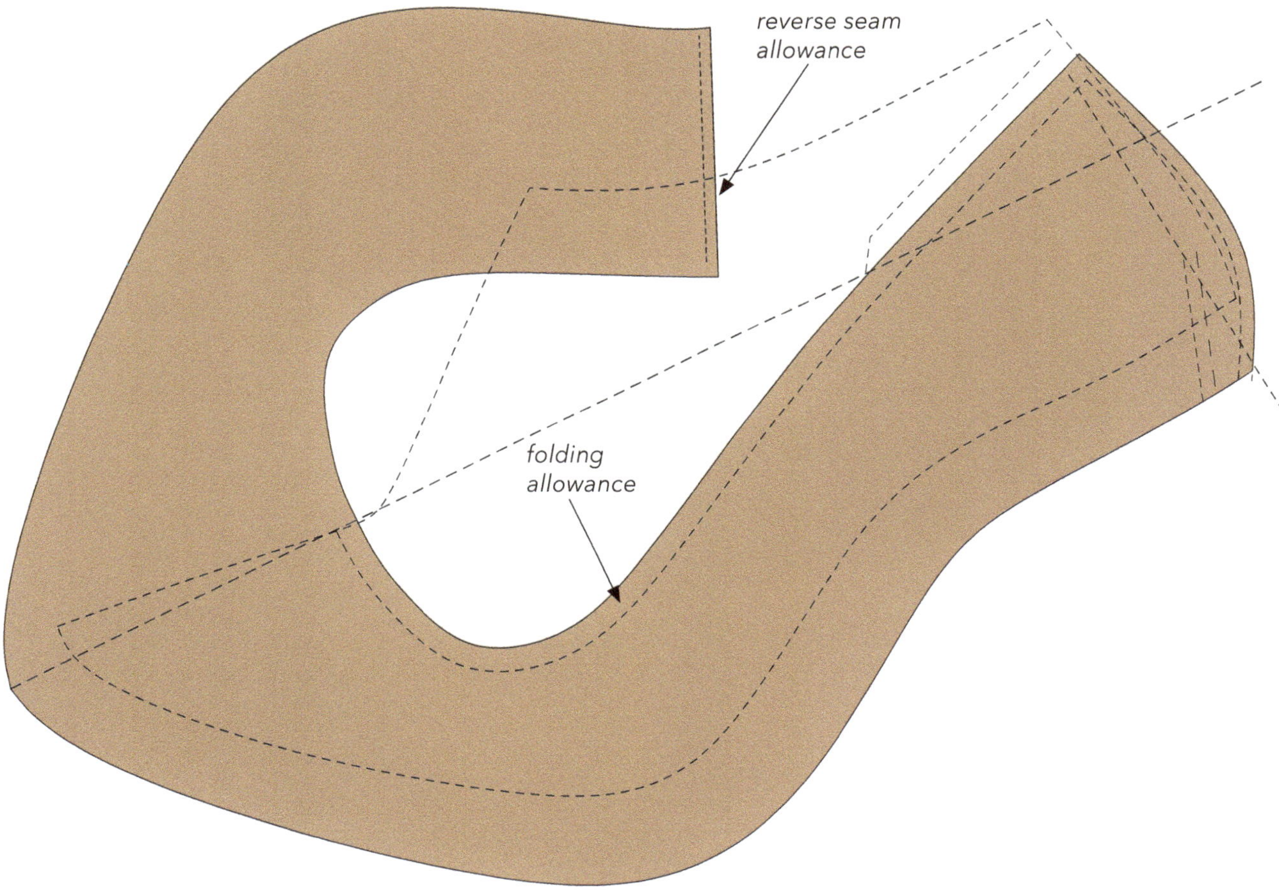

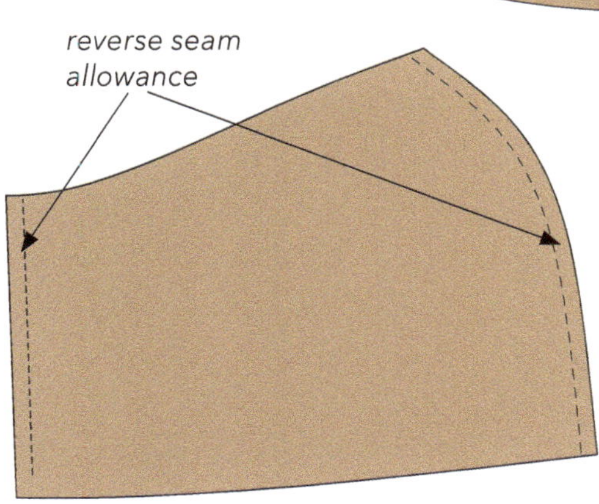

The medial piece – referred to here as a counter, although it is not a counter in the classical sense – is essential to making this pattern possible. It also receives its allowances. Reverse seam allowances are added at both the front and the back, using the same values discussed previously.

The Complete Pattern

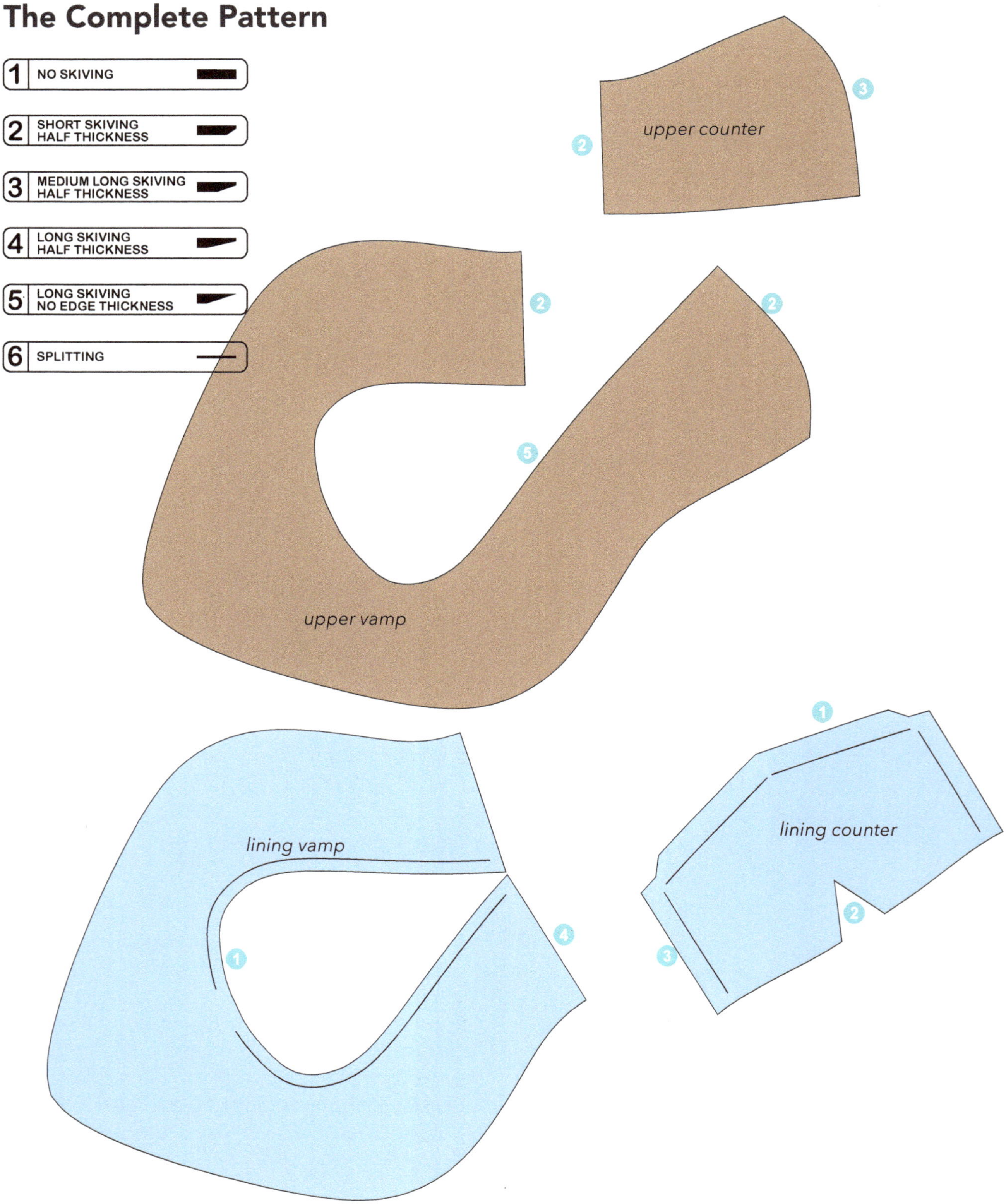

A More Elegant Solution: Pivoting

Pivoting offers another way to create a high-heel pattern while addressing the two-inch heel problem, allowing for the construction of a one-piece vamp. There are clear advantages to this approach, starting with a reduction in labor; the assembly line has one fewer seam to manage, which minimizes sewing, rubbing, and taping. This results in a significant decrease in the overall workload.

Simply rotating the main axis upward, using the topline as the pivot point, may appear similar to the correct pivoting method, but it is an ineffective solution. An over-rotated axis removes excessive material from the vamp area. This loss of volume and surface area can lead to significant tension during lasting and may prevent the leather from properly conforming to the last.

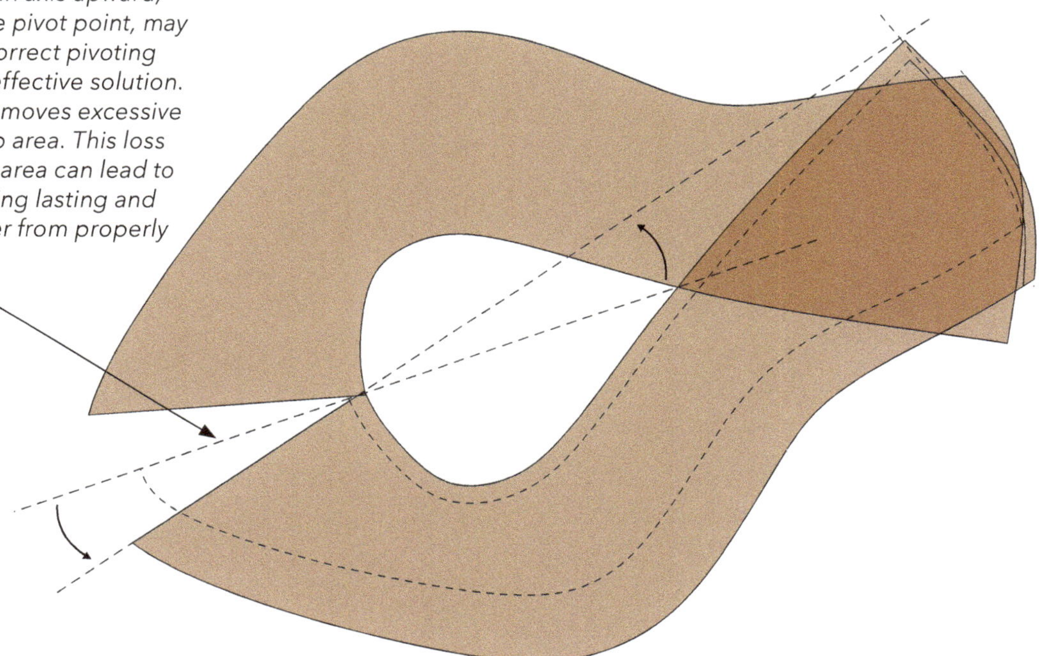

Simply rotating the axis upward–using the lower end near the feather edge as the pivot point–is also a poor solution. This creates excess topline length and adds material to the middle of the vamp, which will distort the lines after lasting. In practical terms, this approach lowers the carefully positioned topline far too much.

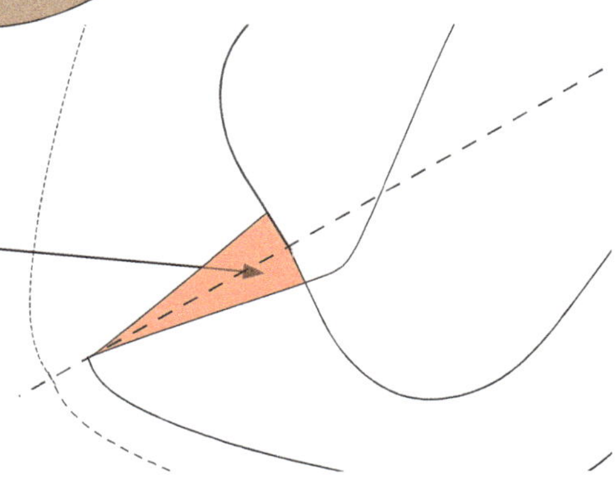

Material considerations provide a second reason for this method, particularly when working with exotic leathers like alligator. In such cases, a two-piece upper makes scale matching extremely difficult and complicates the positioning of pieces during cutting. Pivoting solves these issues by keeping the pattern undivided and preventing axis-related problems. Before examining the correct technique, it is helpful to review how oversimplifying this process can lead to an unsatisfactory result.

Pivoting the Pattern

Here is an advanced yet simple solution. Pivoting the entire vamp downward keeps the front section completely intact, so lasting is not affected. The topline changes slightly, but only in the curved portion. Straightening this curve is not a concern; this type of adjustment is routine when crimping or pivoting pattern pieces.

The major change occurs at the bottom edge, which inevitably becomes shorter. This is a natural consequence of pivoting. What matters is making a conscious decision about this trade-off while keeping the topline length completely unchanged. If the topline length were altered, the upper would no longer fit the last correctly.

This solution causes two major changes to the pattern. The bottom length becomes tighter, which is generally beneficial and only potentially problematic at the medial shank during lasting. The second change may require additional attention: during hand lasting, the topline can slide downward. This can be addressed easily with counter shapers, and a tacking allowance can also be used to correct the issue.

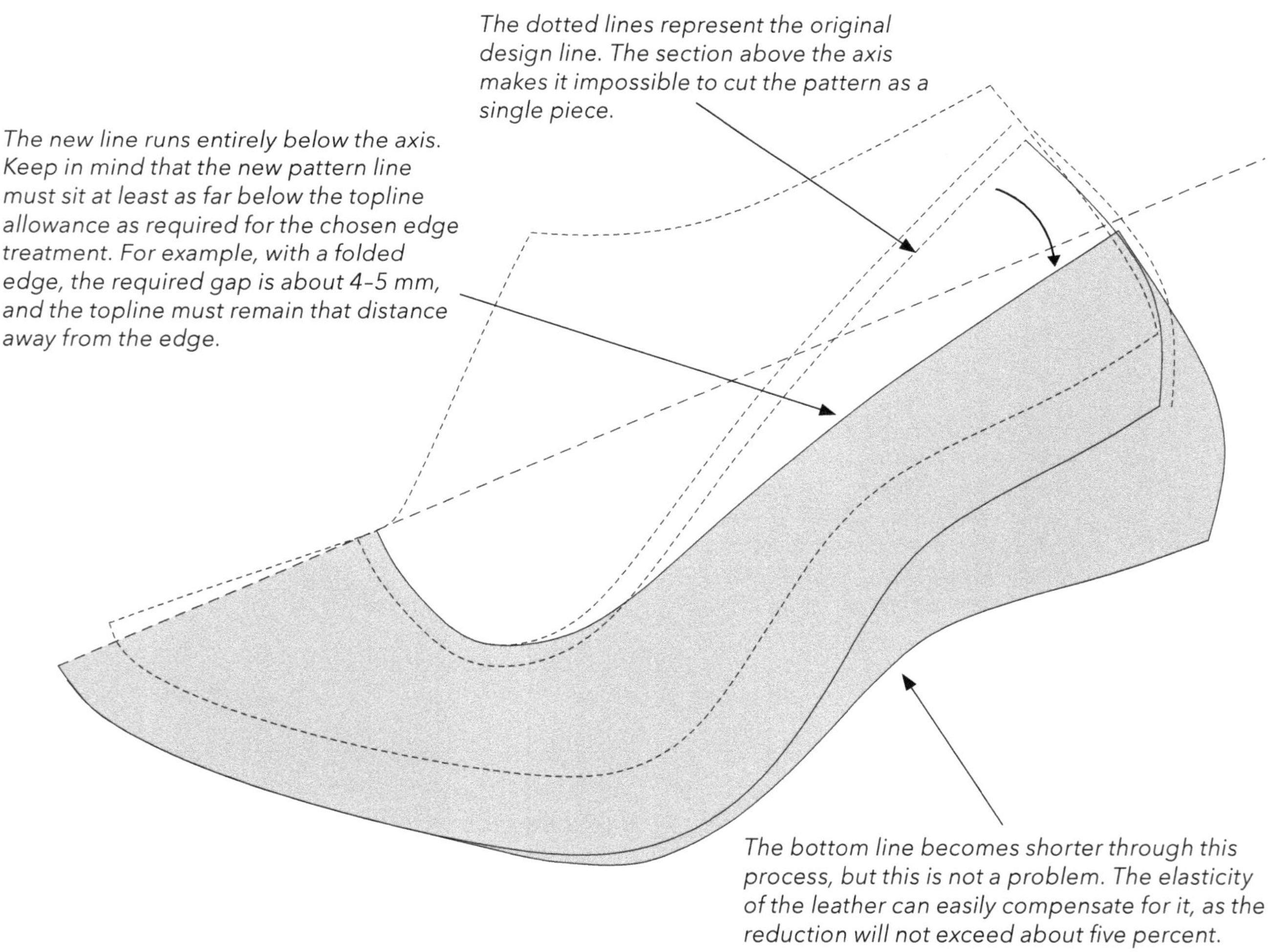

The Pivoted Vamp

The vamp can be unfolded along the main folding axis without cutting away any material from the front or the counter area. With a 5 mm folding allowance, this pattern is fully workable and technically feasible. The real question is whether the leather will tolerate this adjustment – and in most cases, the answer is yes. This solution does not affect the lining structure, apart from the new axis, which no longer restricts the lap seam position.

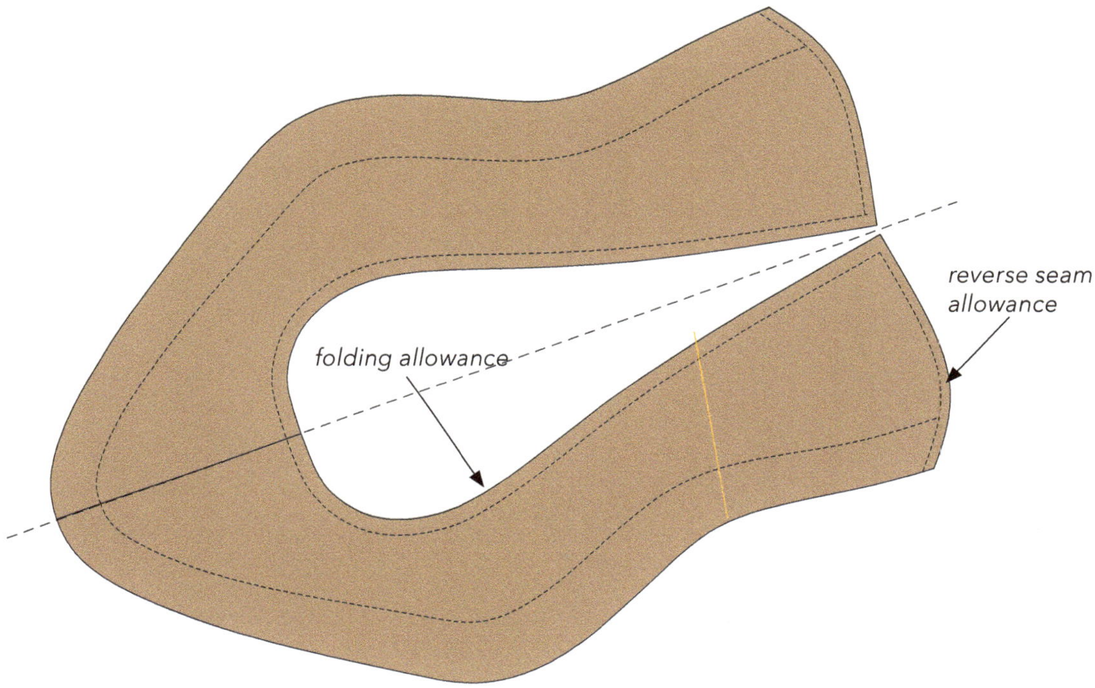

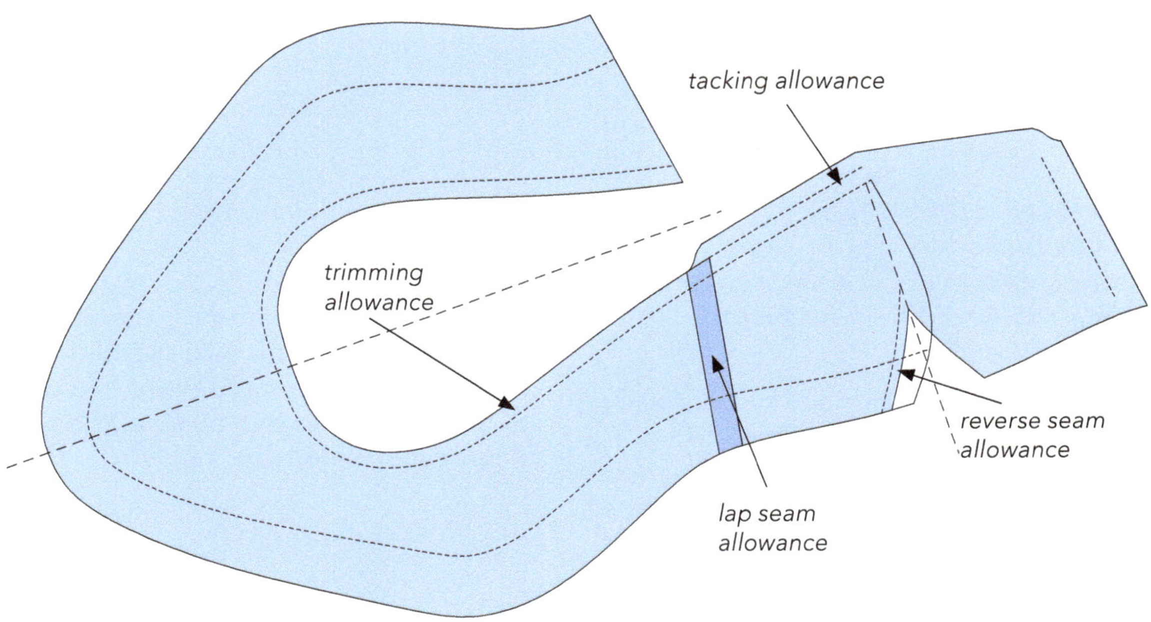

The Asymmetry of the High Heel Pattern

The geometry of the last creates an effect that becomes clearly visible on a low topline, as seen on a pump. To achieve a visually symmetrical result, the pattern itself must be asymmetrical, even though this may sound counterintuitive. This is, in fact, very common in pattern making; most patterns can be asymmetrical to some degree.

The counter portion is almost always symmetrical for a regular pump.

The asymmetrical topline will result in a symmetrical looking vamp, since this asymmetry is just compensating the last shape.

The difference is subtle. Typically, the medial side of the topline sits slightly higher, while the lateral side is lower. The variation is usually only a few millimeters, or whatever the last geometry dictates.

Rather than relying on formulas or geometrical pattern rules, the most efficient way to determine this difference is to draw the pattern directly on the taped last. This approach often eliminates the need for a full test run.

Not all pattern pieces need to reflect this asymmetry. The vamp and vamp lining must be adjusted accordingly, but the lining counter does not necessarily require modification. Even the lasting allowance may differ slightly; in practice, the more generous allowance can be used, while the topline should be corrected to appear symmetrical.

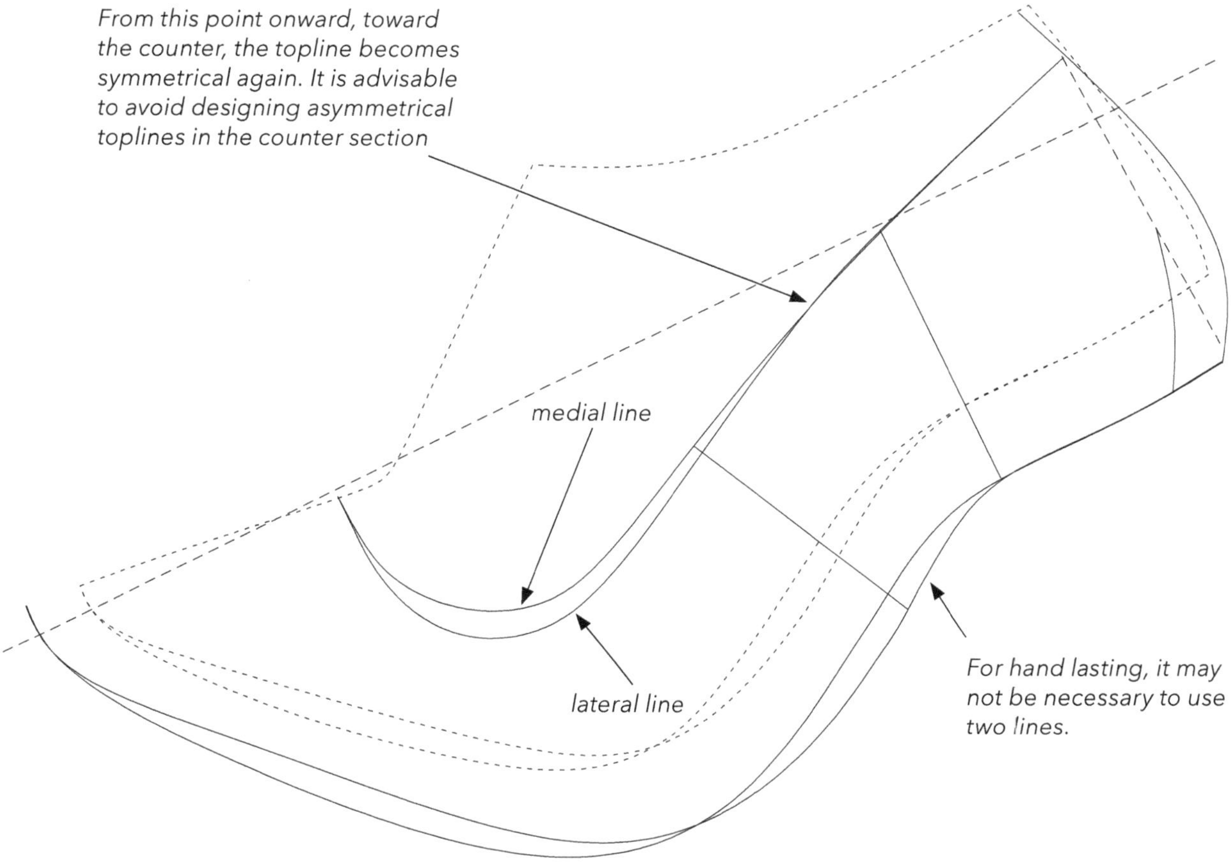

The lasting allowance behaves somewhat differently. The medial and lateral lines can cross each other, which complicates the pattern; therefore, it is advisable to clearly mark the medial and lateral sides to avoid confusion later.

When dealing with two different lasting lines, there are four possible approaches:

1. use only one of them (typically the one that provides more allowance), ensuring the upper will not be too small;
2. use the average of the two;
3. use both lines;
4. Use whichever line provides more allowance - switch if necessary.

These three options must be considered whenever asymmetry is involved.

To determine which approach is most appropriate, the following guidelines can be used:

- if the difference is less than 5 mm, using both lines may not be necessary;
- for slip-lasting, there is no ambiguity - asymmetry must be respected and both lines should be used;
- certain models and pattern elements, especially in the front part of the vamp, commonly require different medial and lateral design lines;
- asymmetry is rarely comfort-related, but when it is, the more comfortable option (for example, a lower topline) should be used, or both lines applied;
- for factory production, asymmetry is almost always the preferred solution.

7 THE LOAFER SHOE

The loafer is a slip-on shoe defined by its absence of lacing, low profile, and relatively shallow topline, combining ease of wear with a wide range of stylistic expressions. Its origins are commonly traced to early 20th-century Europe, with strong influences from Norwegian moccasin-type shoes and informal house shoes worn by the British upper class. These early models emphasized comfort and practicality, featuring apron seams and flexible constructions. By the 1930s, the loafer began its transition from casual footwear to a more codified shoe type, particularly after American manufacturers refined the form and introduced it to a broader market. From a design and construction standpoint, the loafer presents a distinct set of technical challenges. The absence of lacing eliminates a primary adjustment mechanism, placing greater importance on precise last selection, accurate pattern making, and careful control of girth. Small deviations in vamp length, topline shape, or counter position can significantly affect fit and wearability.

The Loafer Pattern Using The Geometric Formulas

The loafer has a few key rules that must be kept in mind to create truly functional footwear. In this context, "functional" has two meanings: the wearer can slip into the shoe without excessive effort, and the shoe remains securely on the foot without heel lift or slippage during walking.

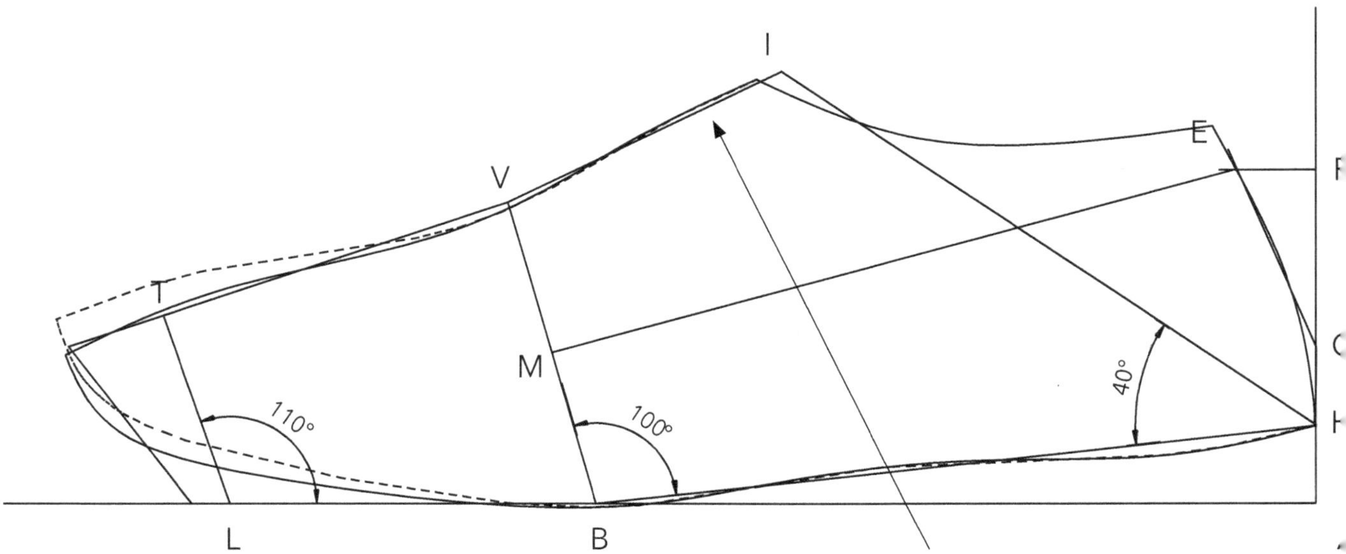

The difference is so minor and, in practice, may not need to be considered during pattern construction, as it occurs along a single axis and therefore has minimal overall impact. However, if it ends up taken into account, a 5 mm adjustment downward from the instep point is sufficient.

A proper pair of loafer should be made on a loafer last. This last is almost identical to a standard dress-shoe last, but with one important difference: the instep is slightly smaller. This tighter waistline compensates for the open-throat construction, ensuring the geometry of the vamp opening works with the foot rather than against it. In other words, the last itself accounts for the lack of lacing and helps maintain a firm, reliable fit.

To further reduce heel slippage, a suede side-in counter lining is used. This is a standard solution for footwear with three-quarter coverage, providing friction and grip where it is most needed to keep the heel stable.

The Waistline Problem

In order to slip into a three-quarter, non-opening shoe such as a loafer, the topline must be positioned accordingly. In practice it means that in a standard medium-wide last, the waist girth is fairly close to the ball girth, so the ball of the foot can pass through a relatively tight waistline. If the foot is wider, this requires more effort. Fortunately for shoemakers, the human foot is not a rigid object; it can compress and adapt enough to pass through a snug opening. It may not be comfortable, but it is possible.

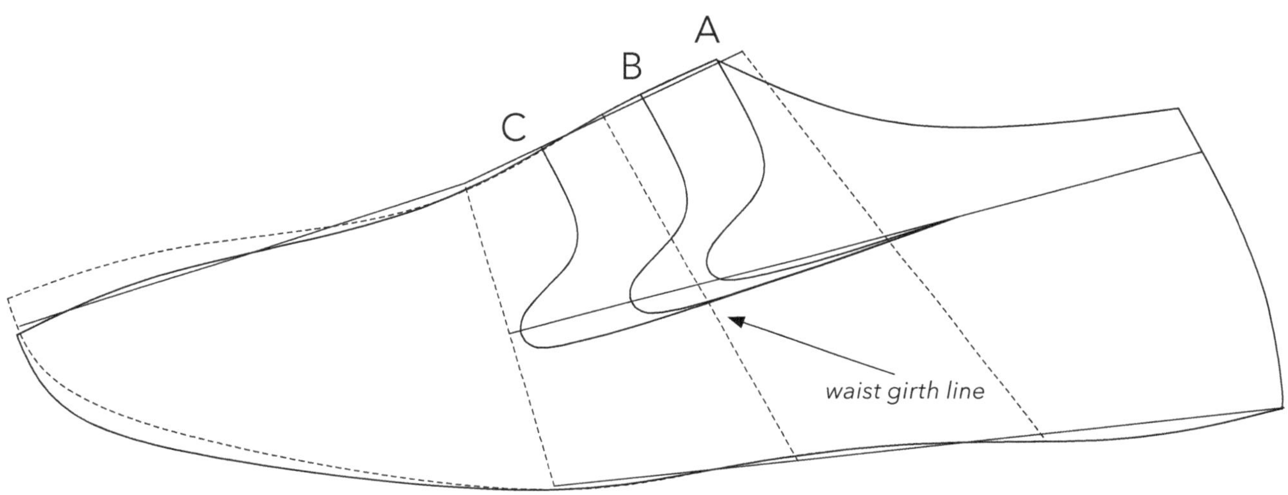

This momentary discomfort is something customers usually tolerate in boots, where difficult entry is common and somewhat expected. In a loafer, however, it becomes a persistent issue – not only during entry, but also through ongoing pressure and irritation at the topline. Because the shoe has no lacing and relies entirely on the geometry of the opening, poor topline placement can lead to constant discomfort.

This is why the waist girth must not be closed down too tightly. It needs to be able to widen temporarily – just enough for the foot to pass through. Once the foot is inside, the waist should return to its intended shape and tension, providing the secure fit the loafer relies on.

The three solutions – A, B, and C –represent different topline placements.

Line A closes the waist girth line entirely, making it a very risky option. It may be wearable for people with narrow feet, but for most wearers it is impractical and difficult to slip into.

Both B and C are functional, workable solutions. B provides an ideal starting point for a classic loafer design, offering a balanced topline height that maintains fit without restricting entry. C is a lower, more refined cut that works well for short-vamp loafers, court shoes, and similar elegant styles.

The Main Folding Axis Problem

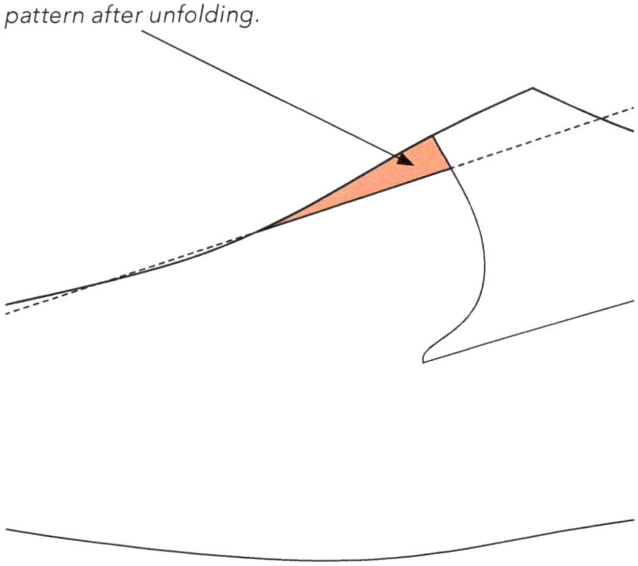

For an apron-seam model, the ideal axis should run close to the last-copy line. If it does not, the apron design line may become distorted. This means the upper part of the vamp – referred here as the tongue – will be directly affected by the axis. A noticeable portion of this area would be cut away during the correction process, and that missing material would compromise the pattern.

Adjusting the axis is not a viable solution for this type of model, and pivoting the entire vamp would be excessive and unnecessary. Instead, a far simpler and more precise approach is to pivot only the tongue and the topline. This preserves the apron design line, avoids unwanted distortion, and keeps the correction localized to the area where it is actually required.

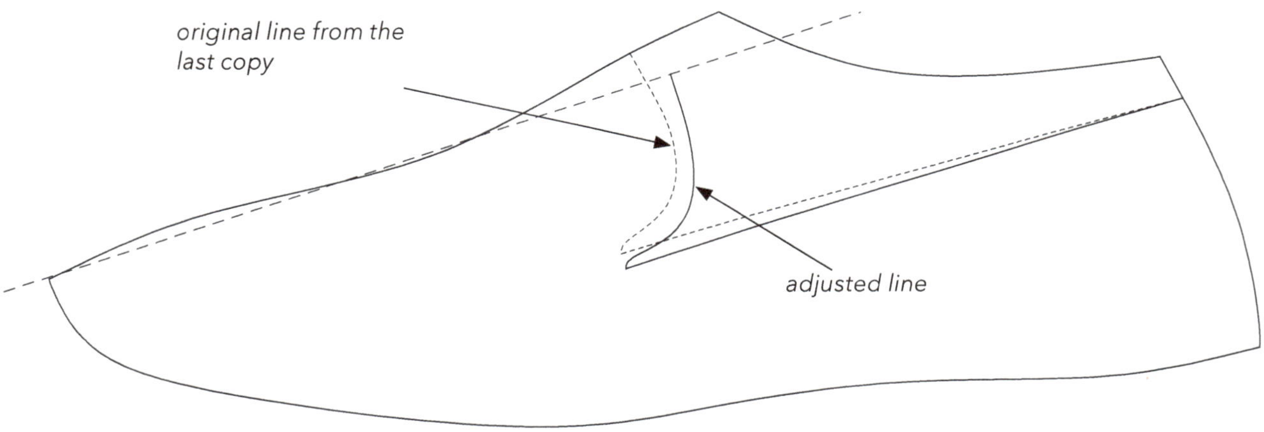

The adjusted line runs slightly lower than the original, which preserves both the width and the shape of the tongue. Although the topline now sits lower on the pattern, this is not a problem. During lasting, the geometry of the front cone profile will naturally pull the topline back into its intended position, restoring the correct height and silhouette.

The Apron Seam

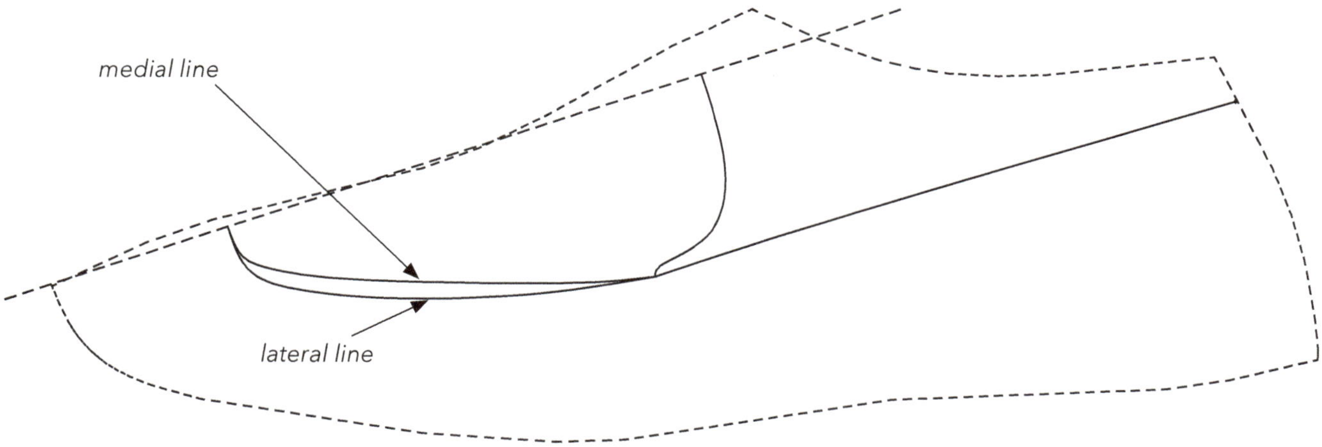

The apron line can be somewhat tricky to draw. While it can be made perfectly symmetrical, the result is often more visually appealing if a slight asymmetry is introduced. On the medial side, the line should run a few millimeters higher and follow a slightly straighter path (generally speaking, medial-side lines usually sit higher because of the geometry of the last).

This subtle difference creates a more balanced look once the shoe is lasted and viewed on the foot.

To determine the ideal shape, it is very helpful to draw the apron line directly on the last. This allows the true 3D surface to be followed and ensures the final pattern aligns with the geometry of the last.

The Asymmetry of the Apron

The apron seam, or, if it is a separate piece, the pattern piece, is probably the best example of why asymmetry is the correct way to place this line. Interestingly, a truly symmetrical line would appear asymmetrical, which is an optical illusion caused by the shape of the last. Regardless, it must be corrected, as the pattern maker aims for a visually balanced result rather than perfect geometry. Raising and straightening the apron line on the medial side will make it appear symmetrical.

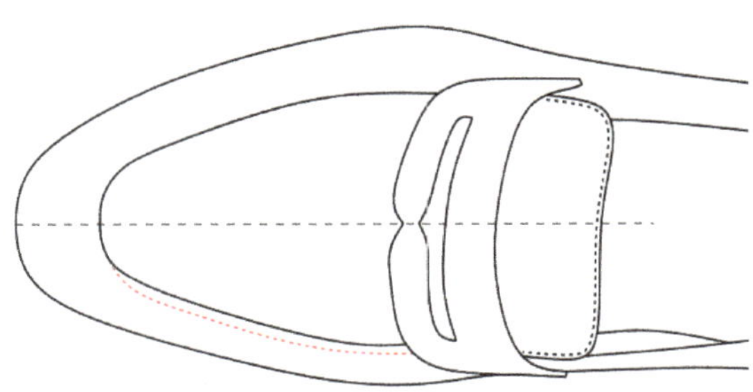

Simply mirroring the design line from the lateral side (the red, stacked line in the illustration) would appear asymmetrical because of the geometry of the last.

The Open Pattern

When working with asymmetrical patterns, many pattern makers - especially those who work primarily on paper - prefer to unfold the pattern and work with both sides laid out flat. This method requires copying the relevant design lines onto their corresponding sides.

In this example, the apron line that sits higher is mirrored onto the medial side. The dotted line on the medial side is shown only for comparison.

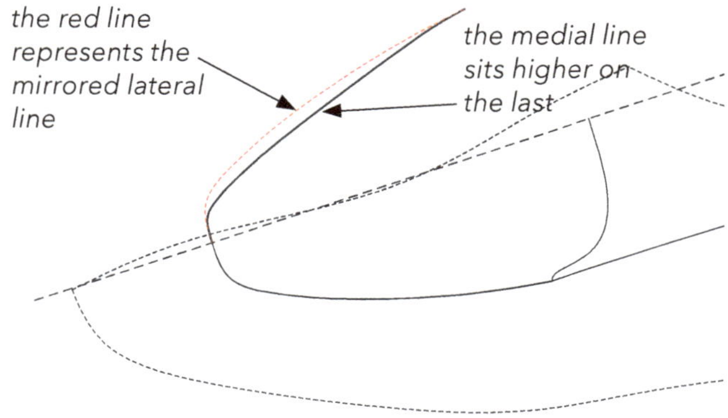

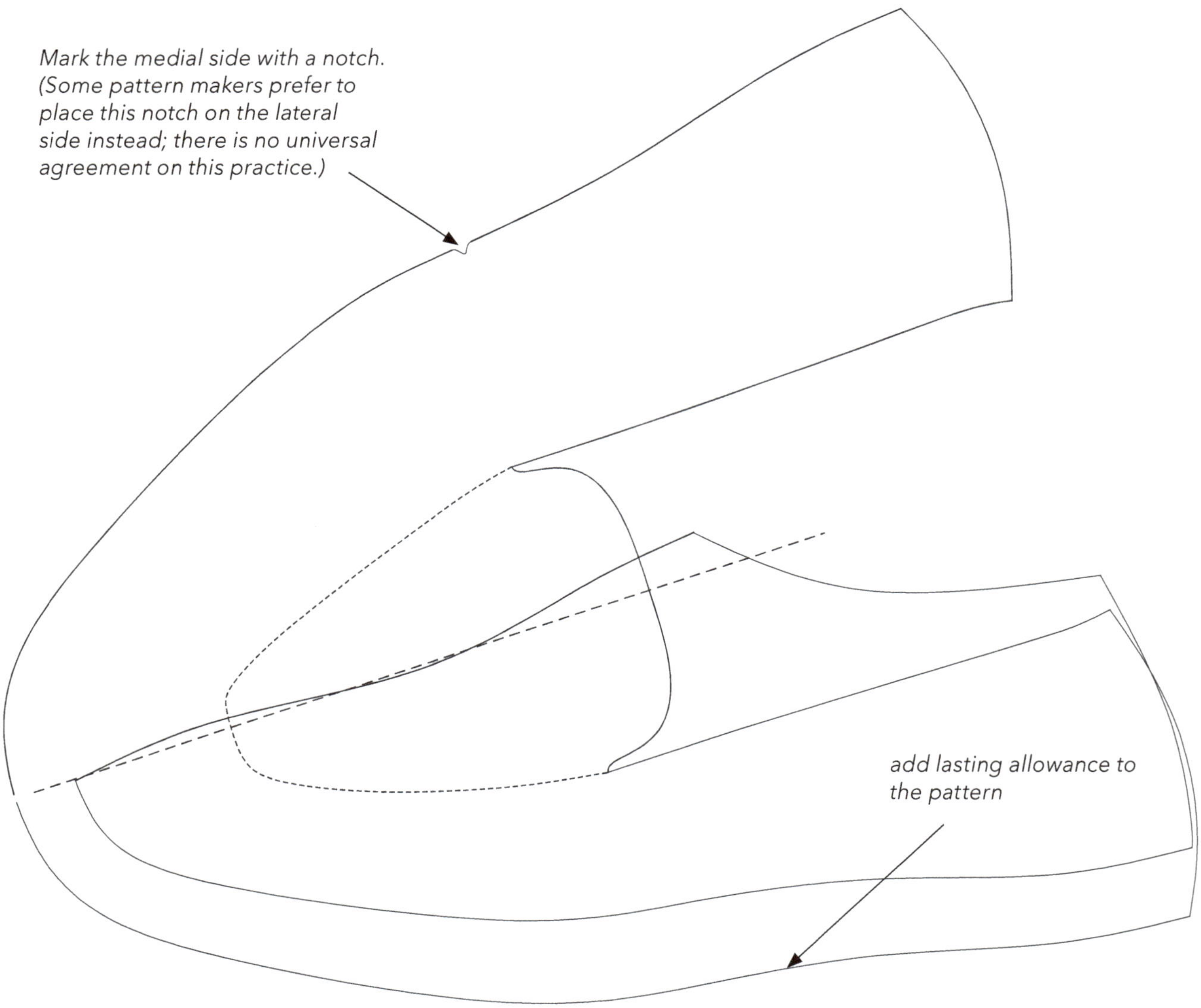

The Strap Construction

The strap is a design element, so there is no strictly "right" or "wrong" outcome. It can follow a classic layout, as in this example, or take on a completely new, creative form. The choice depends entirely on the intended style and aesthetic.

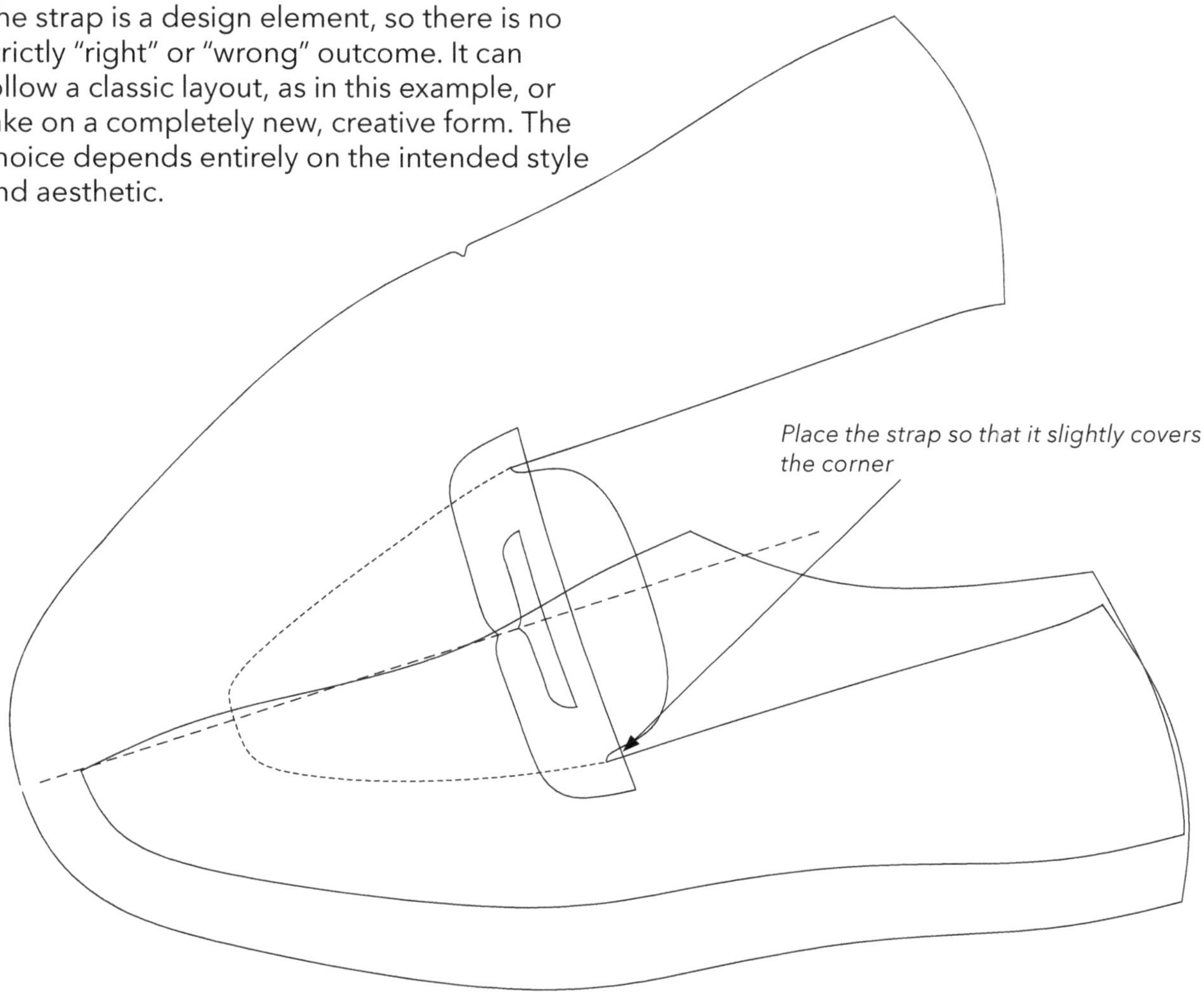

This is the construction of a very classic strap design. Most pattern makers have these as a template and reuse them instead of creating them again. This can result in substantial savings in tooling, as no new die-cutting knife is needed, and helps maintain consistency in a brand's style.

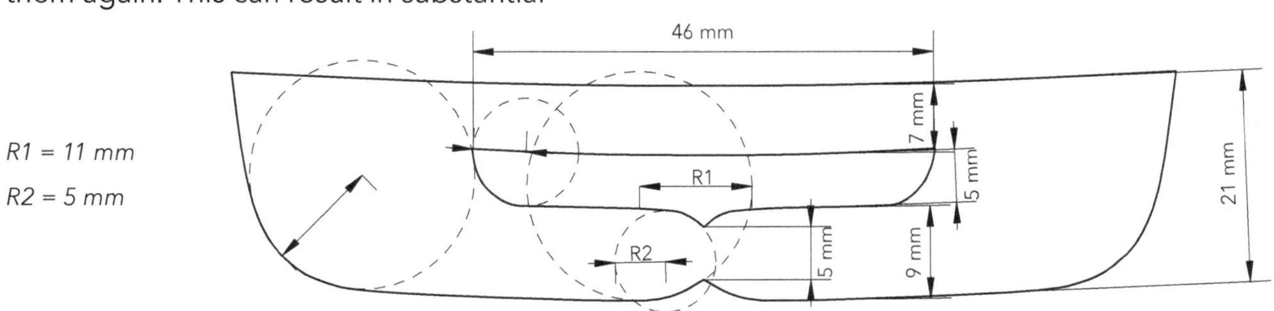

The Stitchhole Problem

For obvious reasons, the stitch holes must sit in pairs on the apron front; otherwise, the apron stitch would drift noticeably. This approach produces two matching stitch lines, and the area between them will be slightly raised, contributing to the signature aesthetics of the shoe.

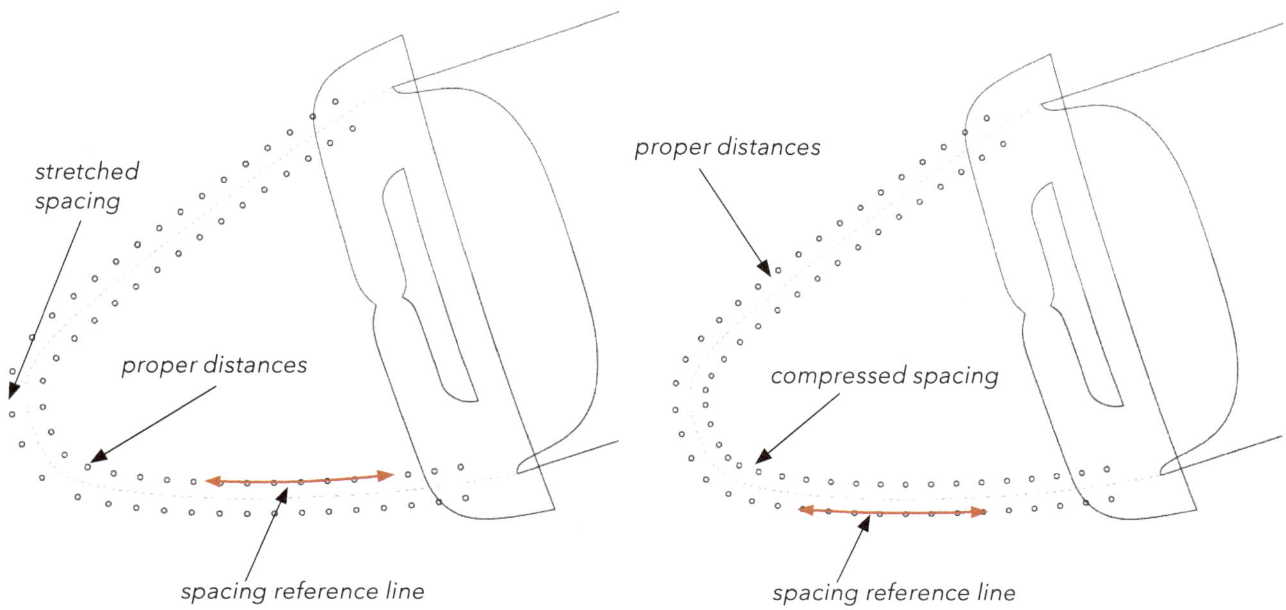

The challenge is that the two lines of holes follow very different curves, so decisions must be made about spacing. If a 5 mm gap between holes is chosen, it can be applied to the inner line and matched on the outer line, or vice versa. In either case, one of the lines will not strictly follow the 5 mm spacing rule.

The illustration on the left shows the result of measuring the distance along the inner curve. The holes themselves are placed evenly, but the outer curve appears less than optimal – the gaps become noticeably elongated as they follow the curve.

The illustration on the right shows the result when the distance is measured along the outer curve. The distances become noticeably reduced, creating a real risk: some of the gaps may end up too small to hold the stitch properly. Under tension, these small sections of leather can easily be damaged or even tear through.

The obvious solution to the problem is putting the reference line between the two hole lines: the holes sit slightly closer together on the inner curve and are gently elongated on the outer curve. The actual feasibility depends on several factors, such as the type of leather and the desired overall effect. The distance between the two stitch lines can be reduced, which decreases the difference between the two radii and minimizes distortion. Another option is to increase the overall stitch length; with longer stitches, the variation between the inner and outer curves becomes far less noticeable, regardless of where the measurement is taken.

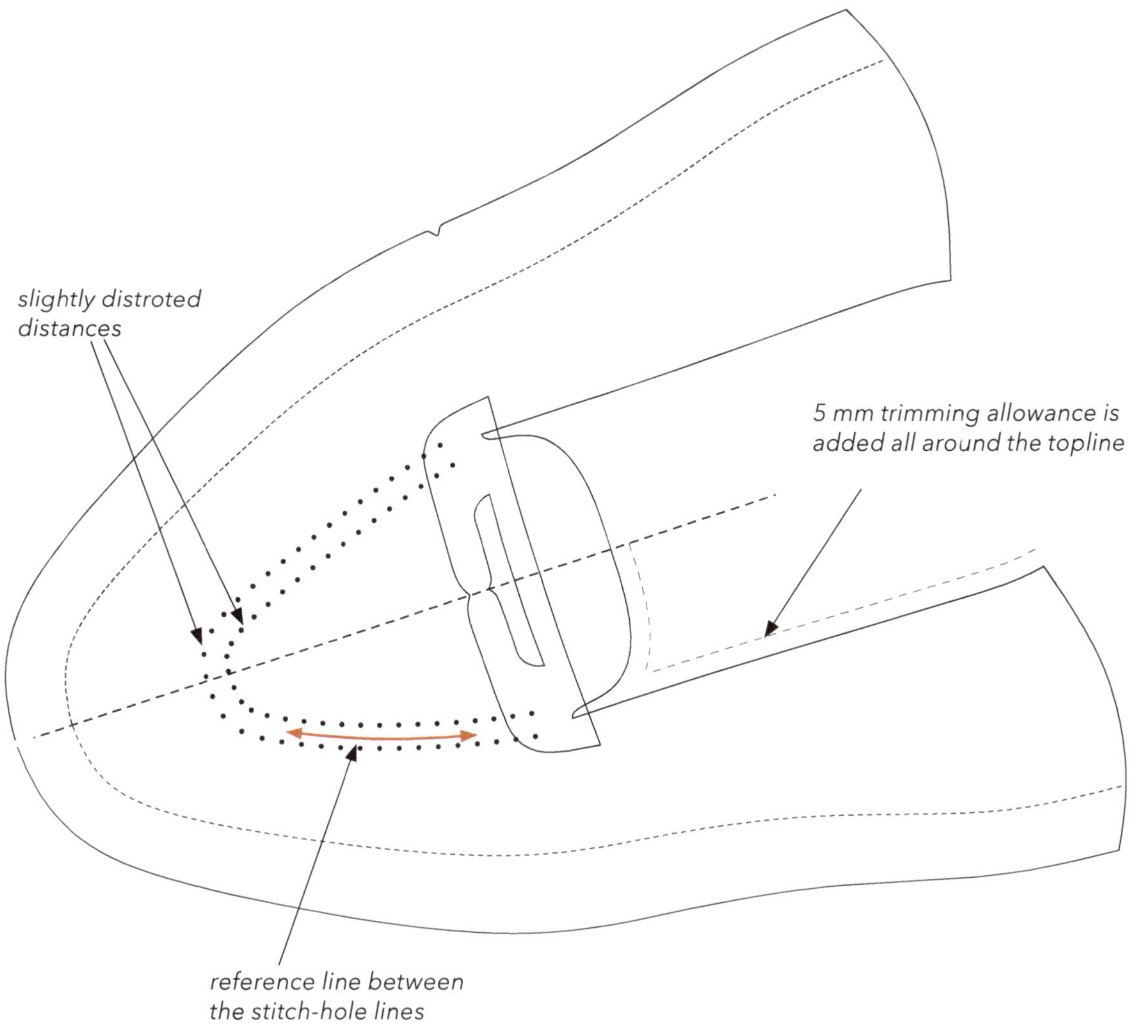

Upper Pieces

The upper pattern is complete. The position of the strap can be marked for accuracy.

In case of a prototype or bespoke product, the marking can be done later and of course the placement should be documented on the master pattern.

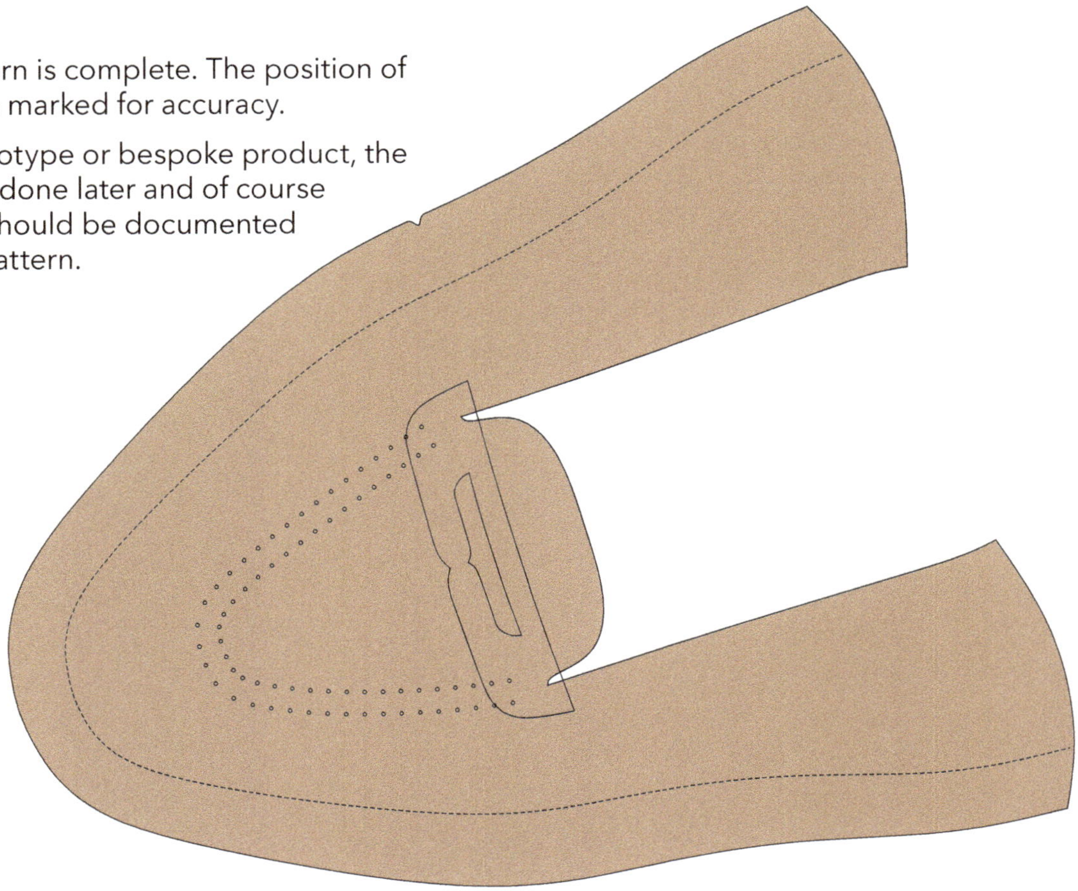

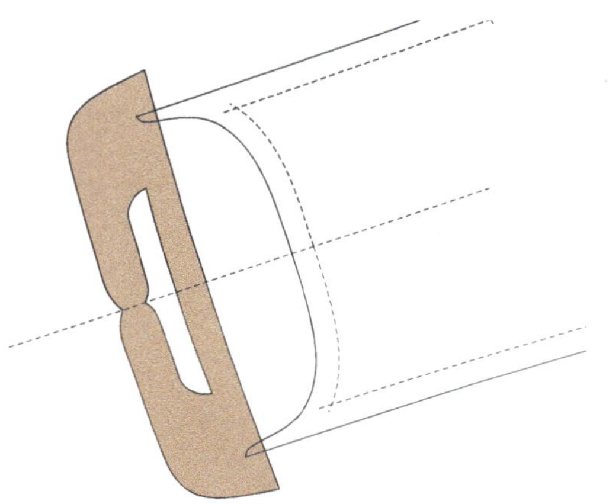

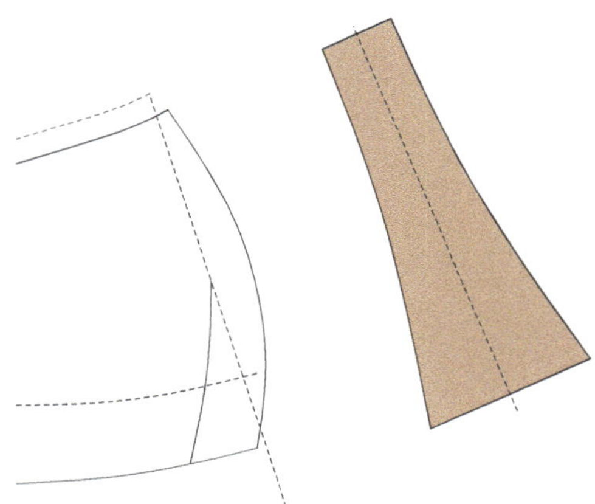

The strap pattern is a relatively difficult piece to cut by hand, so using a die-cutting knife is a very good choice.

The backstrap is also a very general shape; it simply needs to be wide enough to cover the zigzag seam and long enough to span the backline.

The Counter Lining

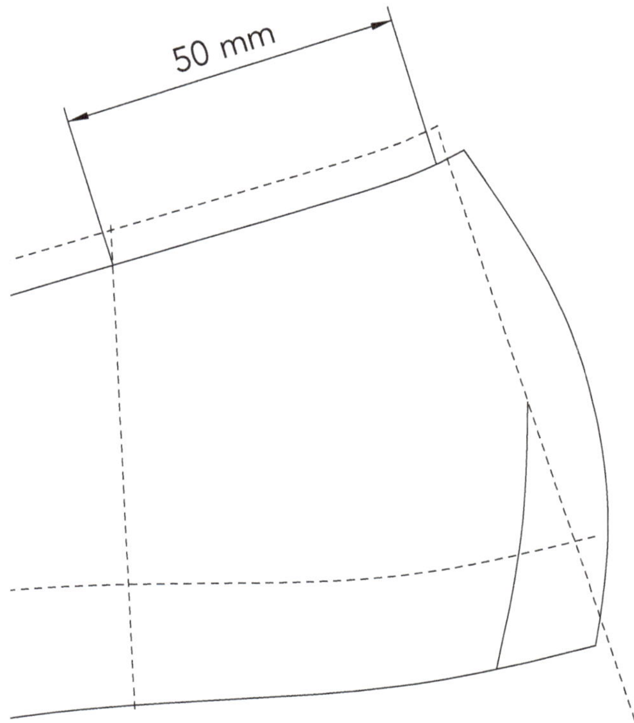

The counter lining pattern and its construction are almost identical to those in the previous models. The main difference is that, in this case, the placement of the lap seam allowance between the counter and the vamp lining is flexible, and it can be assigned to either the counter piece or the vamp lining piece.

The counter axis is identical to the previous models: a 3 mm gap at the top and a 10 mm gap along the lower third of the backline (this is explained in detail in the ballerina shoe chapter).

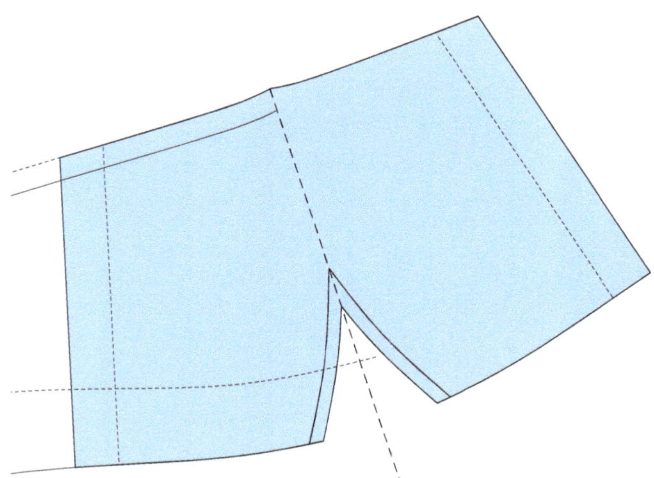

The pattern piece receives a small reverse seam allowance at the bottom, no allowance on the front edge (so it will overlap the vamp lining piece), and a small trimming allowance at the top. In the ballerina shoe, this top allowance was much larger to allow for tacking, but that is not possible here because this pattern uses binding along the topline.

The Lining Pieces

The lining pieces are relatively simple. The vamp receives a 5 mm trimming allowance and an overlap allowance toward the counter. The counter has only a reverse seam allowance and a small trimming allowance along the top.

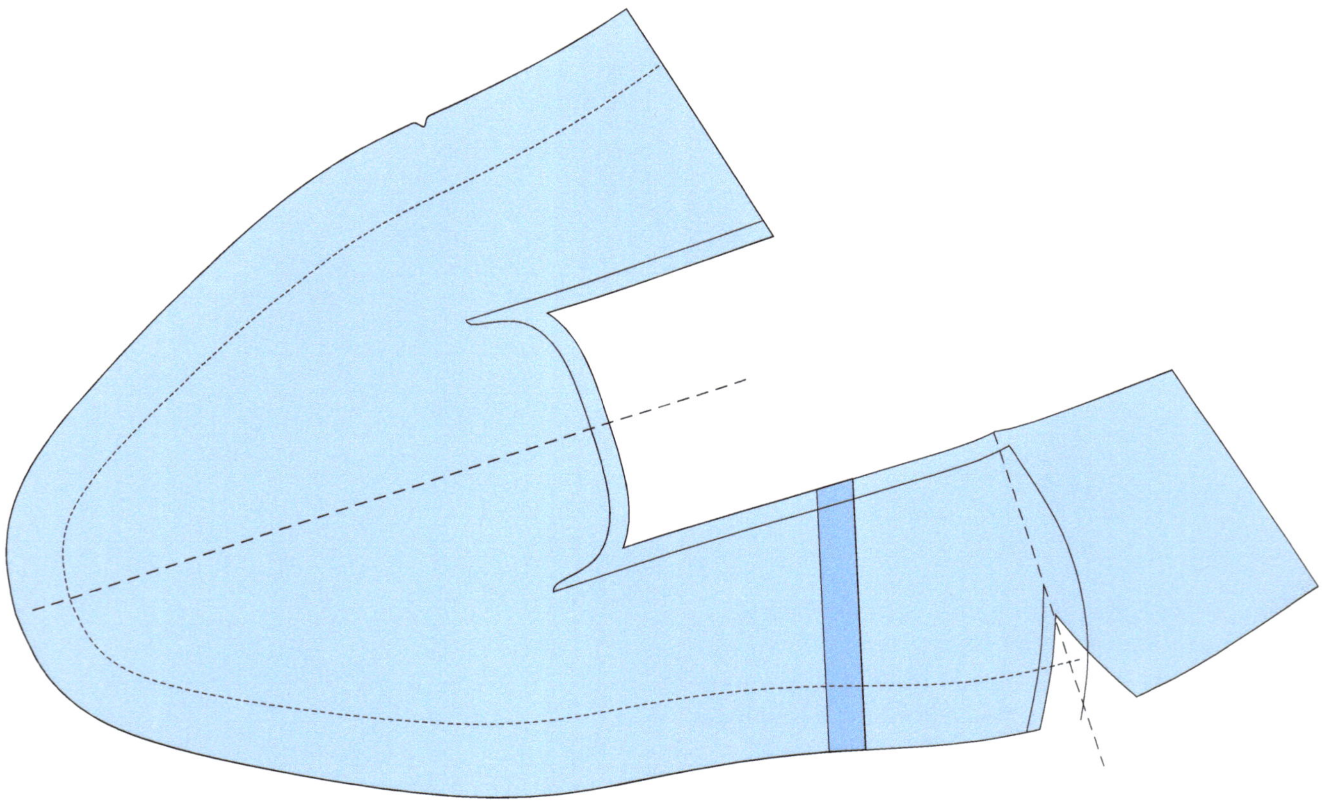

The Complete Pattern

1 NO SKIVING

2 SHORT SKIVING HALF THICKNESS

3 MEDIUM LONG SKIVING HALF THICKNESS

4 LONG SKIVING HALF THICKNESS

5 LONG SKIVING NO EDGE THICKNESS

6 SPLITTING

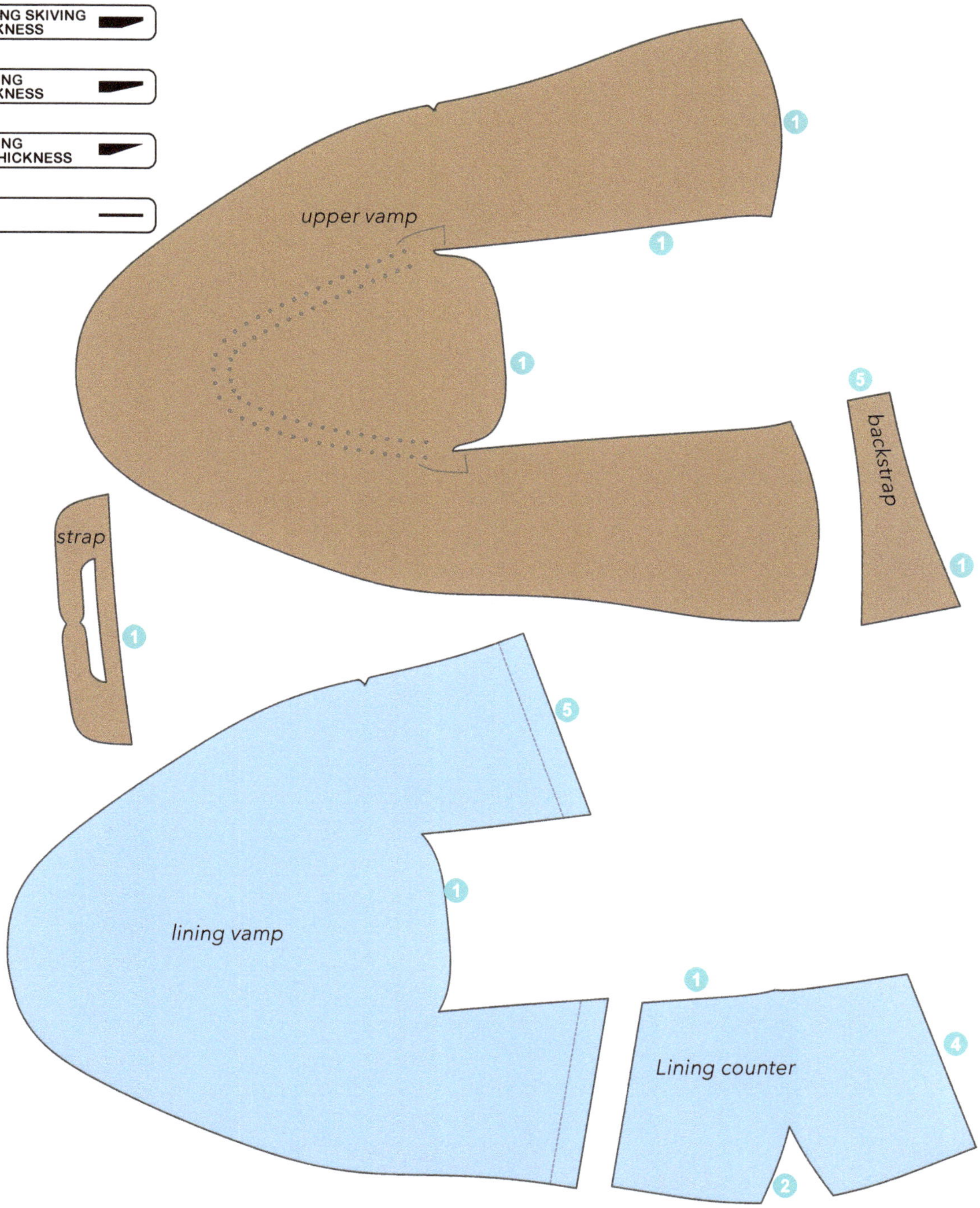

Assembly order

1. Clicking, skiving, marking if necessary
2. Hand-sew the apron seam
3. Zigzag seam the backline
4. Add the backstrap and sew it
5. Add the tongue reinforcements
6. Add piping to the tongue
7. Attach the lining pieces
8. Sew them together
9. Add the lining to the upper
10. Sew the topline around the tongue
11. Trim the lining trimming allowance
12. Sew the binding strip from the inside
13. Turn the binding strip and sew the outer edge
14. Add the front strap and sew

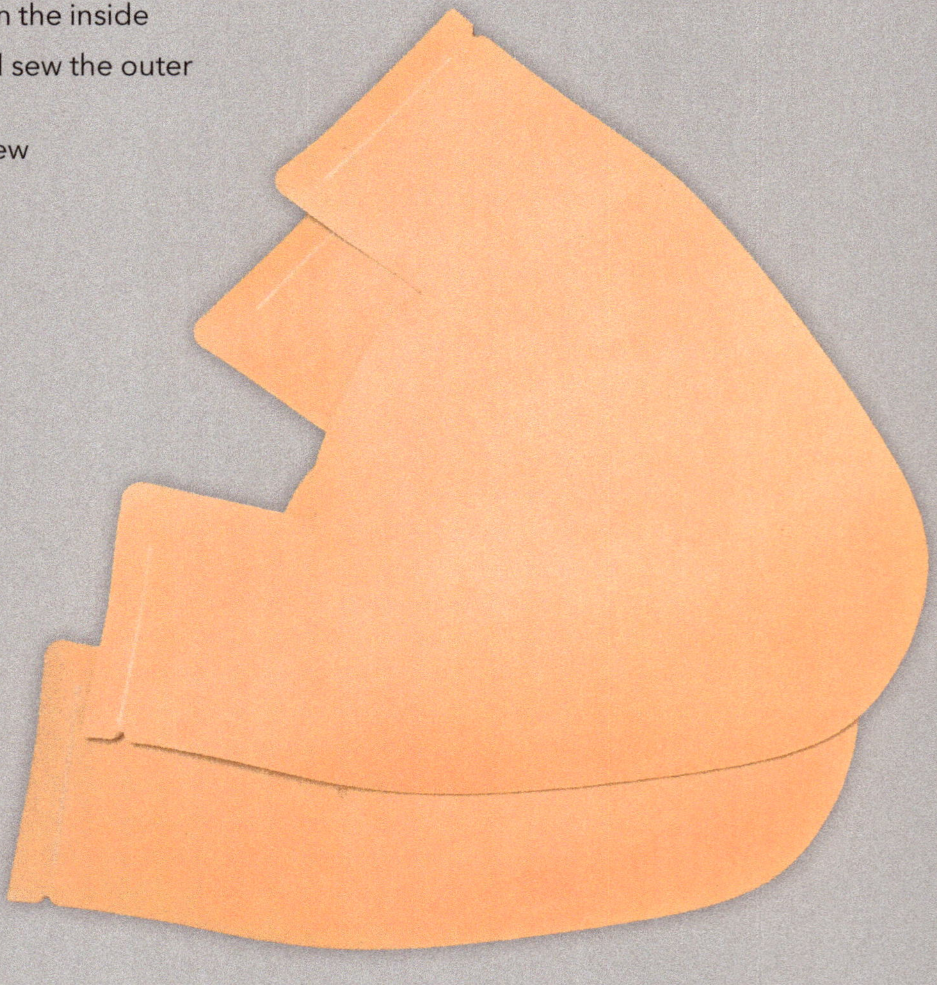

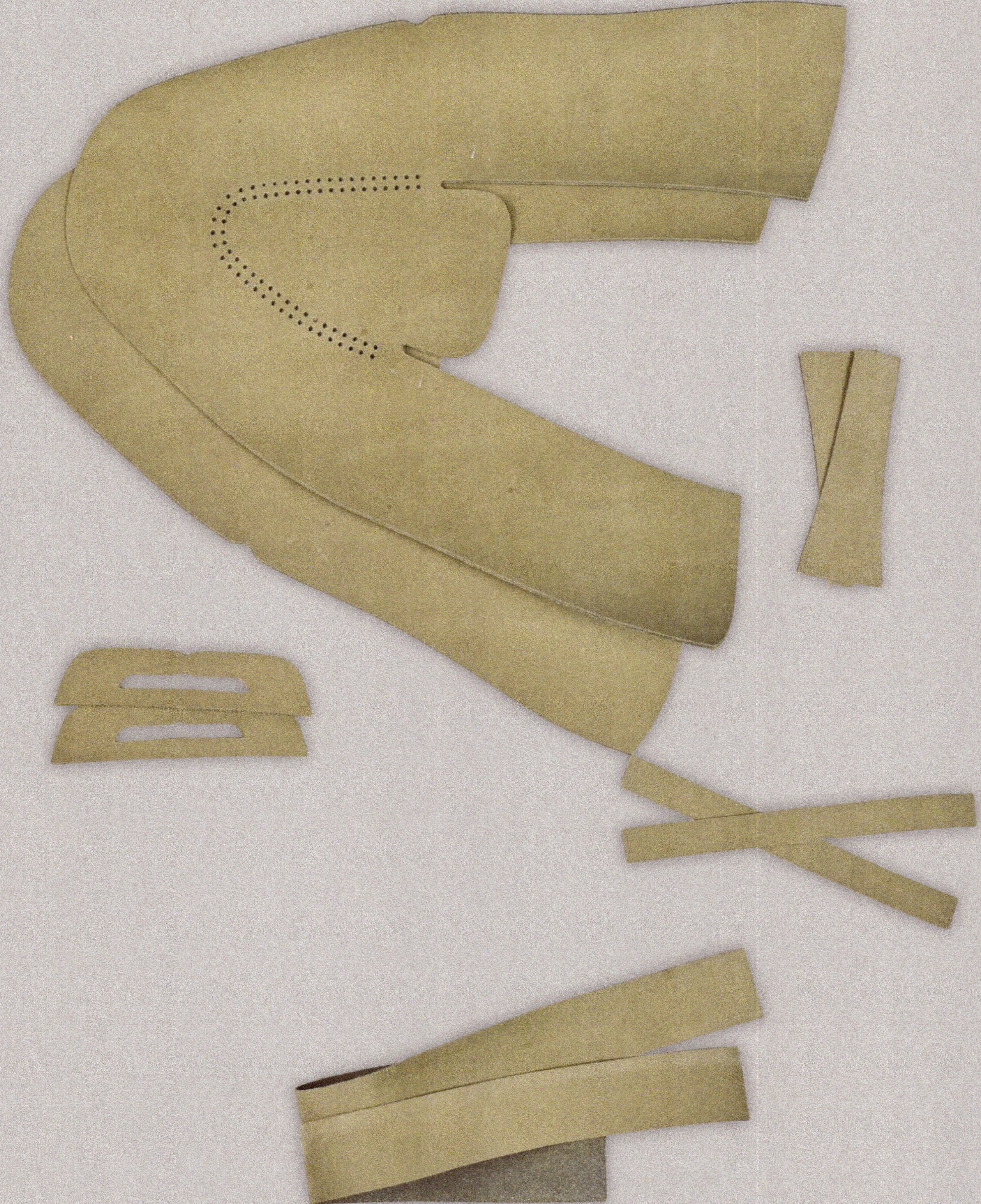

Skiving in the case of loafers is not done solely for ease of assembly, but also to enhance the appearance of the design. All edge treatments, such as reverse French binding, depend on thinning the materials to achieve a smooth, clean finish. Therefore, skiving the topline and the binding strips to the correct thickness can make or break the final appearance.

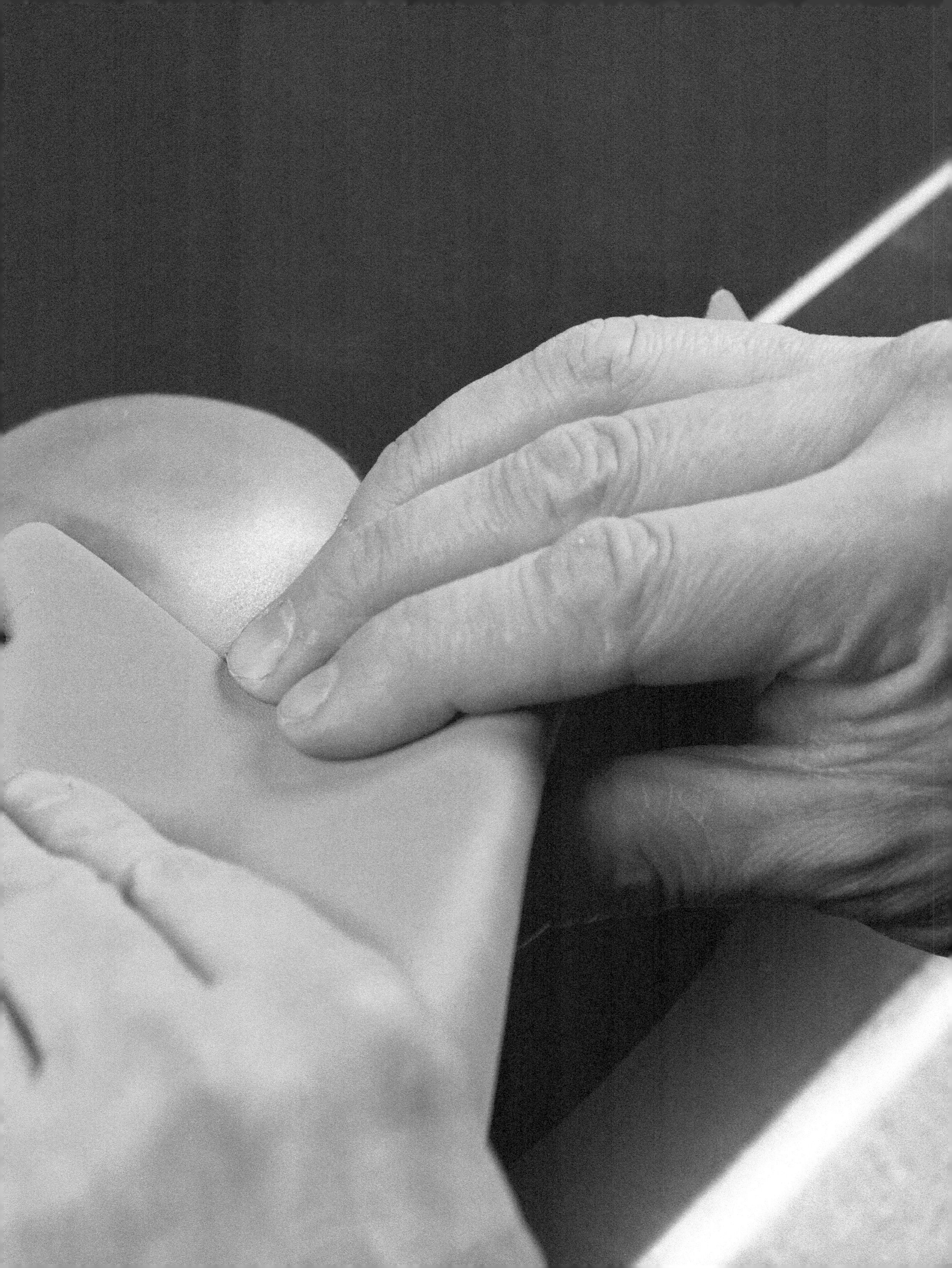

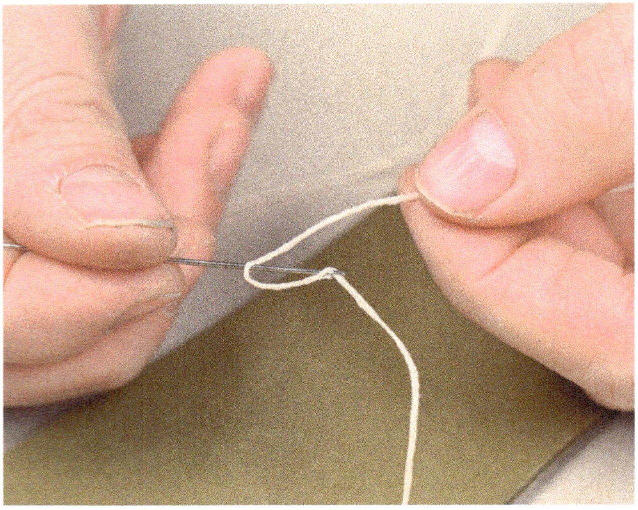

Sewing the vamp seam is the first step in the assembly process. To determine the thread length, use this rule of thumb: three times the length of the seam, plus 4 in (10 cm) on each side to finish comfortably. Thread the needles onto each end of the thread, additionally securing them by piercing the thread as shown in the picture, ensuring they do not come off during the process. Use a relatively thicker thread for this stitch, preferably in an accent color, although this is a design decision.

The placement of the strap should be clearly marked. The decision to incorporate the strap into the stitching also falls under aesthetics. In this example, the simpler approach is shown, where the stitching ends are hidden under the strap. This is the preferred choice for a slim loafer such as this one. In the case of a moccasin, incorporating the strap into the stitching is usually the better option.

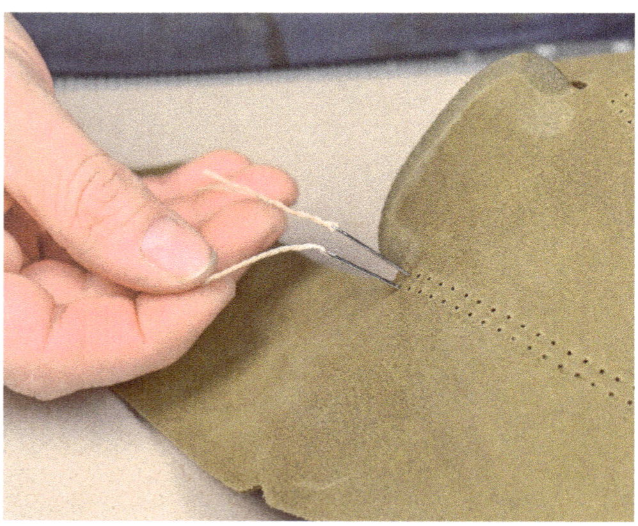

To start, both needles should pass through from the flesh side, which is the opposite side of the seam itself. As mentioned earlier, this stitch does not contribute to the structural integrity of the upper; therefore, reinforcing the ends of the seam is optional and primarily decorative.

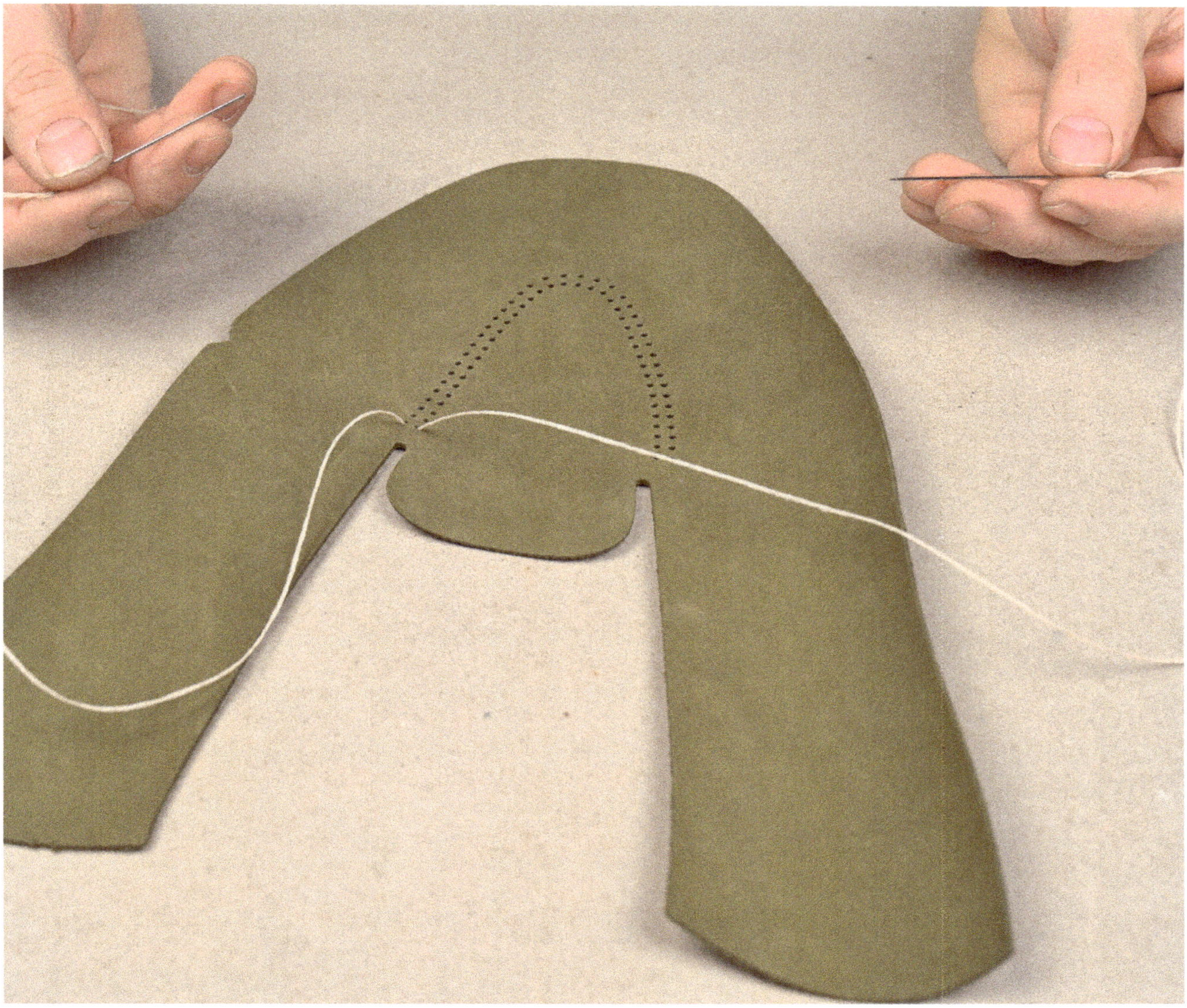

This seam is probably one of the easiest stitches in upper making, and attention to a few minor details will help achieve a neat-looking stitch line. Once the needles are on the top side (the grain side), make sure both have the same length of thread; otherwise, one side may run out before reaching the end. At this point, the stitch can begin.

The simplest stitch used in leather craft is the saddle stitch. By default, the saddle stitch uses the same hole, alternating the needles and allowing the thread to move up and down, forming a continuous line on both sides.

Ideally, these two lines appear identical on each side. In this case, however, the saddle stitch is slightly different. Two holes are used, so both needles must pass through both holes at the same time, changing positions. Since the holes are pre-made, the likelihood of poorly formed stitches is low; nevertheless, the same movements must be repeated consistently to ensure an even and clean stitch line.

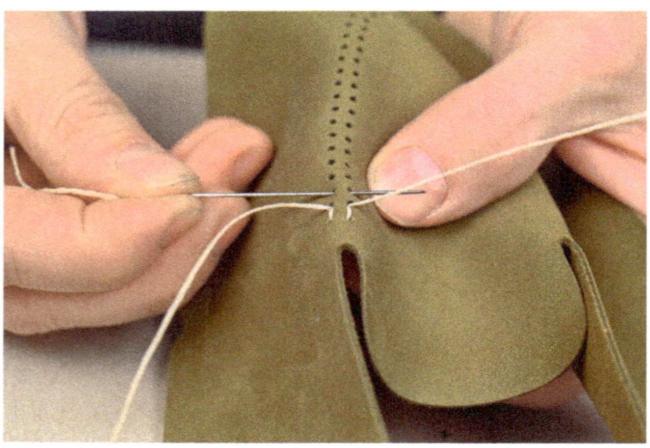

A few practical techniques will make this stitch easier and the final result more polished. First, there is no need to push the needle through both holes at once; it is sufficient to work one hole at a time. This is especially important when working with heavier leather, and in some cases it may be the only practical approach.

The other critical point is tension. The threads should be pulled with moderate and, most importantly, consistent tension. Over- or under-tensioning the thread will affect the appearance of the stitch. Once this step is completed, there is little opportunity to correct tension-related issues.

Repeat the same sequence all the way to the end. Do not worry about the curve; it does not affect the technique.

At the end of the seam, pull through both threads and tie a knot. Avoid pulling too hard in order to keep the stitches even.

Cut and burn the thread ends. Avoid cutting too close to the knot; otherwise, it may later open due to hammering or stress.

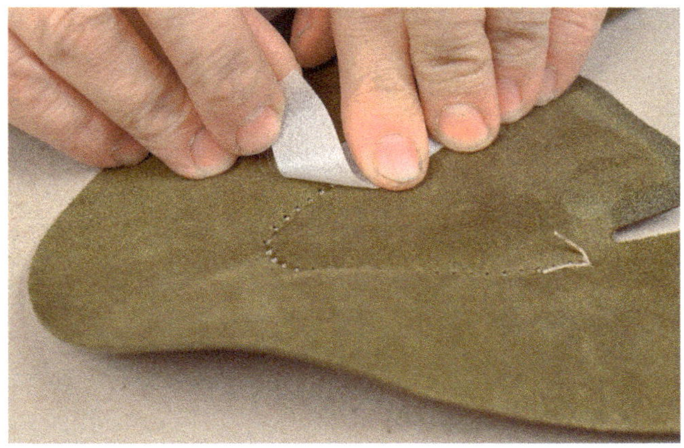

Cover the stitches with a reinforcement tape to protect the thread and insulate the holes from the inside.

Make sure to cover all the holes. When the leather gets wet during wear, exposed holes can cause leakage.

Hammer the seam lightly. This will make it appear more even and also press down the material.

Pro Tip!

The backseam is covered with a backstrap, which makes the zigzag machine an excellent option. In a zigzag seam, the pieces lie side by side without overlapping, resulting in a joint that is perfectly flat apart from the thread. This seam should be started at the top so the topline remains level. Even if the two pieces are perfectly matched, the machine can shift the alignment by the end of the seam. To help prevent this, adding a piece of reinforcement tape is recommended.

Marcell Mrsan

The seam is complete. Note the backstitches at both ends. Even though this seam has only a temporary role in holding the two sides together, if the thread were to unravel, it would create a major problem.

Hammering this seam is crucial. It helps flatten the thread and minimize the profile so it does not show through the backstrap later. This can also be done at a later stage, in which case the backstrap will be affected as well.

To attach the backstrap to the backseam, double-sided tape is used. Regular cement also works, but since this upper is made of nubuck leather, a clean solution is preferred—in this case, double-sided tape, which adheres well to most materials.

Place the backstrap on the seam. Lightly pulling the piece taut while doing so helps the strap follow the curvature and create a natural 3D shape, as discussed in the chapter on pivoting. These narrow pieces do not need to be pre-shaped.

Sewing the backstrap poses a serious challenge. If it is sewn by following the curve of the counter, the strap may be pushed aside unintentionally. A much better method is to temporarily turn the upper the other way by caving it inward, almost as if starting to turn it inside out. This way, the strap sits securely in a natural recess.

Start sewing at the top, moving toward the bottom, then turn back upward. Sewing the backstrap in this manner pushes the sides together even more, which is beneficial. There is no need to sew the topline as well.

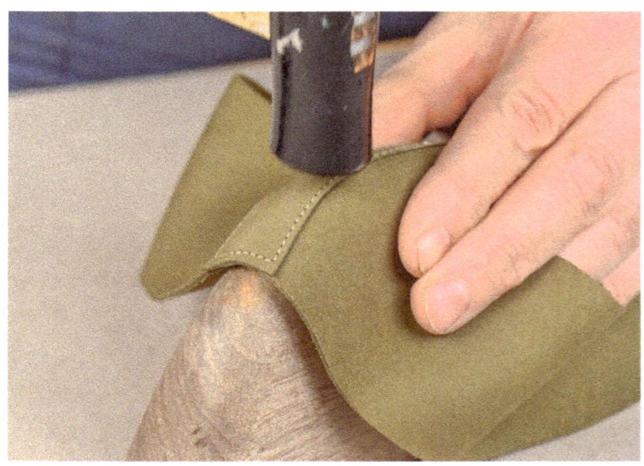

Hammer the seam lightly. This not only flattens the stitches, but also helps it take its final, proper shape, with the counter curving outward. If the zigzag seam was properly hammered in the previous step, there should be no thread lines showing through.

Add the tongue reinforcement. A strong press cement is used for this step, even though it will be secured by seams later. This operation can be done earlier, but grouping the process by relevant shoe components can make the sequence easier to follow.

Prepare the piping for the tongue. Instead of folding the two skived edges exactly at the center, fold them slightly off-center. An offset of about 1/8 in (3 mm) is sufficient and helps prevent the edge of the piping from showing through on the grain side. Use double-sided tape or glue.

To make the ends even thinner, the ends of the piping strip are skived off. This is a relatively easy skiving operation and can be done on either a flat or a curved surface. It is one of the many techniques used in bespoke work that are not strictly necessary, but together contribute to a more polished result. The skiving is minimal and can be done with almost any blade, even a clicking knife.

Bend the piping strip and hammer it down, which "teaches" the material to hold its shape. This reduces the likelihood of the piece opening up when it is attached.

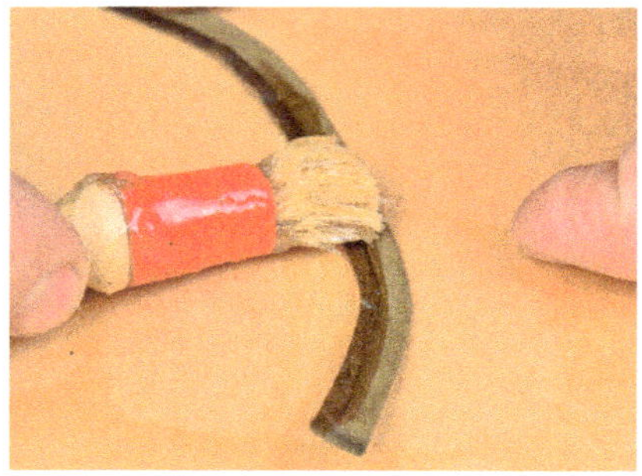

Add glue to the strip, but make sure to keep the edge clean. As a reminder, nubuck leather is very difficult to clean. Since this is a temporary bond, this narrow strip of glue is sufficient.

Do not forget to apply glue on the other side. Be very careful during this process. Hold the upper firmly, do not let it slide, and avoid accidentally placing it onto a fresh glue spot on the work surface.

Attaching the pieces is not particularly difficult if the pre-shaping was done thoroughly. Piping comes in different styles. Some makers prefer piping that almost resembles a folded edge, aligning closely with the thin, skived edge. Another approach is used here, leaving the piping to extend over the edge with the raw edge clearly visible. This reverse French binding provides a more robust edge, which fits the style in this case.

The narrow strip requires only minimal adjustment and light hammering with a folding hammer. If the skiving was sufficient, there should not be excessive material buildup. As a comfort feature, a thin foam layer can be used to level the gap before adding the lining. This is optional but may help improve fit to some extent.

At this point, work can begin on the lining. Since it consists of only two pieces, there is very little to do: apply glue to the counter-vamp lap seam. When the seam is straight, as in this upper, use the counter to mask the vamp from glue. Make sure to apply the glue to the grain side of the counter.

Add the pieces together. Sticking to the pattern markings is very important here, as even small deviations in overlap can lead to bunching and poor fit. While a shorter lining (too much overlap) can be managed, a longer one (too little overlap) is very difficult to correct.

Gently hammer the overlapping areas. Ensuring good adhesion is only part of the reason for this step; wet glue should not enter the sewing machine. If waiting 30 minutes before sewing is not desired, this step helps speed up the process so sewing can proceed immediately.

Start sewing on the bottom side; this way, once one side is finished, it can be followed immediately by the other. This so-called "chain sewing" method is very common in the industry, as it speeds up workflow and reduces thread waste. This is not primarily about the cost of the thread, but about the limited capacity of the bobbin. Changing the bobbin slows down the process, and for this short seam, the time saving can be around 50%.

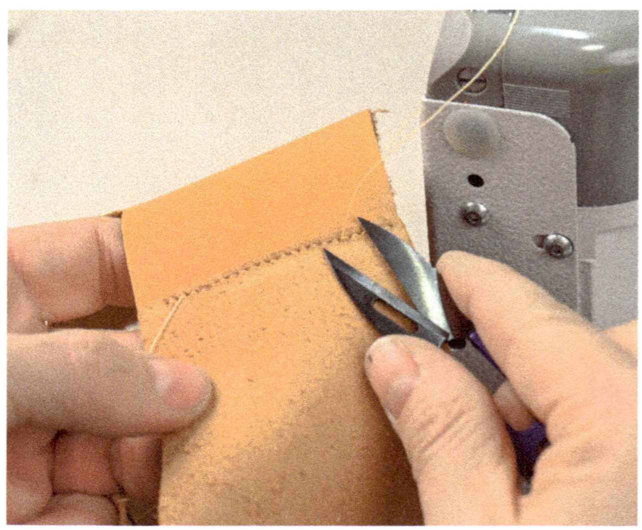

Make sure to add ample backstitches in the lasting allowance.

Cut and burn the loose thread ends. If a higher level of finish is desired, they can be pulled through to the inner side and burned there. Use extreme caution when burning threads on lining leather, as heat can leave dark marks that affect the quality of the final result.

Adding the lining is not a particularly difficult task, but even slight misplacement can complicate the process. It is preferable to start at the center of the counter. To achieve the necessary accuracy, make a visible center mark. If there is any chance this marking might remain visible later, use a pen with removable ink, a barely visible punch mark, or a fold line. In this particular case, the reverse French binding will cover everything, so a regular black pen is acceptable, as there is no risk of the mark showing later.

To adhere the pieces, apply glue only around the tongue, for the reasons we've covered earlier. Stay away from the edge here, so no cleaning will be necessary. Around the top edge there's little chance for errors, and there is no cement necessary on the trimming allowance.

After attaching the pieces, sewing follows. Leaving the piece only glued or sewing it are both valid solutions. The distance of the stitch line from the edge is also a stylistic choice. Since this design has wider piping, a greater distance for the stitch line is a natural choice.

The advantage of the post-bed machine is apparent, but most of the process can also be done on a flatbed machine, including this seam. When sewing a "sandwich" structure such as this, the main challenge is that the layers can slide apart at corners and tight curves. Since this area is highly visible, slow the machine down or sew stitch by stitch if necessary.

I am again, using the corner of the workspace for trimming the lining. Good news is, since the topline will have a binding, which will cover this edge later. Just let the raw edge guide the knife and other than cutting into the upper leather, there are no major concerns here. The tight corner can also be cut by knife, or a hole punch of the right size can also be helpful. This portion will be covered by the binding strap, therefore making this a very forgiving detail.

This is a reverse Italian binding. Sew it on the inner side, with the grain side of the strip facing the lining. Start the seam at the corner of the tongue and finish on the opposite side at the same position. This seam is not particularly difficult, apart from one key requirement that determines the final appearance: a constant, precise distance from the edge. To achieve this, hold the binding strip slightly away from the edge so its position remains clearly visible throughout the process. If the strip is applied unevenly, achieving a clean and appealing binding becomes very difficult.

Add glue to the top of the strip. Make sure it is a very thin layer so there is no need to wait long. Apply the glue only 6-7 mm wide, since a different type of adhesion will be used on the other side.

Gently pull the strip upward, closing the cemented layers. The tension must be consistent not to curl the upper outward. As the strip turns toward the seam, press the layers together gently.

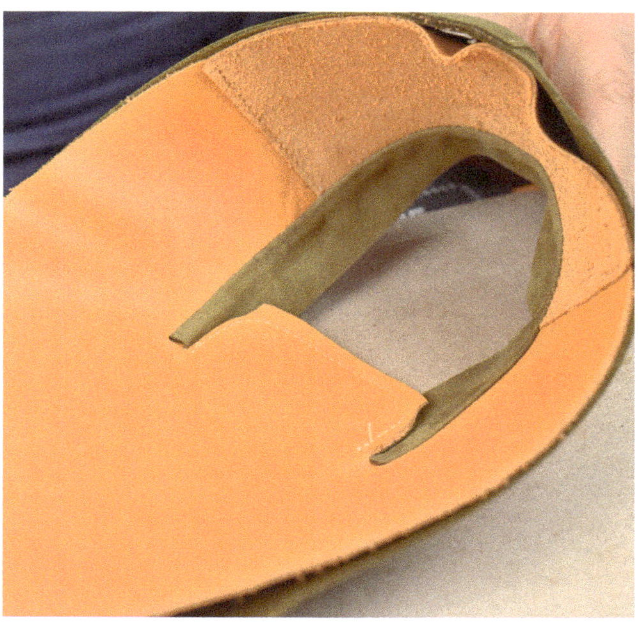

The strip is turned up and ready to be glued down on the other side. Hammering at this point is very important; otherwise, the added thickness of the binding can affect the top-stitching.

Here, the appropriate size of double-sided tape can be used. It is preferable not to apply the tape across the full width, leaving 2-3 mm clean for practical reasons.

Sew the edge of the strip down to the upper, through all layers. This is not very difficult, since the layers are even (except around the backstrap). Make sure to adjust the upper as you go; this may be the only challenge here.

The seam is complete. This is close to the minimum safe distance from the edge, as the turned binding strip on the other side can push the roller foot to the right and potentially break the needle. Using this width ensures sufficient clearance from the raised edge.

Position the strap on the upper. In factory production, the placement is properly marked; for one-of-a-kind work, it may be advisable to last the upper temporarily with a few nails to ensure correct positioning.

It is advisable to mark the turning points in advance to achieve precise turns. Since the area will be visible only at approximately a 45° angle during sewing, marking the correct turning points beforehand ensures accuracy.

This seam can be challenging, depending on the machine setup. The underlying layers are very uneven, so it must be sewn with great care and at low speed. Start from the right side (holding the upper in this position) and do it slowly. The binding is very close, so it may force the upper to sit at an angle.

The seam is relatively easy around the middle. Make sure to maintain an even distance from the edge, which can also be carried through the corners. Keep an eye on the pre-marked turning point so the last stitch can be adjusted slightly longer or shorter to hit the exact mark. This is a highly exposed part of the upper, so the extra care is worthwhile.

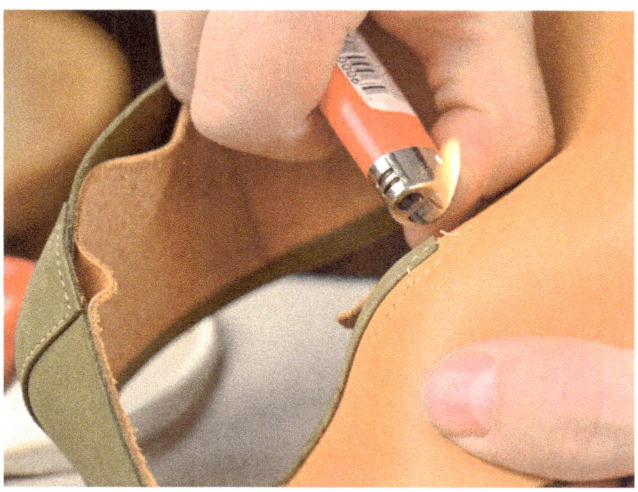

Finish the upper by burning the loose threads. Here is a simple technique to avoid burn marks on the lining: hold the lighter upside down at an angle so the flame itself does not contact the leather. The heat from the flame will still be sufficient, while providing much better control and allowing the thread ends to be melted in a controlled manner.

The upper is complete and ready for lasting or pre-lasting. As can be seen, the top of the tongue still has a small amount of excess lining, which can be used as a tacking allowance if necessary. This is difficult to judge at this stage, so it is better to leave it in place. Leaving this extra material helps avoid nail marks on the upper, which is preferable.

8 THE DERBY SHOE

The derby shoe is technically more complex than the models discussed earlier, yet it is still considered a relatively simple style. Its technical complexity comes mainly from the lining structure, but the advantage is that the same structure – and the same assembly order – can be reused for many other patterns.

Not all possible variations are discussed here; only the core construction of the model is covered, featuring a reverse seam at the back and a folded topline edge.

The practicality of this model lies in its wide opening, which makes it easier to wear for people with wider feet. The pattern itself has a long history and can be found in historical footwear and boots dating back several hundred years.

In the 19th century, this simple open-lacing system was rediscovered and reintroduced as the derby, or the closely related Blucher shoe. The style has changed little since then. Until relatively recently, it was considered less elegant than the oxford. The key difference between an oxford and a derby is simple: if the quarter overlaps the vamp, it is a derby; if the vamp overlaps the quarter, it is an oxford.

The Derby Pattern with The Geometric Method

The derby shoe pattern introduces us to new parts to consider which adds new layers of complications, but learning it properly offers a major advantage: most similar patterns with a comparable structure will follow this kind of lining structure. Its uniqueness comes from the "scissor" construction, in which the scissor cut on the quarter pattern wraps around all the vamp layers, passing over the tongue from the inside.

The pattern can be made from five or more pieces, depending on the design – a toe cap can be added, the quarter can be divided into additional pieces, the tongue can be made as a separate component, and so on.

Pivoting the Last Copy

Starting with the geometric standard, the toe line (TL) gives a clear indication of where the last copy should be pivoted. To be fair, it is entirely possible to create a correct derby pattern without pivoting, but using it makes axis positioning much easier, so it will be applied here. The remaining reference lines will be used as well.

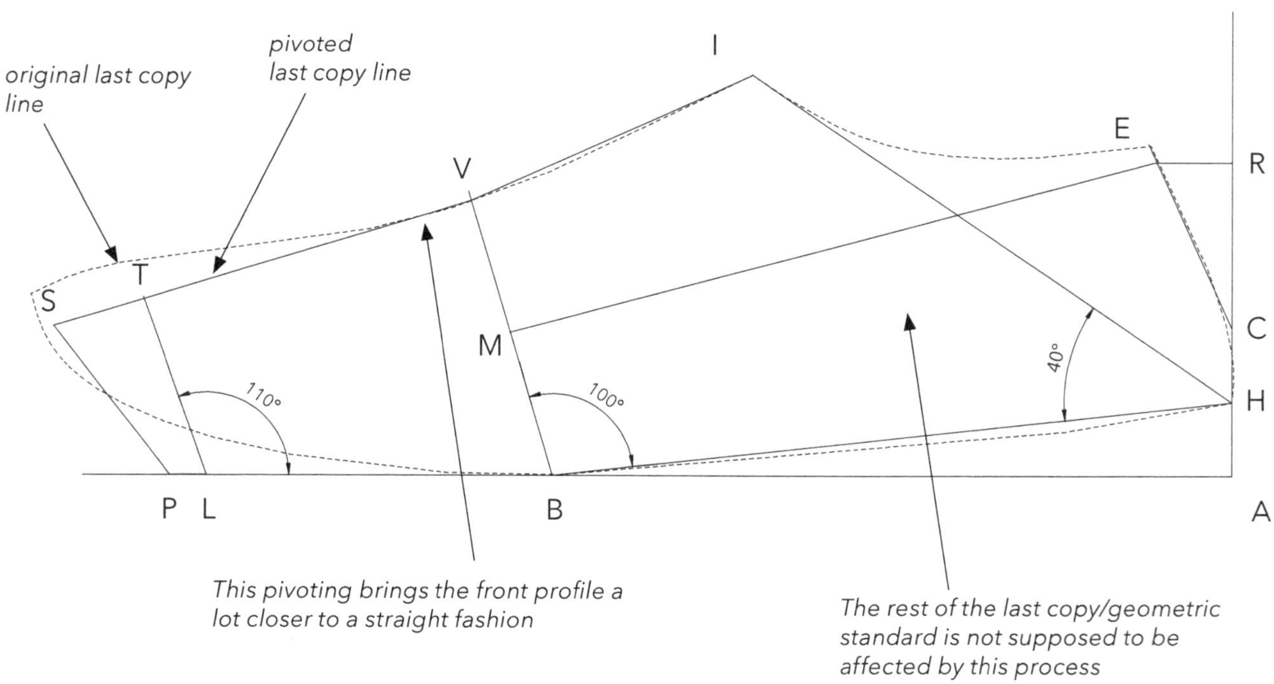

About Toplines

Toplines are signature elements and a highly recognizable indicator of a pattern maker's approach, almost like a personal mark. By definition, a derby typically has a relatively low topline, as it is considered casual footwear, but several factors must be taken into account.

The topline is a defining element of a shoe's visual identity. In casual footwear, the topline typically follows a smoother, softer curve, creating a relaxed and approachable appearance. In contrast, more elegant designs – especially those with elongated toes – often feature a deeper, more dramatic cut through the midsection.

Topline height alone does not determine how easy the shoe is to put on; the length of the lacing is also important, and in many cases even more so. Extending the lacing close to the ball girth line, or in some models even beyond it, creates significantly more room for the foot to pass through the topline. A good example is a minimalist athletic shoe with a short vamp and long lacing: even with a higher topline, this structure allows easy entry.

The topline position is a key design element and must be in harmony with the vamp line, toe decoration, heel counter, and overall proportions. Certain styles even call for a very high topline, such as PNW-style derby shoes.

There is one anatomical limitation that restricts topline placement: the ankle bone.

For the comfort of the wearer, the topline should not cut into the ankle bones. The same consideration applies to the counter height. Exceeding the values determined by the geometric method is risky and can cause significant discomfort or even pain.

In some cases, toplines are not symmetrical, just as ankle bones are not. The lateral ankle bone is typically lower than the medial one. Traditional patterns often account for this by using an asymmetrical topline. Today this is less common, partly because feet have become wider and a slightly lower topline is generally preferred. Using the same topline on both sides is also simpler for practical reasons.

Creating the Topline

Using the original geometric last standard, additional reference lines are introduced, starting with the topline. Follow the instructions shown in the diagram and add the topline to the last standard. There is some flexibility in this step, as a wider last or foot generally requires a lower topline. For the sake of simplicity, a symmetrical model is used here.

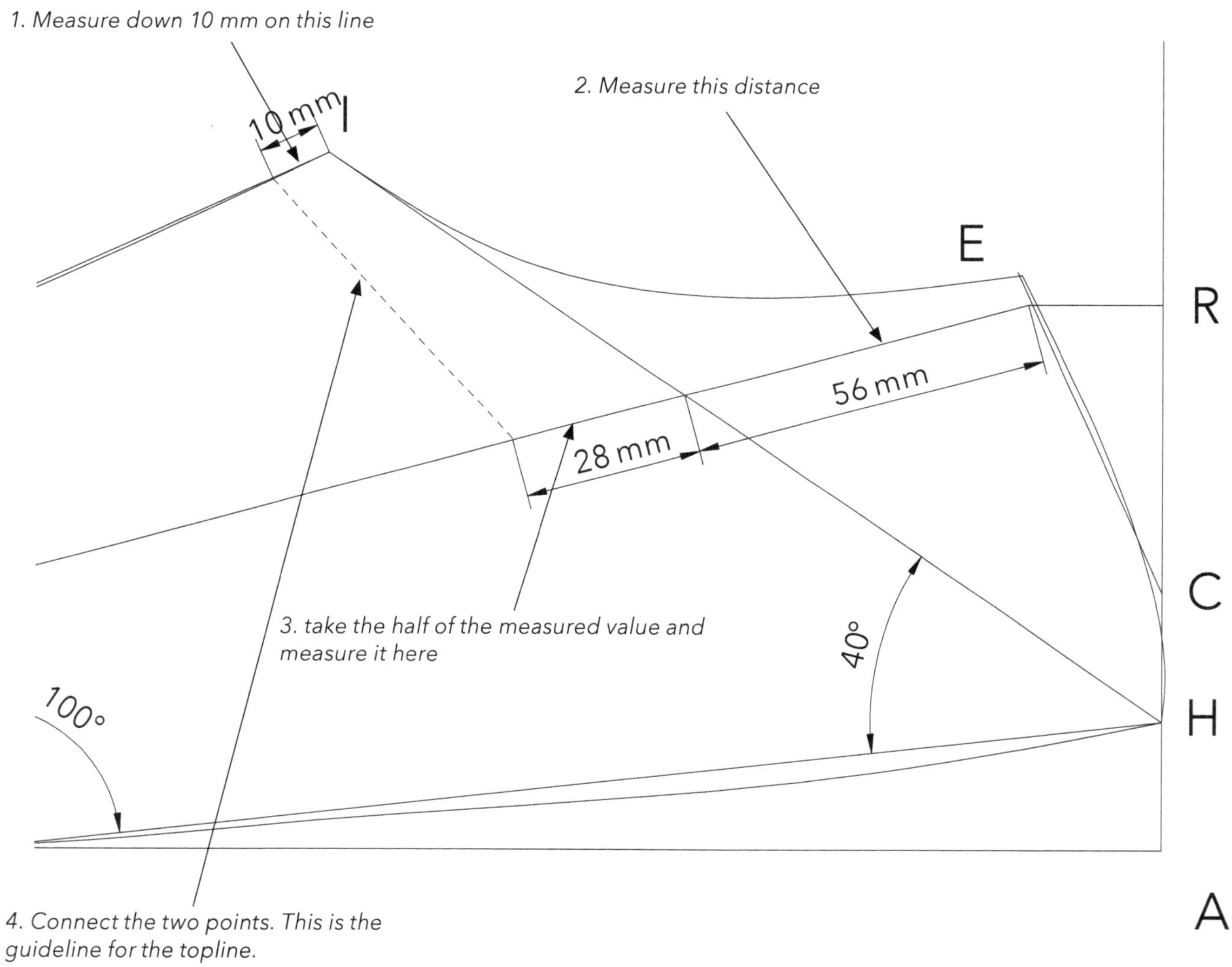

The topline is not only a design choice; comfort and practicality also influence its placement. It must be deep enough to allow easy entry into the shoe, positioned correctly to hold the foot securely, and shaped so it does not interfere with the ankle bones.

If, for any reason, an asymmetrical model is used, the medial side is placed higher, as is typically the case.

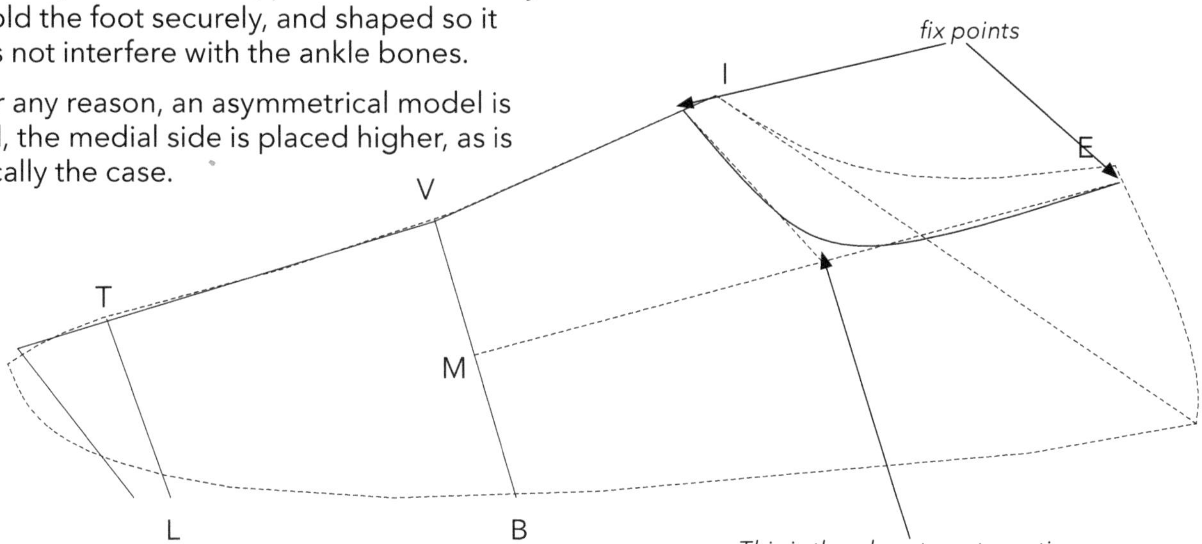

This is the place to get creative. The curve can be less or more smooth.

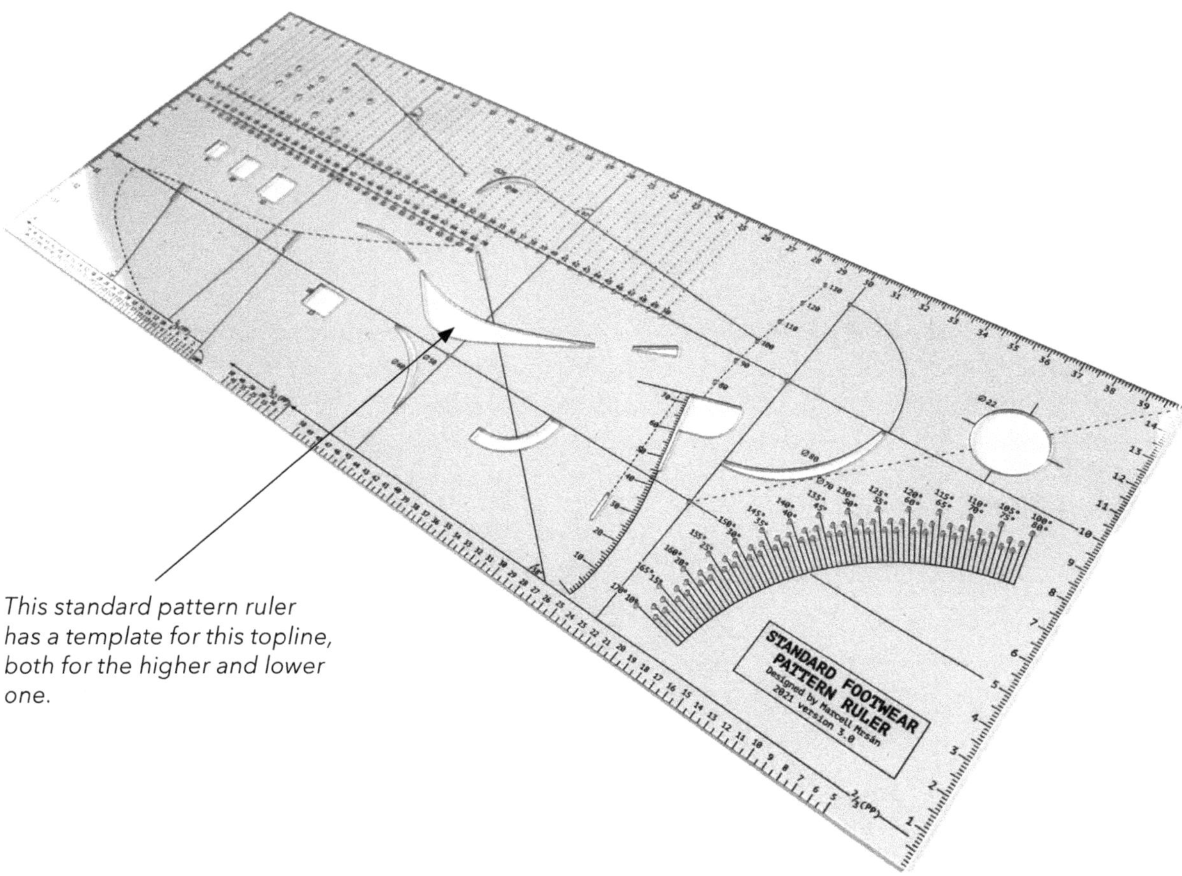

This standard pattern ruler has a template for this topline, both for the higher and lower one.

Additional factors can also be considered when placing the topline. This example illustrates one of them: the brogue line. Even if the topline and the vamp line are technically well spaced, the presence of brogue lines influences how this distance is perceived visually. When brogueing is involved, it must be taken into account during line placement.

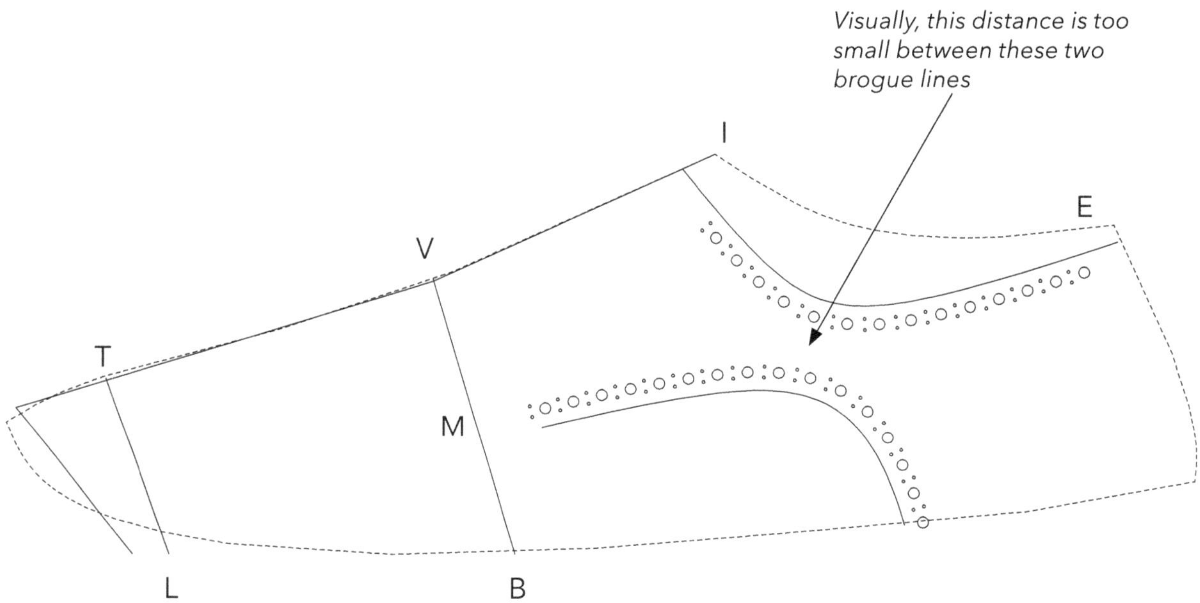

There are also technical reasons to avoid positioning brogue lines too close to one another. Brogue lines require reinforcement underneath, which can result in overlapping layers in this area. More importantly, brogueing weakens the material; even with reinforcement, it effectively behaves like a perforation. Placing two brogue lines too close together can therefore compromise the structural integrity of that section.

Creating the Lacing Section

The lacing area follows a set of practical rules, which will be discussed in the following pages. This part of the upper plays a crucial role: it must accommodate the shoelaces and eyelets while providing sufficient strength for the laces to perform their function – holding the shoe securely on the foot.

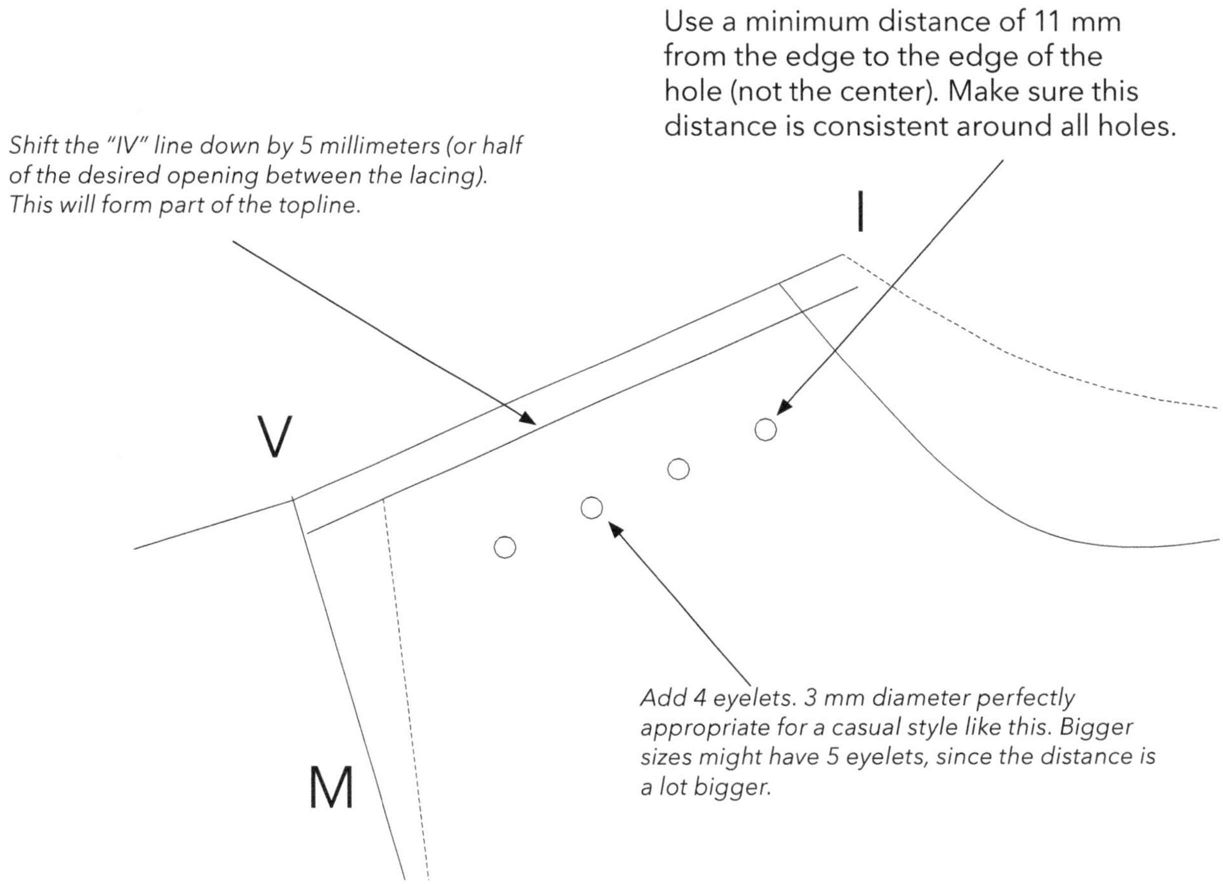

The lacing pattern must ensure adequate structural strength for the eyelets and support comfortable entry into and exit from the shoe.

From a design perspective, this area is one of the most visible parts of the footwear, and its styling should follow established traditions as well as technical requirements. Traditionally, a classic derby has four – occasionally five – eyelets. The opening between the two facings should be around 8-10 millimeters, and both the top and bottom ends of the lacing section should have rounded corners. Depending on the edge treatment, the corner radii should not be less than 12-13 millimeters; otherwise, folding quality may be compromised.

Common Lacing and Lace Hole Errors

There are a few typical errors in lacing design that should be avoided. Unlike other lines in footwear design – where incorrect placement may result only in an aesthetic issue – a poorly positioned set of eyelets can compromise the entire pattern and lead to durability problems.

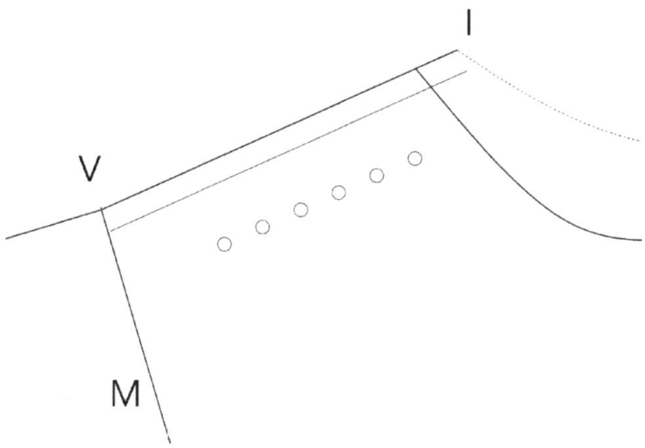

Too many lace holes in a short distance. This results in spacing that is too small between the holes, or, worse, insufficient clearance for eyelet hardware if it is used.

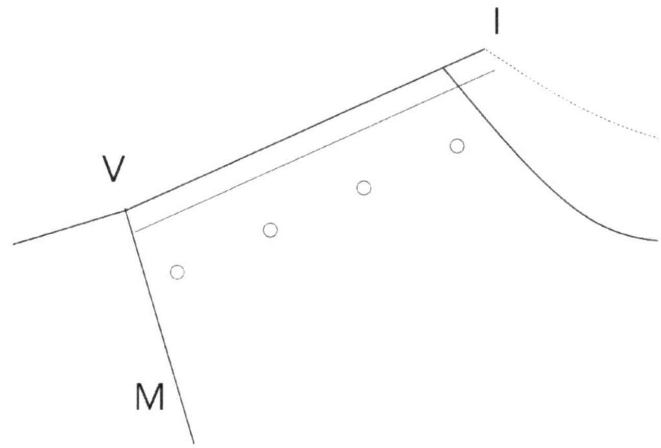

The holes are spread too far apart, leaving insufficient distance at the top. Even worse, the ball girth line is used improperly as a design reference to place the last hole, resulting in it being positioned too close.

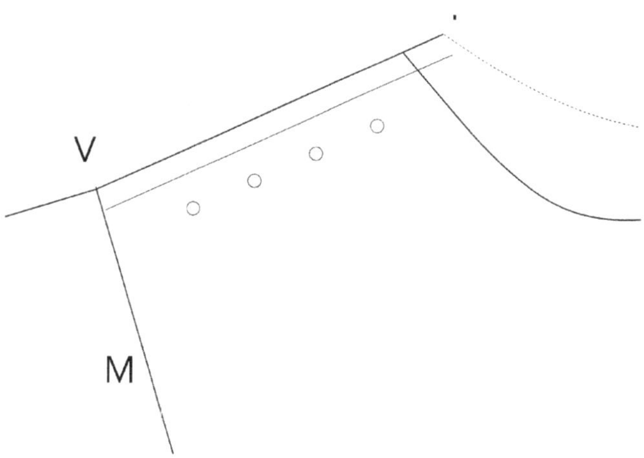

The eyelets are positioned too close to the edge. This creates a vulnerable strip of leather between the hole and the edge, which cannot withstand the continuous stress from the laces and is likely to tear over time.

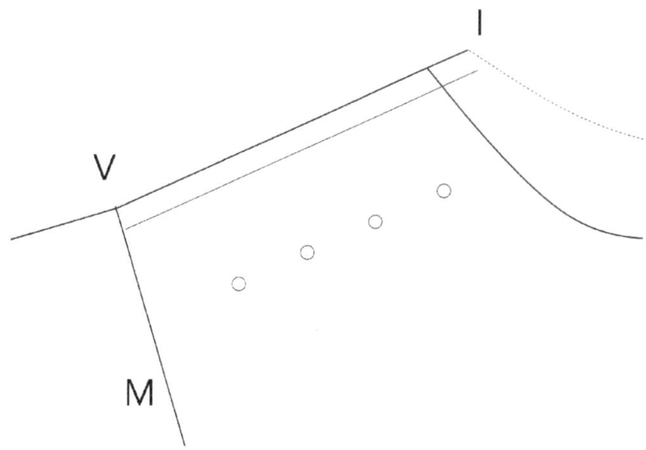

The holes are placed too far from the edge. While they may still function, there is a risk that the tongue will not fully cover them from the inside. Exposed laces against the foot can cause discomfort and are a major concern in terms of wearability.

The Derby Line

The derby line is the front edge of the lacing section. This edge is not sewn down, except for the bottom corner, where it is secured with a reinforcement stitch – hence the term *open lacing*. The placement of this line directly affects how easily the shoe can be put on, making it extremely important. Since a simple derby shoe has relatively few design elements, this line also plays a significant stylistic role. Older pattern books suggest a mildly curved shape, though a straight line is also acceptable; these are aesthetic choices. What truly matters is the position of the bottom corner, where the line transitions into the vamp line.

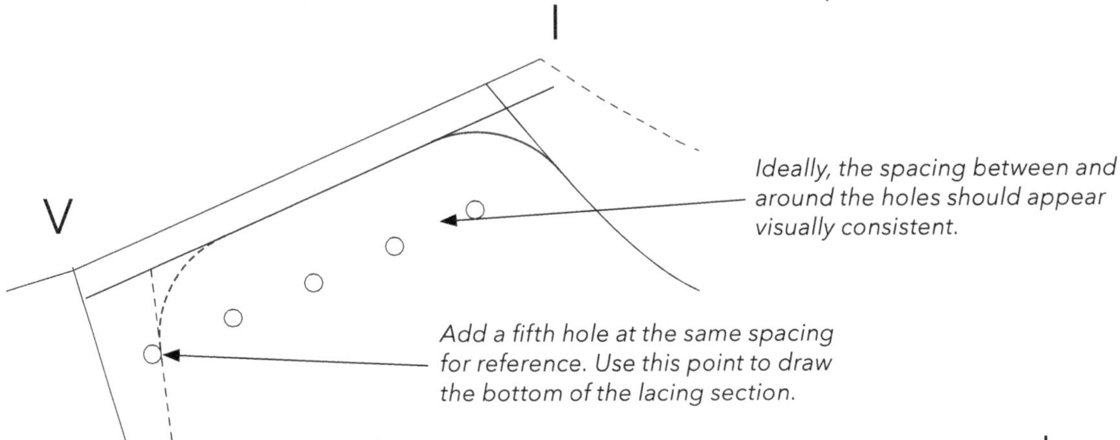

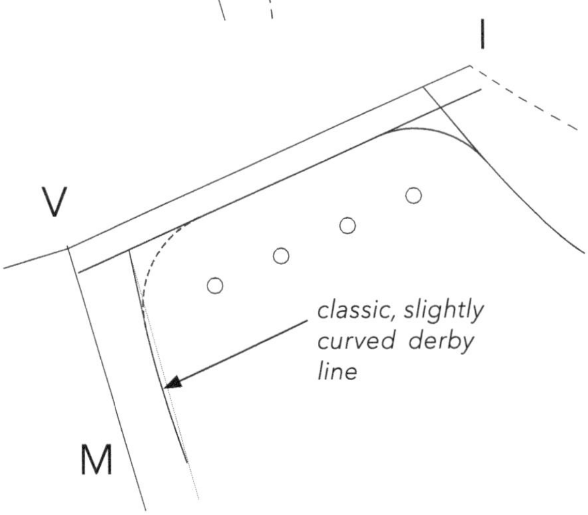

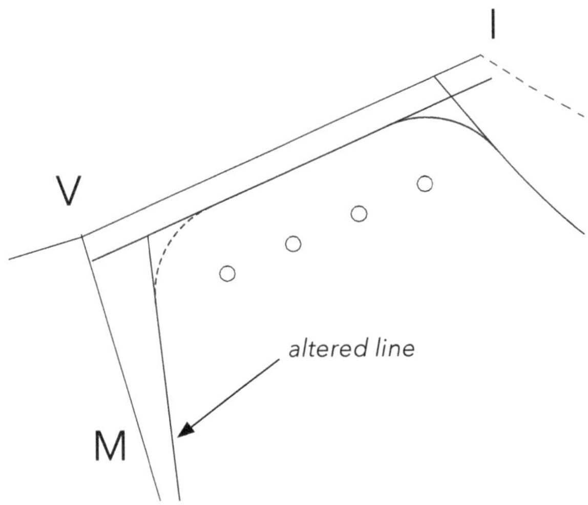

Some older textbooks suggest this solution for the bottom style line, or even a slight curve. This approach is perfectly functional. If the four eyelets are positioned correctly, the opening – defined by the position of the bottom of this line, as discussed later in this chapter – will be sufficient for a regular or slightly wider foot. For much wider feet or a high instep, a different placement may be required.

The altered line is a rotated version of the original, which is parallel to the ball girth line. This rotation makes the top slightly shorter, mainly as a stylistic change, while shifting the bottom portion farther toward the toe. This shift is not merely aesthetic; it has important practical implications for the fit and function of the footwear.

Creating the Vamp Line

The vamp line is the lap seam between the vamp and the quarter – this is where the latter overlaps the former. Its placement is primarily a design decision. The reason for discussing a technical approach is that certain proportions are considered classical and have become widely accepted standards. While there is room for stylistic variation, it is still possible to make mistakes in placing this line, which will be discussed later.

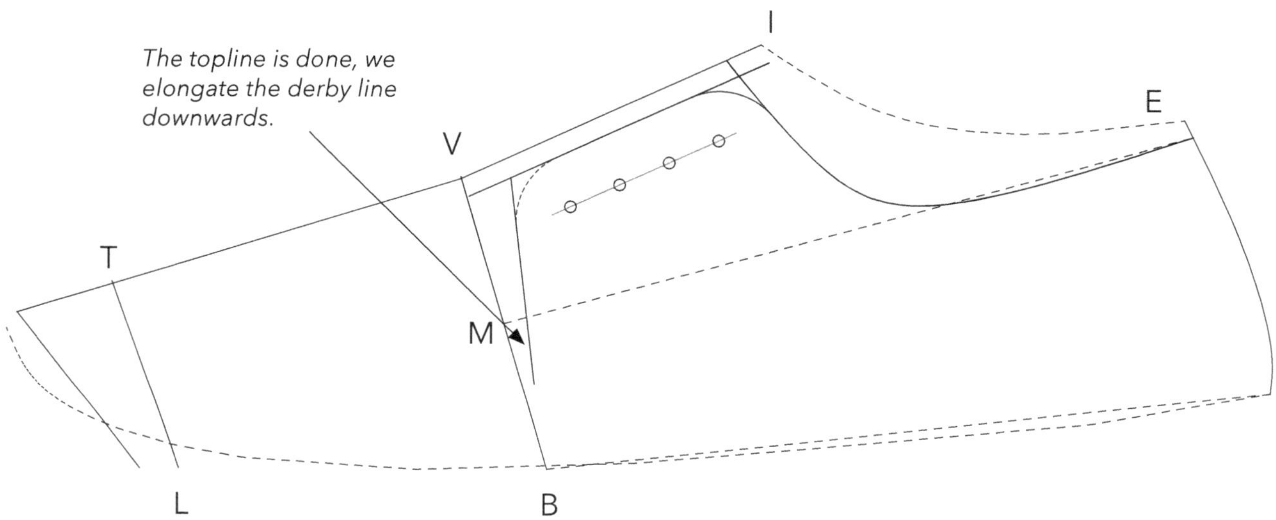

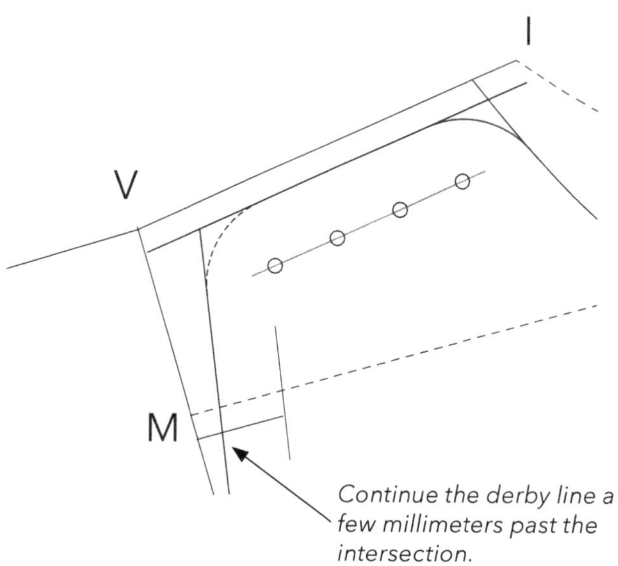

Draw a line pointing slightly backward from a few millimeters below the intersection. This adjustment makes the vamp line more proportional and provides slightly more opening for the shoe, improving comfort when sliding the foot in. Technically, point M is the center of the ball girth line (BG), but because some distance was removed to create the opening, the geometric center is no longer the visual center. Shifting the point downward by 3-4 millimeters results in a more visually balanced placement.

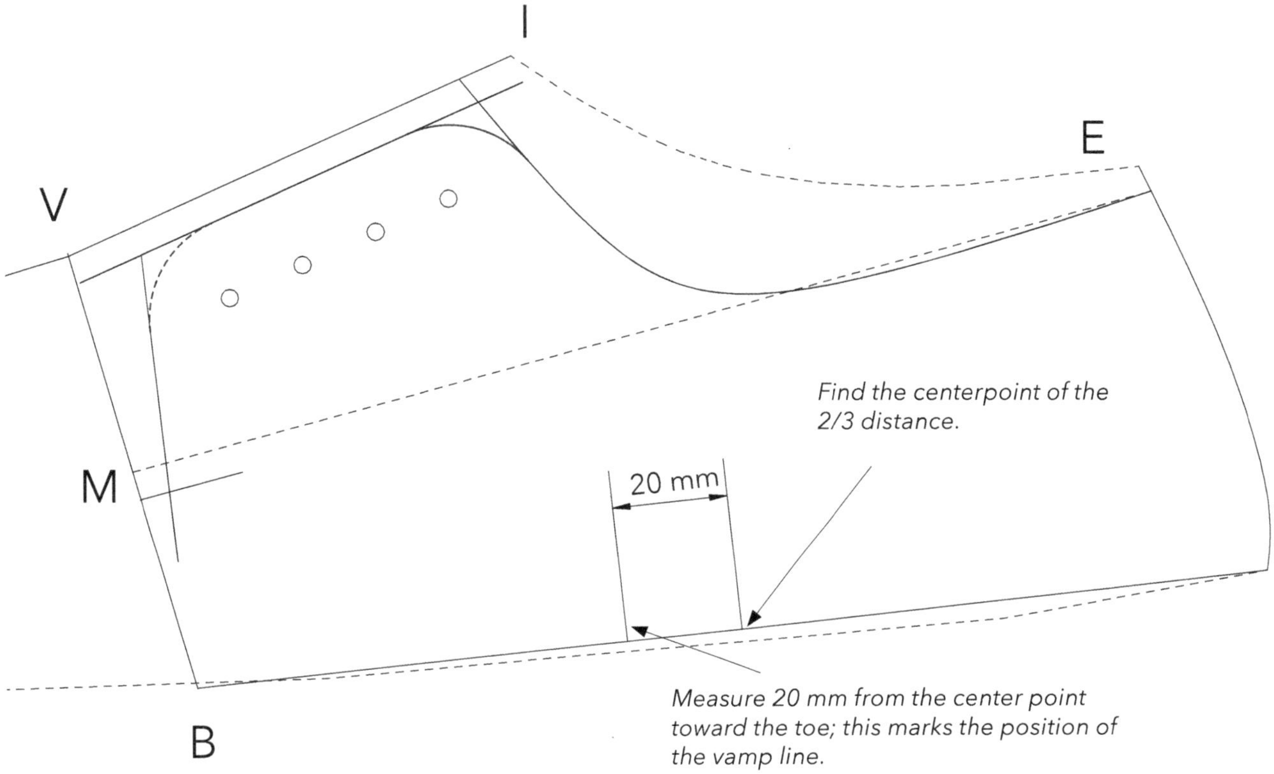
I
E
V
M
B
20 mm
Find the centerpoint of the 2/3 distance.
Measure 20 mm from the center point toward the toe; this marks the position of the vamp line.

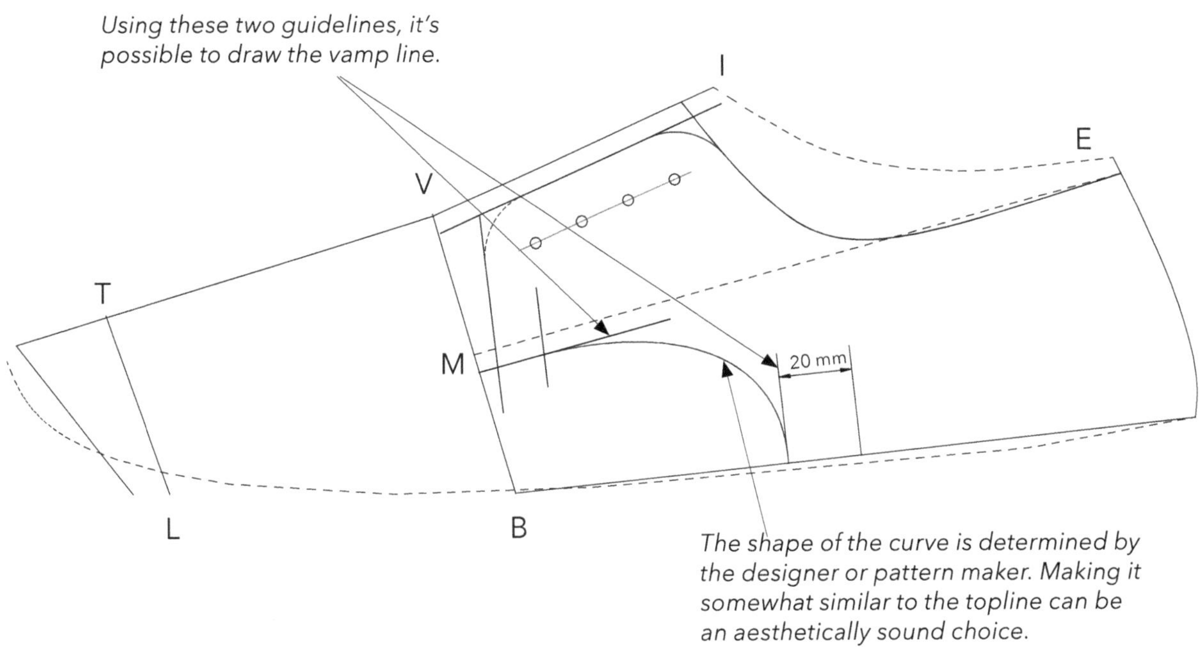
Using these two guidelines, it's possible to draw the vamp line.
I
E
V
T
M
20 mm
L
B
The shape of the curve is determined by the designer or pattern maker. Making it somewhat similar to the topline can be an aesthetically sound choice.

Vamp Line Placement Pitfalls

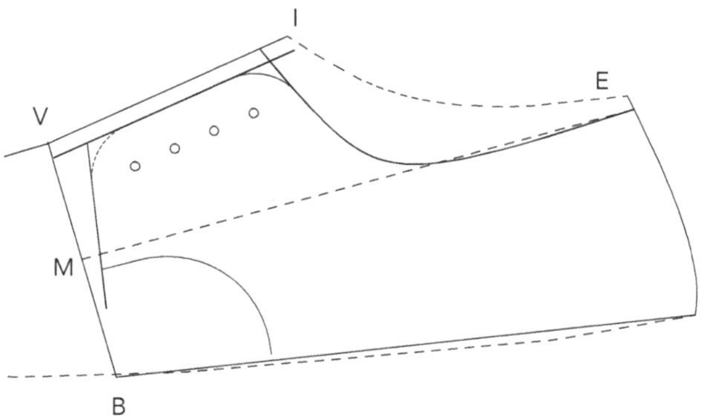

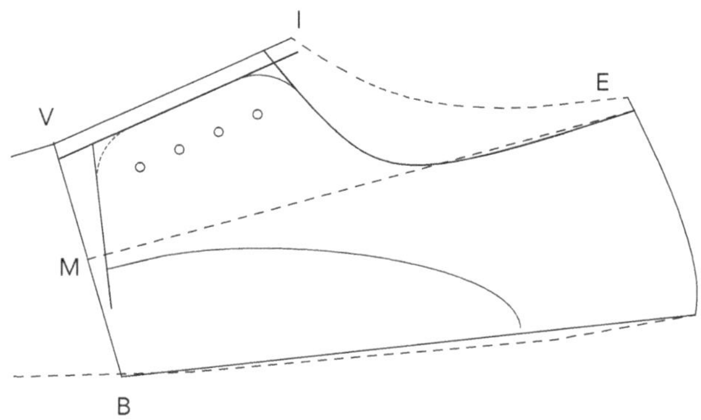

Making the vamp line this short is not only an aesthetic issue; it can also overlap the lining lap seam underneath, resulting in four layers meeting at one point. This is far from ideal and will become visible regardless of skiving. While it is technically possible, using this line may require special arrangements.

Extending the vamp line too far is a design choice and not technically incorrect. However, this approach is not classic and creates a strong dissonance between the topline and the extended vamp line. That said, it is still workable, and variations like this have appeared on the market. It makes a bold stylistic statement that can be divisive and should be used intentionally.

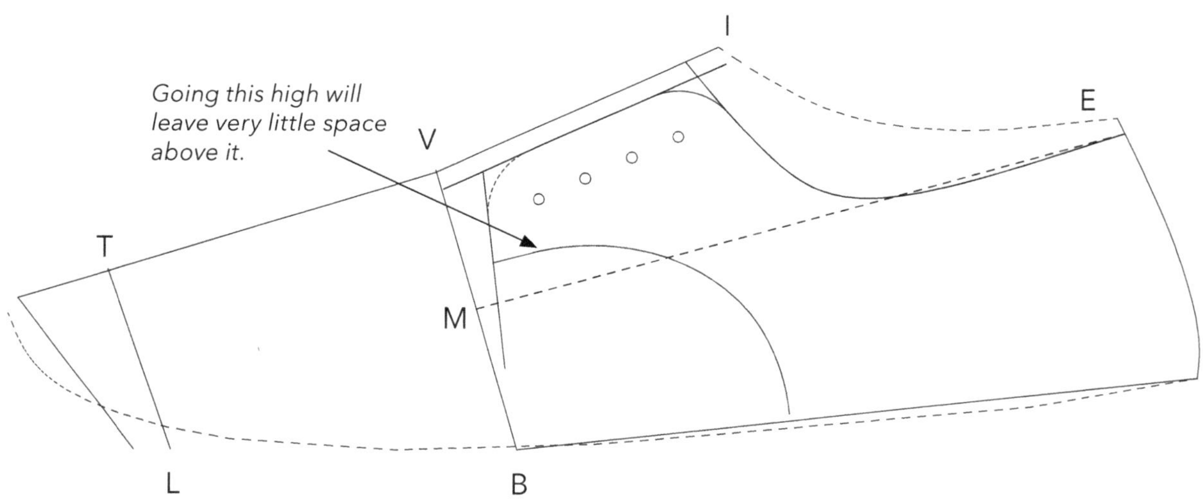

Placing the vamp line this high is incorrect and should be avoided. The reason is the overlap allowance between the vamp and the quarter is around 10-12 mm. Once this is added, the line shifts upward, leaving very little space for the tongue. There is a significant risk that after lasting, the tongue will fail to cover the laces, exposing them to the foot, which is a serious issue. In addition, this placement is not visually pleasing.

The next step is to begin creating the tongue pattern. In a derby, it is very common – especially in bespoke and high-end shoemaking – to cut the tongue as part of the vamp. Factory production is different: since manufacturing focuses on cost efficiency, the best part of the hide is typically not used for a component that is barely visible.

In this example, the more exclusive approach is followed.

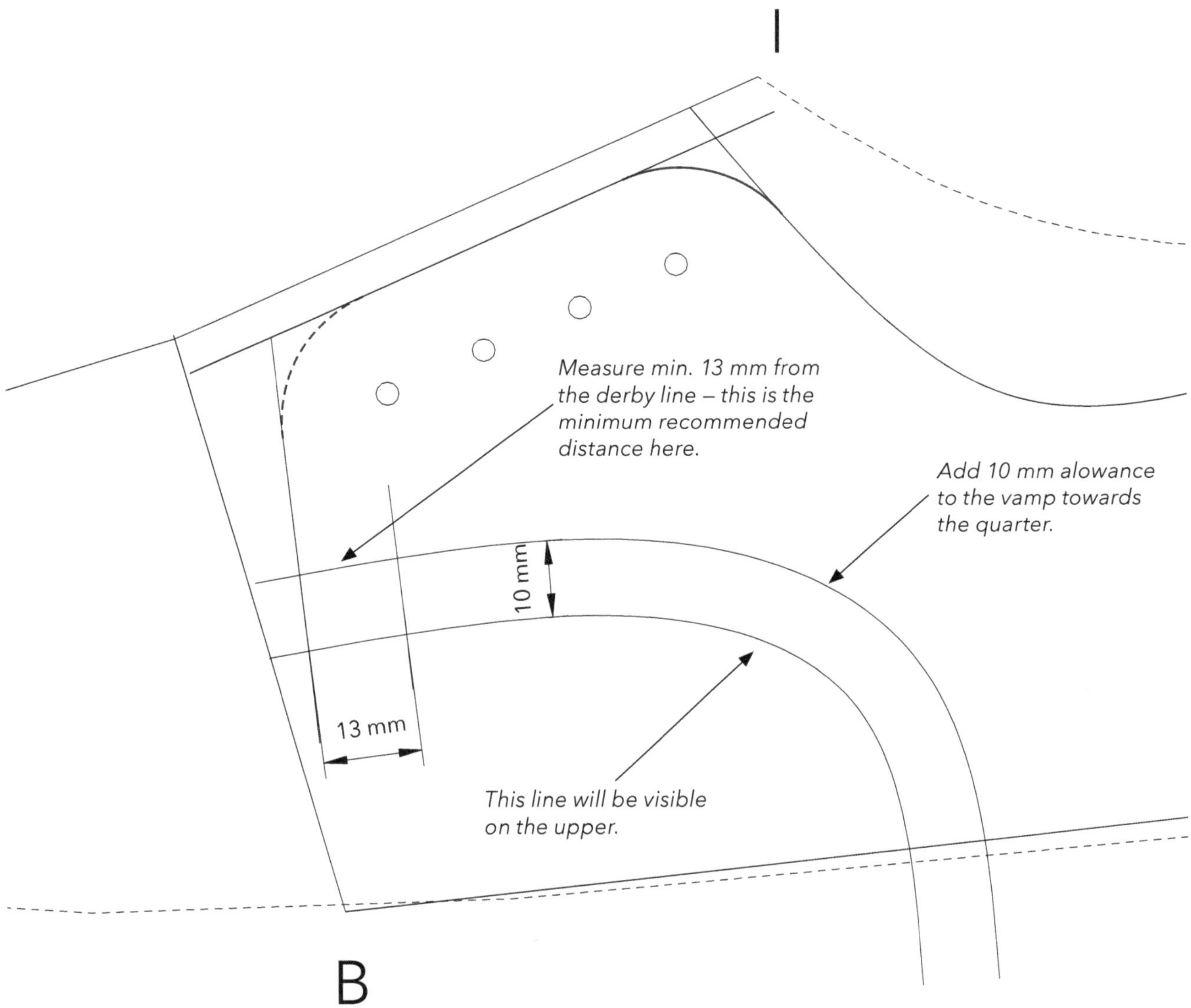

The Tongue

In this example, the tongue is cut from the same piece as the vamp. A standard derby tongue measures approximately 70-80 mm wide. Its primary function is to separate the laces from the foot, preventing irritation and bruising on the instep. It also closes the structure beneath the facings, protecting the foot from dust, dirt, and unwanted cold airflow, and preventing foreign particles from entering the shoe through the lacing area.

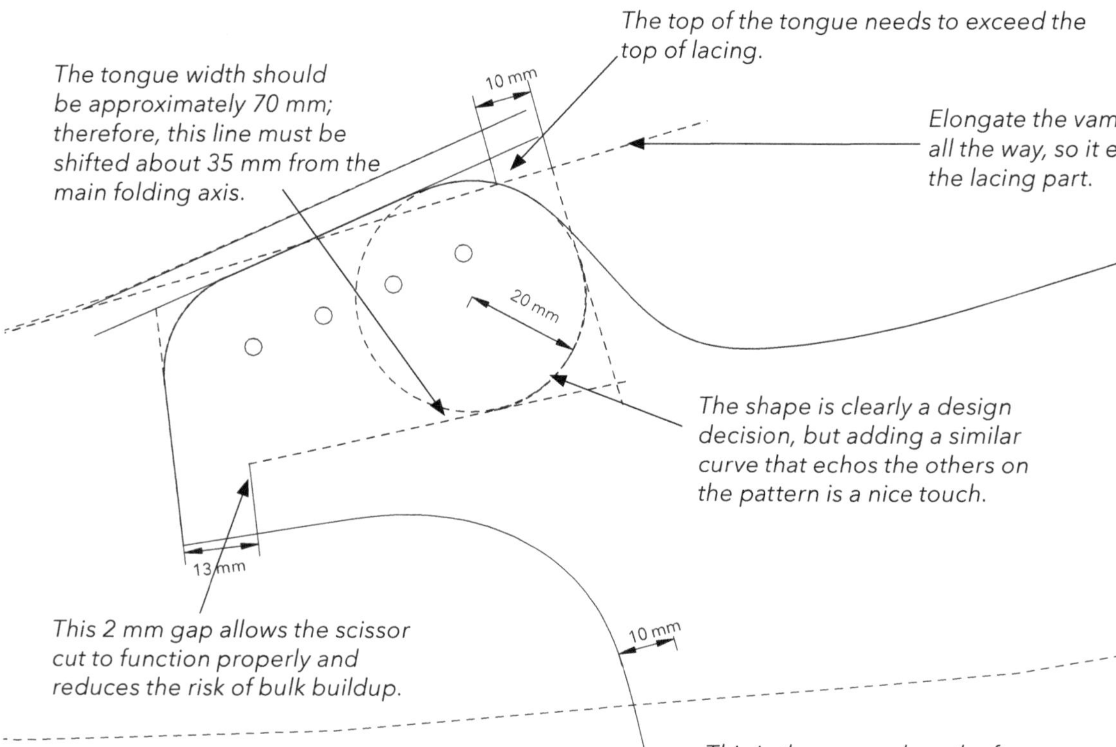

The proper length of the tongue should extend just slightly above the lacing. Too long and it may cause discomfort and look unbalanced; too short and it may not function properly. It must also be considered that the tongue can slip downward during lasting. This is influenced by the materials used, or more specifically by how slippery the lining is. In extreme cases, a tacking allowance can be added to the tongue and temporarily fixed to the last. Tongue slippage is also affected by the number of eyelets: four eyelets usually provide sufficient grip, while fewer eyelets increase the likelihood of slipping.

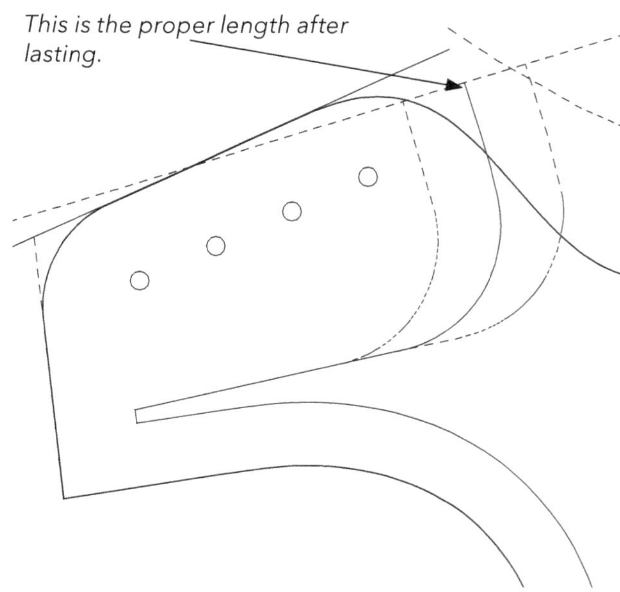

The Backline

> We often see photos of professional makers driving a tack through the counter to hold it in place during lasting – some even present this as a hallmark of true "handmade" craftsmanship. In reality, this is a classic example of the "don't question the old ways" mentality, which is both misleading and potentially does more harm than good.

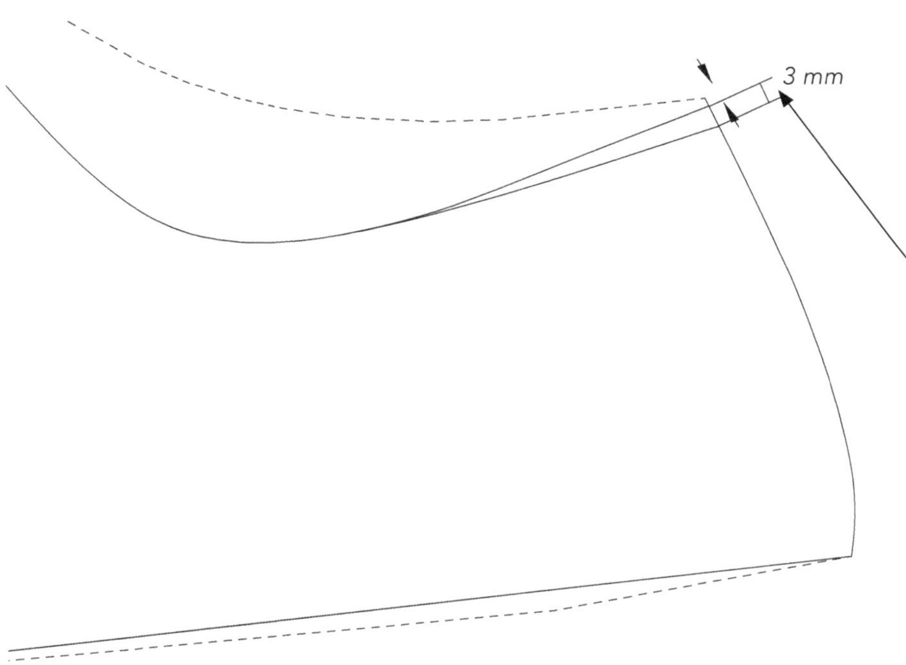

The first step is to raise the topline by 3 mm from its original position, as dictated by the geometric rules.

A standard 1/1 upper, where the upper covers most of the last, does not require such destructive treatment. Proper pattern making ensures that the counter cannot slip below its intended position, eliminating the need to damage the upper. In some cases – most notably loafers or styles with long toplines – a tacking allowance can be left, but only on the lining, so the affected area can be trimmed away afterward.

It is also important to note that hand lasting and machine lasting, especially when combined with a counter-shaping process, are fundamentally different. The method discussed here applies primarily to hand lasting. In machine lasting, a hard, pre-shaped counter is used, making slippage highly unlikely.

The first step of creating a counter that stays firmly in place is raising the topline by 3 mm as demonstrated in the ballerina shoe chapter.

Pivoting the Backline Curve

Pivoting the backline curve is key to achieving a tight topline and a stable counterpoint position, without the need for nails. The counterpoint has already been elevated, which helps compensate for the natural slipping and stretching of the topline during lasting.

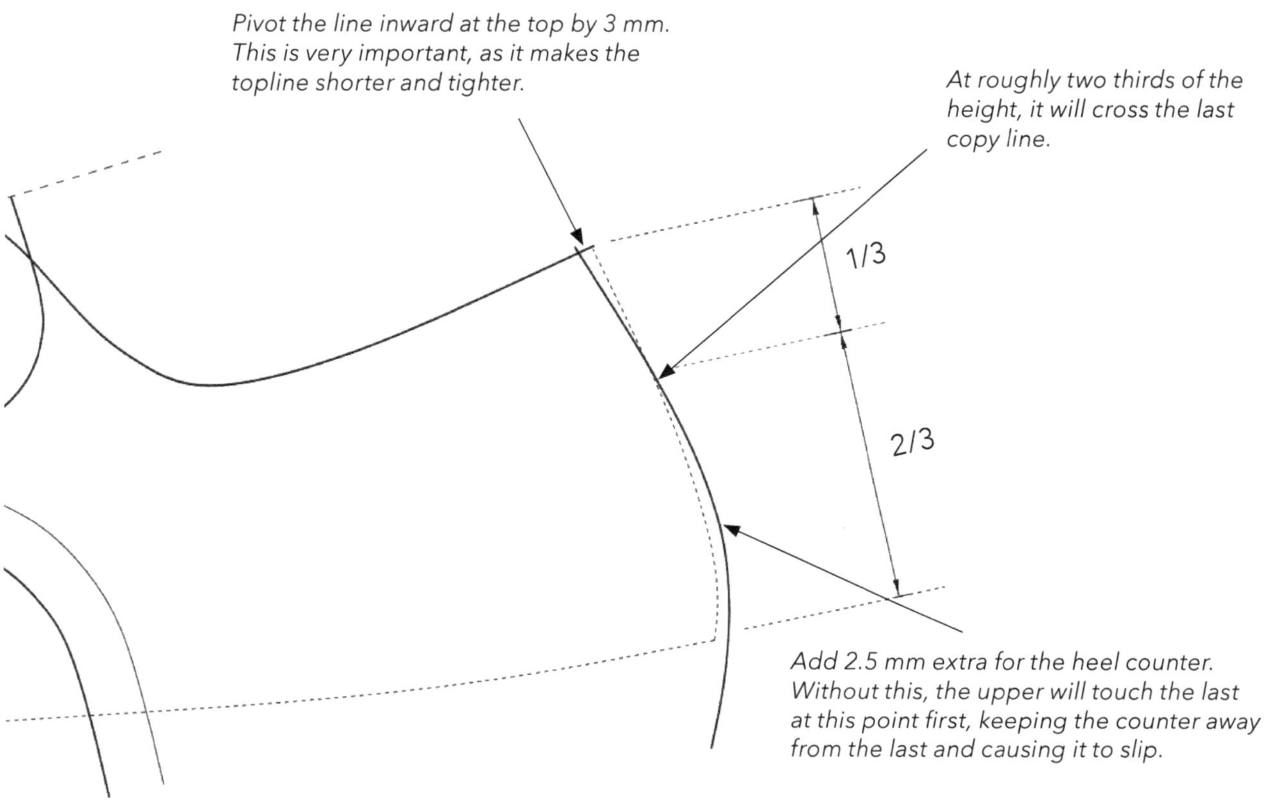

Another important modification is accounting for the thickness of the counter. A typical heel counter for a dress shoe is around 2.5 mm thick. If this is not considered during pattern making, the counter will pull the counter line closer to the last by that amount, but only at the bottom. As a result, the counter line will touch the last at the base while the counterpoint remains offset.

By adding space to the lower third of the pattern, the material thickness can be compensated for and the correct back curve maintained. Although lasting will shape the counter, this shaping occurs during the process, when it may already be too late to prevent unwanted slippage.

These two modifications ensure that the counter follows the last accurately and remains securely in position.

The Dogtail

The shape of the dogtail is purely a design choice, but its purpose is very practical: it reinforces the topline at a critical stress point. Without this reinforcement, the topline would not hold up for long, as this area experiences significant stretching when the foot slips in and out of the shoe. A simple reverse seam would not withstand this level of stress.

This small section of the upper is reinforced to accommodate the increased wear and tear, and named the dogtail because of its shape. Dogtails always run from the lateral side to the medial side. The dimensions and exact shape are only guidelines; some makers use a square version, which is equally functional and, from an assembly standpoint, even easier to work with.

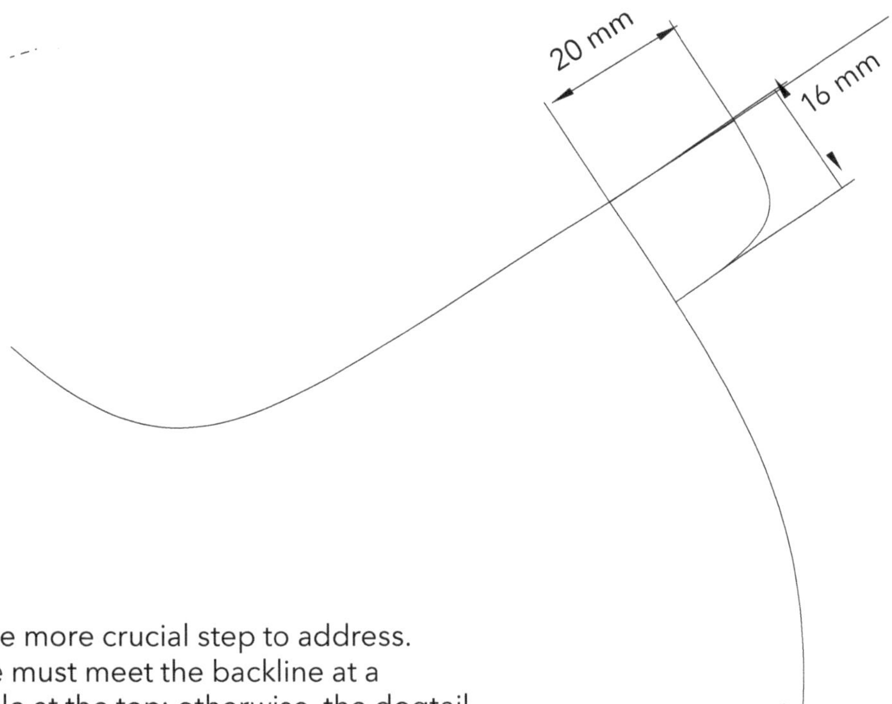

There is one more crucial step to address. The topline must meet the backline at a square angle at the top; otherwise, the dogtail will sit too high (most likely) or too low (less frequently) . This can be corrected by raising the topline, but preferably by adjusting its shape. For the record, this may not require any correction, as leather is a forgiving material and this minor inaccuracy can be easily corrected, resulting in an even tighter topline. A slight curve can be added at the end, or the center can be lowered by introducing more curvature and relying on the lasting process to straighten it slightly. This approach is quite common for certain models. This exact shape is included on the standard pattern ruler.

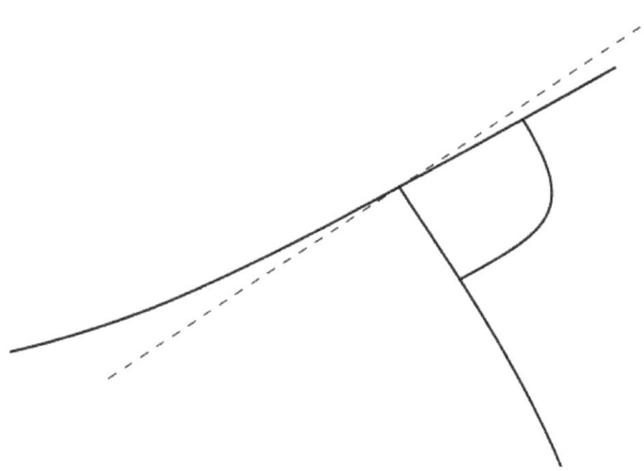

The Lasting Allowance

The lasting allowance is the extra margin of material added around the feather edge to allow the shoemaker to stretch and secure the upper over the last, particularly in traditional hand-lasting or comparable machine-lasting processes. However, it is not a universal or fixed measurement that applies to every shoe.

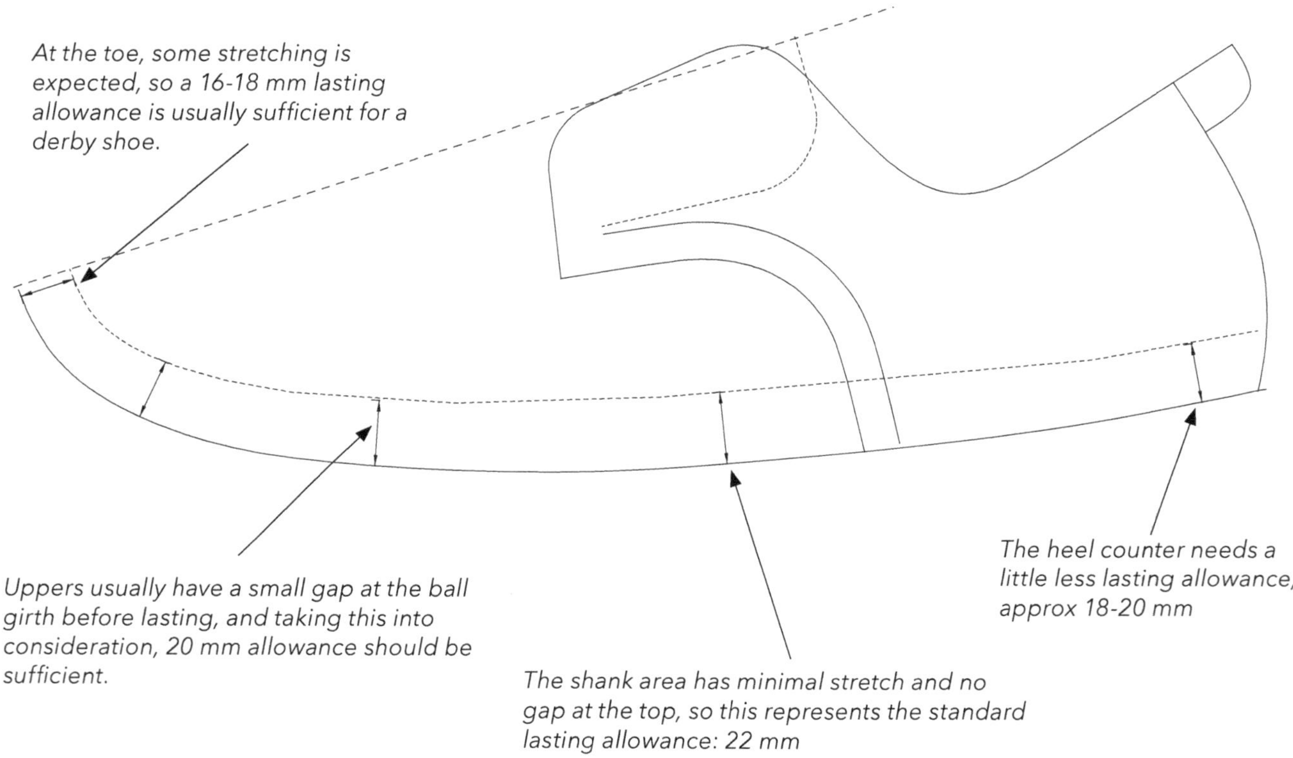

Different construction methods shape the upper in different ways. In some technologies – such as many athletic footwear constructions, moccasins, or other pre-formed uppers – the upper is engineered very precisely to its final shape on the last. In these cases, the lasting allowance may vary significantly or may not be required at all.

For this reason, the pattern maker must always consider the specific manufacturing technology when designing the pattern.

Material choice is the second critical factor. This involves not only how much the material can stretch, but also its thickness and stiffness. An 18 mm lasting allowance at the counter may work well with a thin upper and a light insole, yet be insufficient for a heavy work boot made from thick leather, combined with a robust insole and a bulky heel counter.

A poorly designed lasting allowance can compromise the work, create distortions, or even make proper lasting impossible.

The Scissor Cut

The scissor cut is the V-shaped cutout at the front of the quarter lining. It allows the lining, starting from the very bottom of the layered structure and overlapping the vamp lining from the inside, to rise to the top of the upper tongue and wrap around the vamp layers. This illustration clearly shows how the cut opens and embraces the layers, resembling the action of scissors, which is where the name comes from.

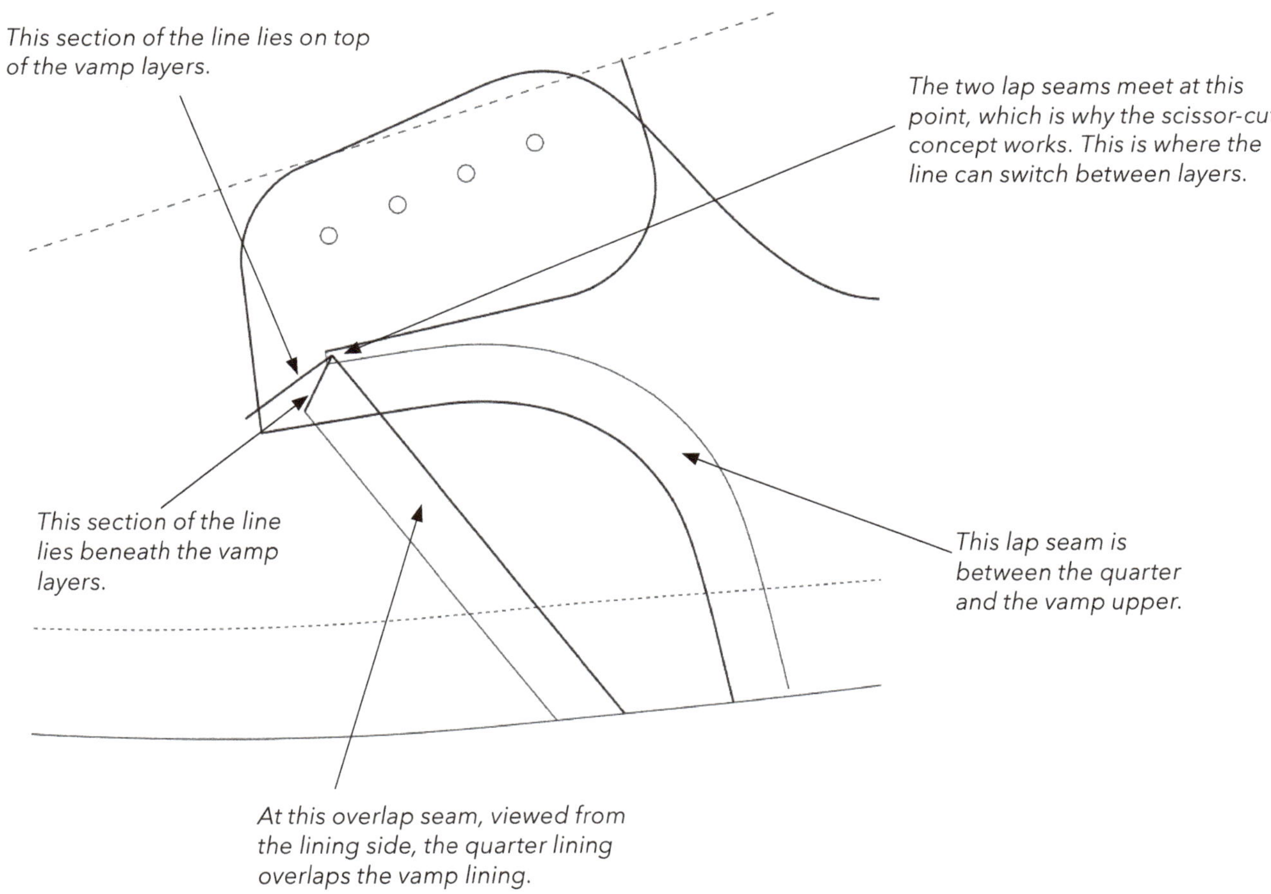

Upper Pieces

The Quarters

The quarter pieces have two raw edges, and allowances must be added for the topline and the backline. The reason for the asymmetry is the dogtail.

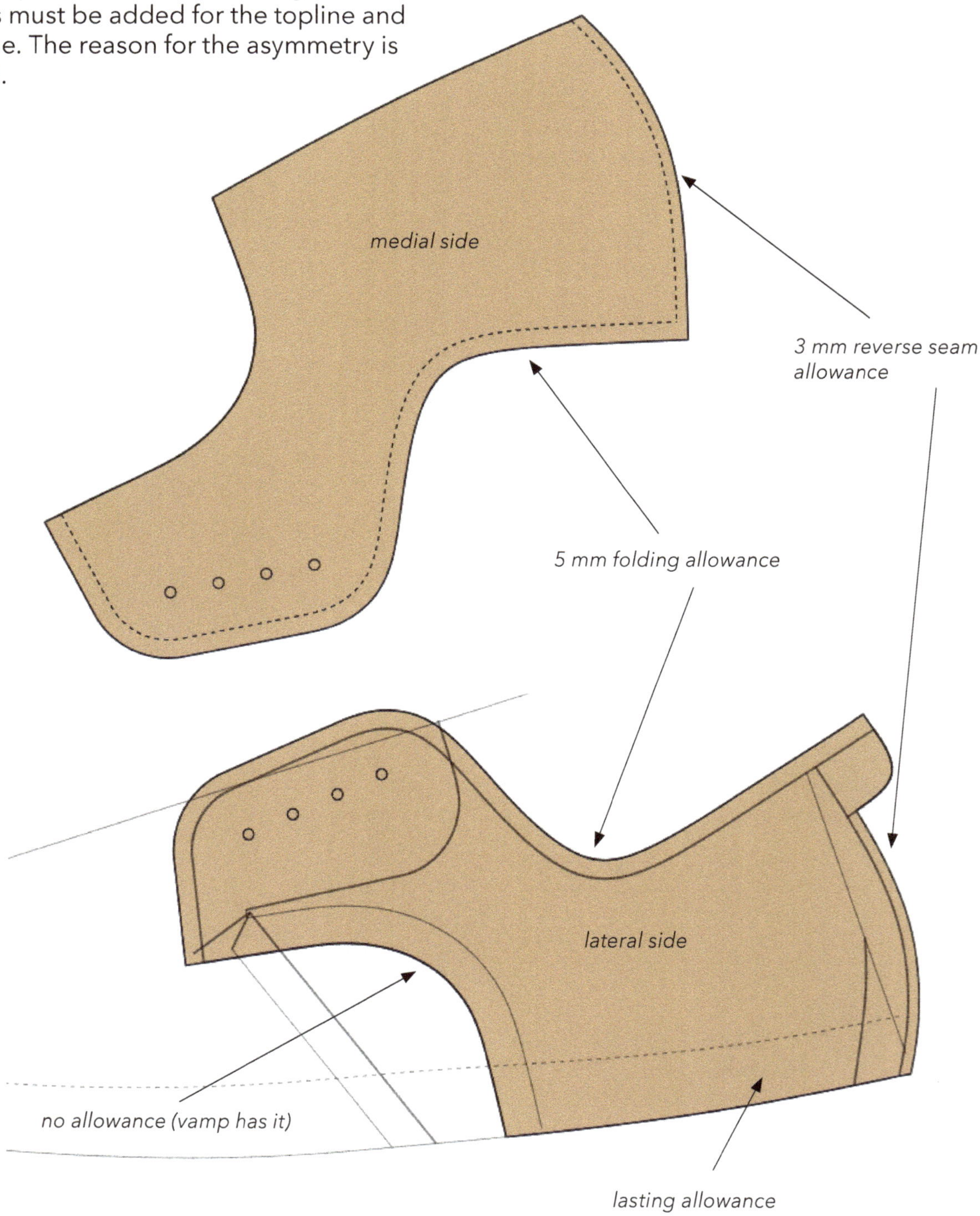

The Vamp

The vamp piece has no exposed edge, except for the small section at the top of the tongue. Lap seam allowances must be added along the vamp lines.

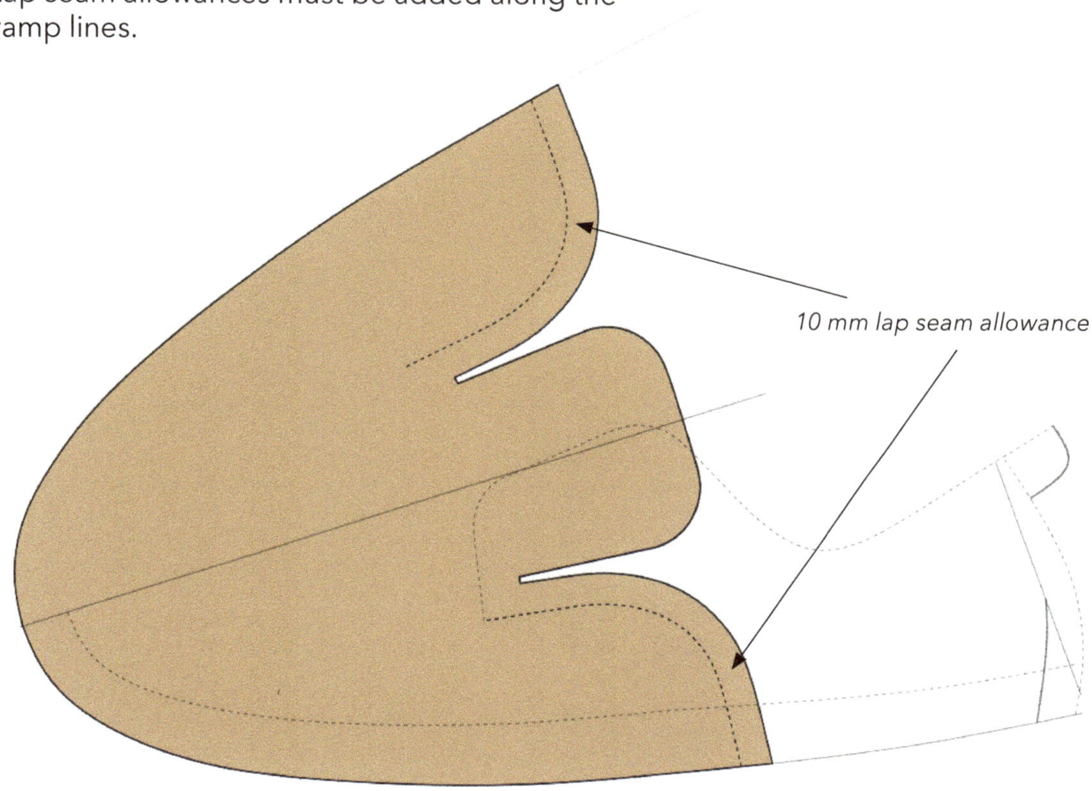

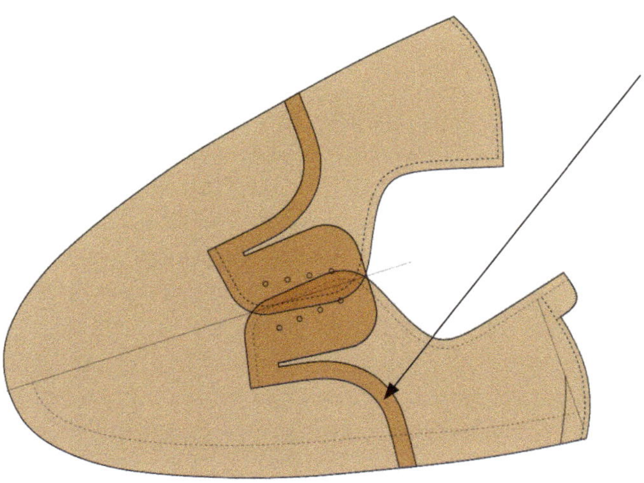

This illustration demonstrates how the lap seam allowance joins the vamp and the quarters. This junction is where the actual seam line sits. The overlap on the tongue, however, is not considered an allowance.

The lap seam allowance can vary from the typical 10 mm. Thicker materials may require a larger allowance, while thinner materials can work with less. In general, this range falls between 6-12 mm.

There is one notable exception: when the quarters feature brogueing and the allowance needs to act as the bottom, visible layer. In that case, the allowance can be increased to 15 mm or more to achieve the desired visual and structural effect.

Lining pieces

The Vamp Lining

The vamp lining has a trimming allowance at the tongue. This can be 3-5 mm. No additional allowances are necessary.

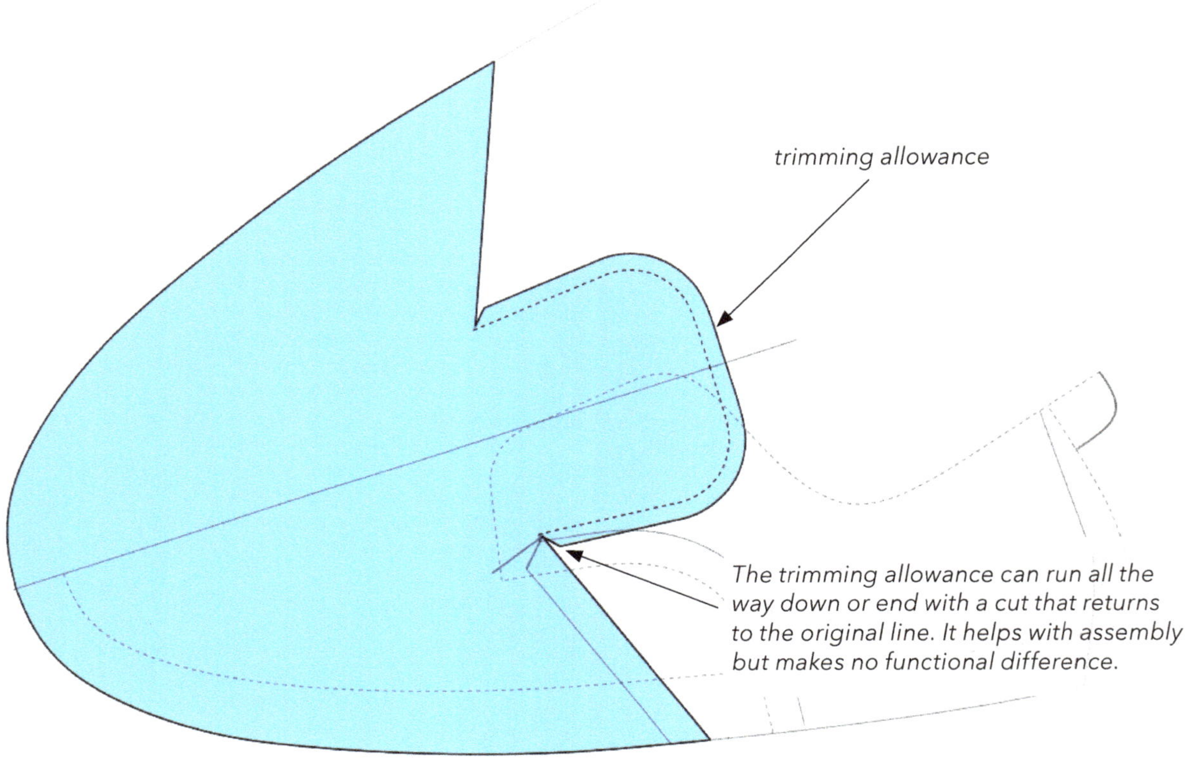

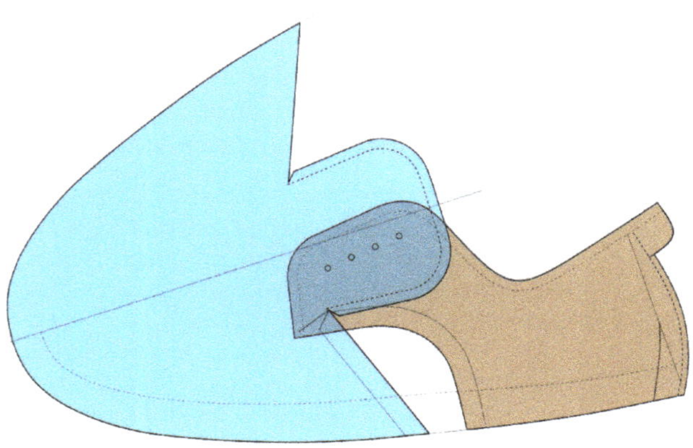

It is worth pausing briefly to discuss the relationship between the quarter upper and the vamp lining. The two lap seams are positioned differently, so there is no point where four layers of leather stack on top of each other. This is especially important at the feather edge, where excess bulk would be problematic. This structure allows the upper maker (closer) to access the lining lap seam without interfering with the lap seam of the upper.

The Quarter Lining

A single-piece quarter lining is a slightly more high-end solution. For manufacturing, however, it may be somewhat less cost-effective than a multi-piece version. Make sure to use the vamp axis when unfolding this piece.

Next, establish the counter folding axis while accounting for the orange-peel effect. This is done in exactly the same way as described earlier in the ballerina shoe chapter.

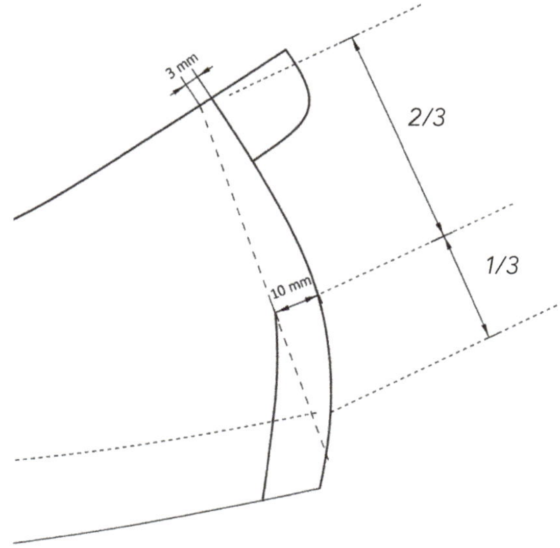

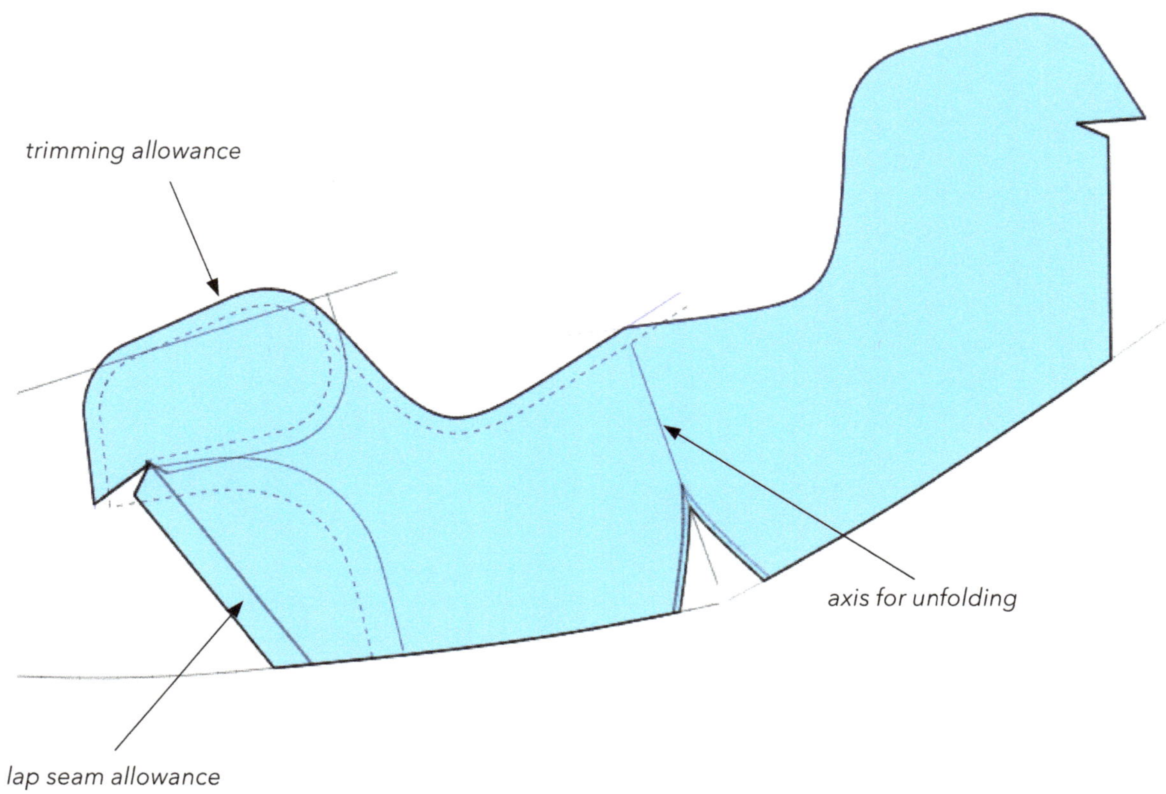

This is a far more common solution for the quarter and counter lining structure: dividing the lining into three pieces allows for more cost-effective nesting, as the components are significantly smaller. A lap seam is used between the counter and the quarters, and it is up to the maker to decide which piece sits on top and which underneath. The counter piece can be placed suede side in, but in practice this makes very little difference and is more a matter of design and appearance than function.

A hybrid solution is also possible, offering a slightly higher-end look while remaining efficient in nesting: using only one lap seam, resulting in a two-piece quarter lining structure.

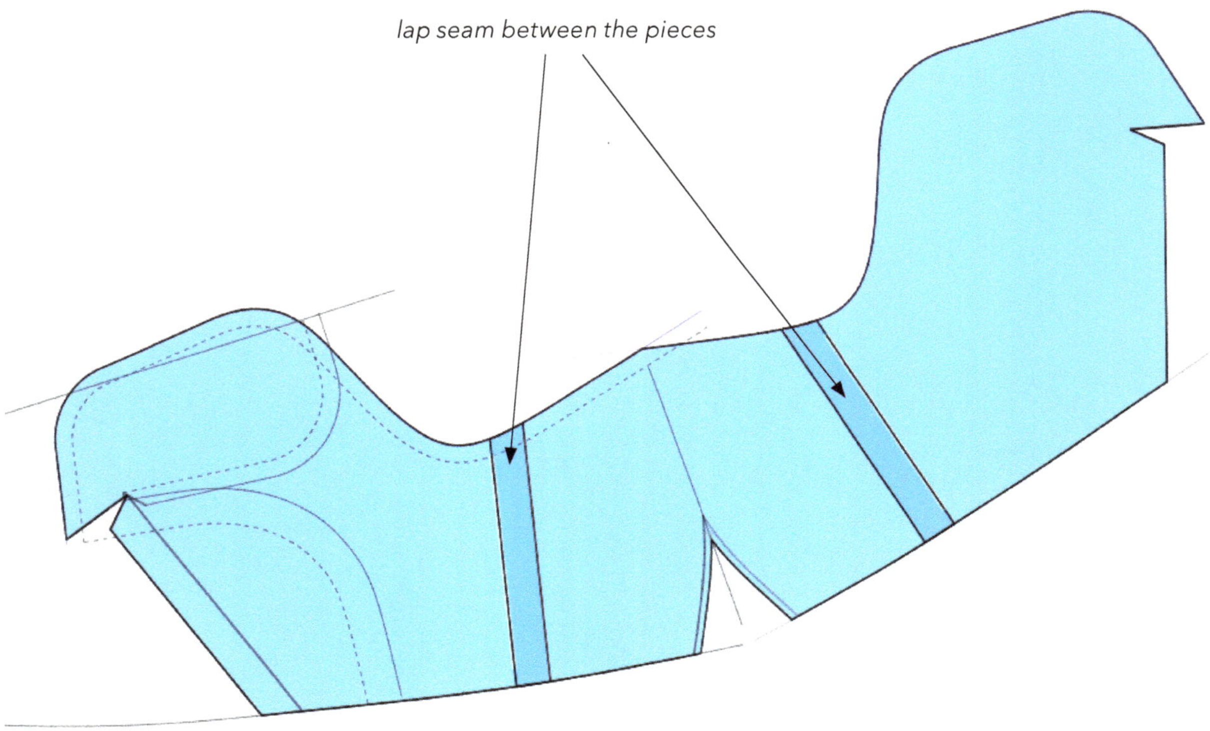

The Complete Pattern

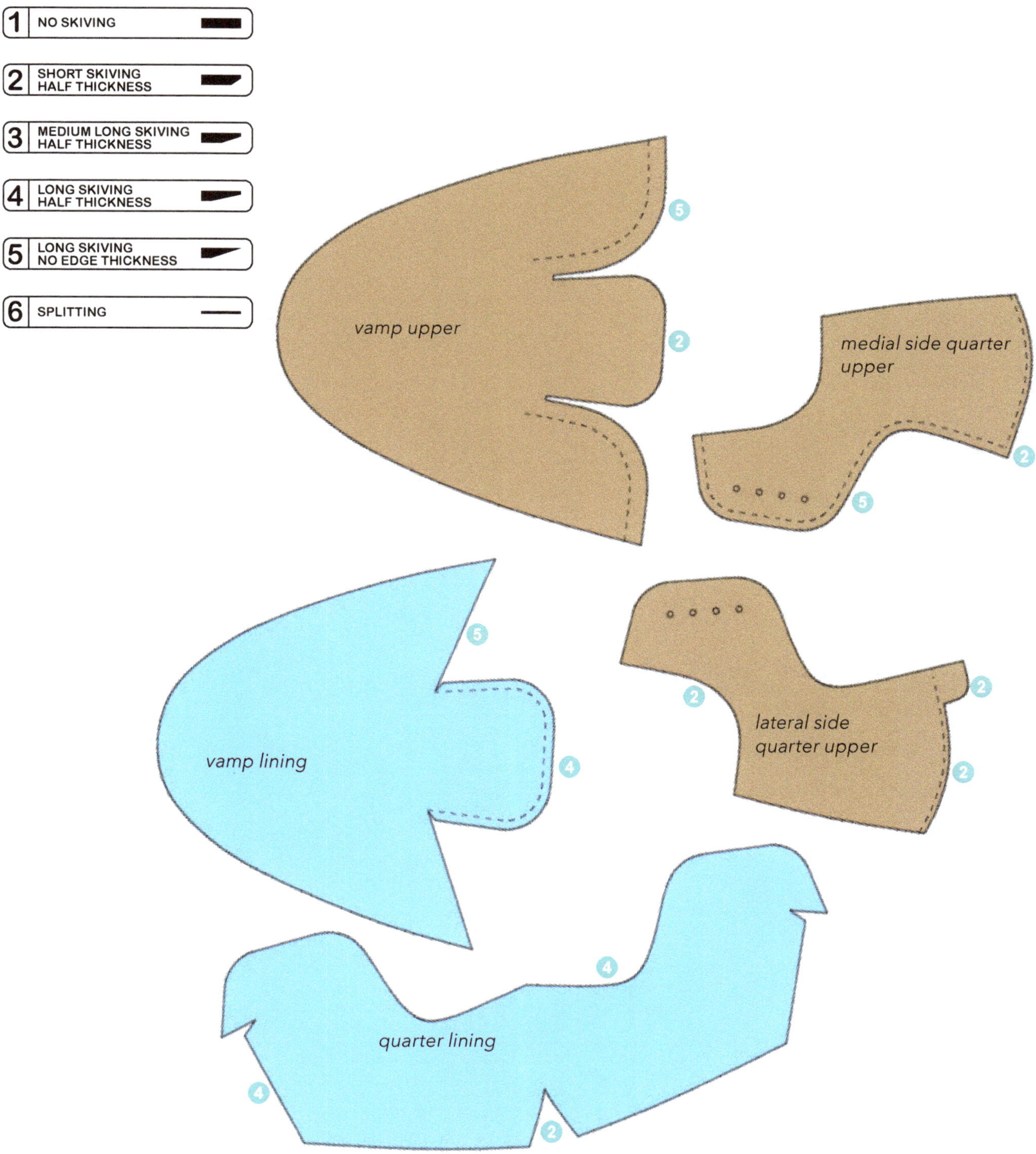

Assembly order

1. Reinforce the lacing and the topline
2. Fold the topline
3. Sew the back seam, trim, and rub
4. Add double-sided tape to the dogtail and attach it to the medial side
5. Finish folding
6. Sew the backline topstitch
7. Sew the lining quarter reverse seam
8. Attach the quarter lining to the quarter upper
9. Sew the topline
10. Trim the quarter lining trimming allowance
11. Attach the vamp lining to the vamp upper
12. Sew around the tongue
13. Cement the right side of the quarter to the vamp upper, and do the same with the lining
14. Sew the upper
15. Cement the left-side upper pieces together
16. Sew the upper pieces
17. Sew the lining lap seam
18. Sew the reinforcement seam

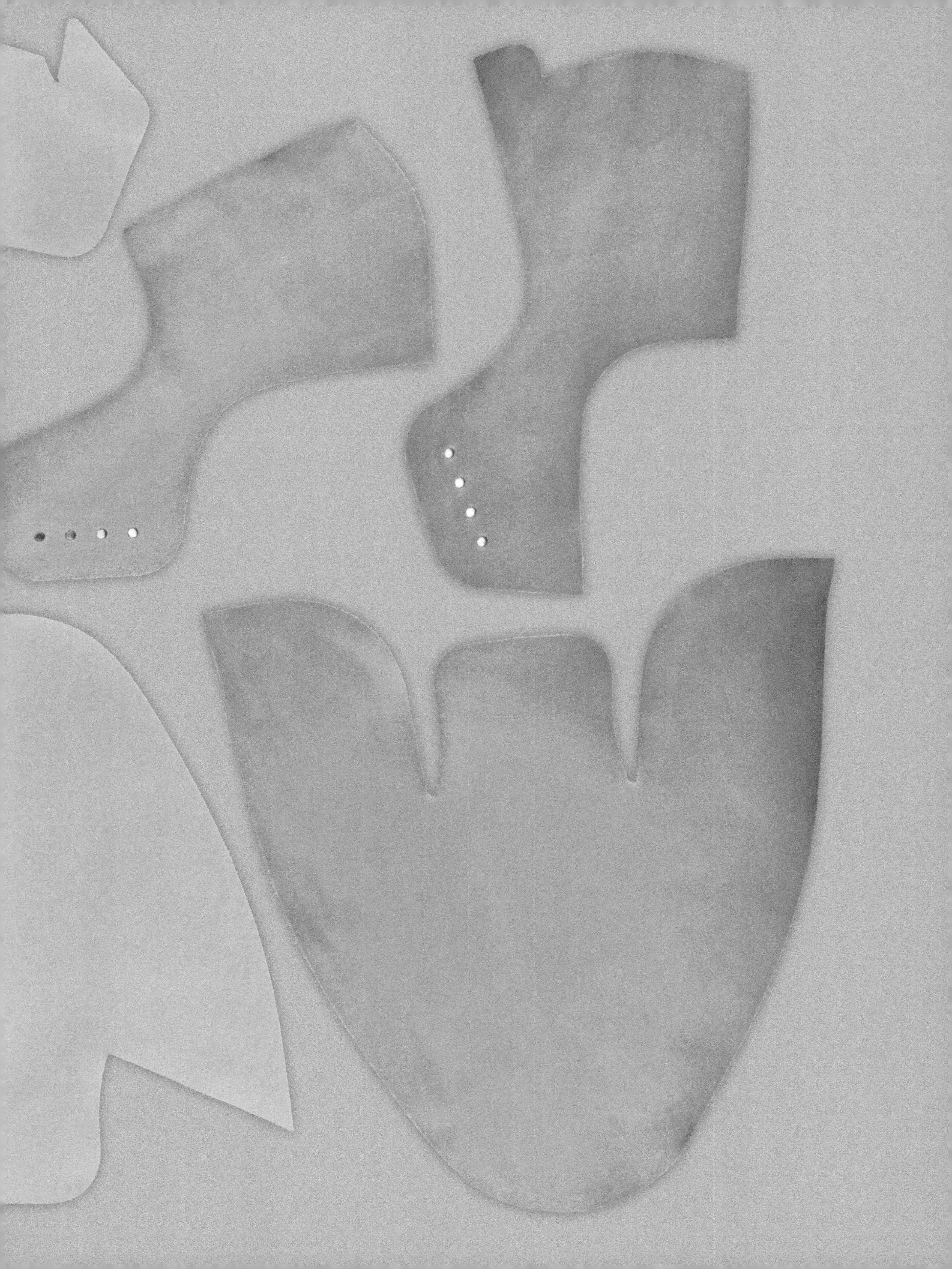

After skiving is completed, reinforcements are added to the upper. Since a relatively thick leather is used, only a few different types of reinforcement are required.

The lacing tape is 25 mm (1") wide and made from a synthetic, non-woven material. This material is practically unbreakable. The tape supports the lace holes, preventing them from overstretching or deforming. In addition to its reinforcing function, the tape is cut to match the shape of the upper, helping to create a clean, crisp, and accurate folded edge.

The topline tape used here is a 3 mm (1/8″) wide woven tape. It provides additional strength under the fold and ensures that the thin, skived edge has a clear guideline during folding. The primary function of this tape is to prevent the topline from overstretching over the life of the footwear. After attaching the tape, the folding allowance is cut where necessary to allow the edge to spread properly.

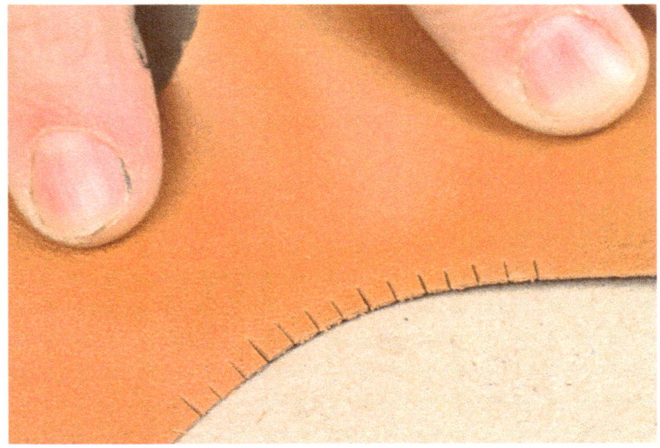

Without these relief cuts, the edge will curl inward or outward. It is very important to make these cuts along concave curves. Some leathers can be folded without them, but the result is usually underwhelming, and the soft edge is likely to become distorted. It only takes a few seconds to make these cuts, and the result justifies the effort.

The folding hammer is a special tool used to make folding easier by hand. The hammer head functions like a regular hammer, and the technique is simple and easy to learn. A thin layer of cement is applied to the folding allowance and the narrow strip where it will be folded. After allowing the cement to dry slightly, the allowance is folded down. The leading hand lifts the allowance from the table, while the folding hammer follows with controlled tapping.

Folding along a curve is also straightforward if proper reinforcement is used. To manage the excess material, distribute it as evenly as possible. Create two larger pleats at the corner, then divide them and repeat this process as long as possible. This results in small, even pleats that are easy to hammer flat. As an alternative, notches can be cut, but this requires more effort and does not always produce a better result.

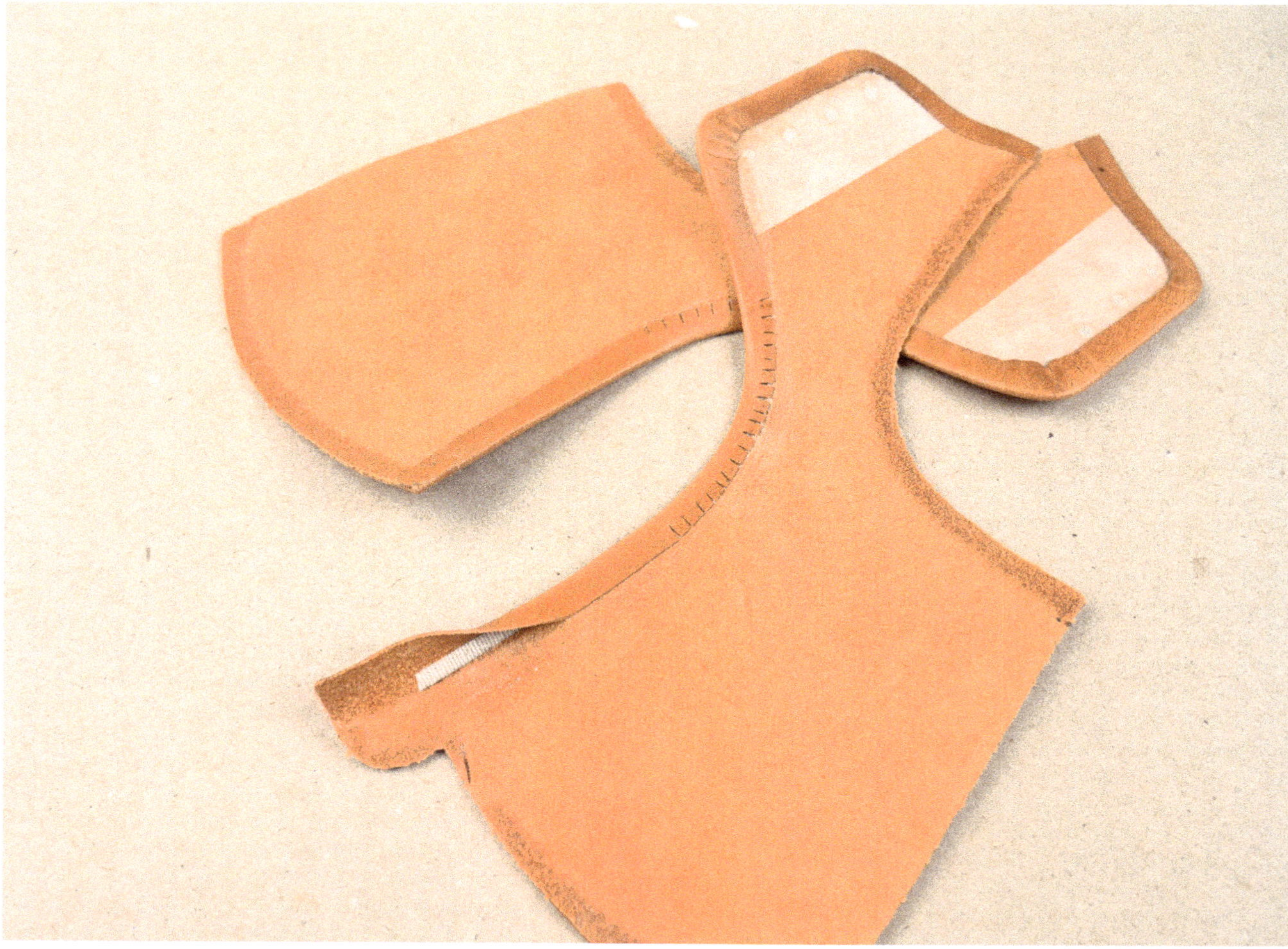

Folding is a typical edge finish for derby shoes and other classic dress shoe designs. The style generally dictates that the topline should have a finished edge, rather than the raw edge seen on some other styles. However, this does not mean that every edge on the upper requires treatment. As shown in the image, the vamp line—the curved line where the quarters overlap the vamp—is left untreated.

It is also possible to choose a coated edge instead of folding, provided the leather is sturdy enough. Selecting the appropriate edge treatment is one of many decisions the pattern maker and shoemaker must consider. Each option has advantages and disadvantages, but none is inherently superior to the others.

The quarter can lose its shape and the curve its definition during the lasting and wearing stages, so additional reinforcement tapes or interfacing materials should be used to provide extra strength.

The skiving is complete–the intact portion of the fold runs under the dogtail folding allowance, then quickly tapers, making this area precise and smooth.

It may seem excessive to skive the small corner where the dogtail overlaps the topline on the medial side, but this detail distinguishes a bespoke-quality upper from a mediocre one. Using a piece of curved glass as a work surface provides a clean and easily accessible angle for this task. Without this extra step, the fold would become bulky, not to mention the additional excess required to fold over an already folded edge. With this step, a 5 mm allowance is perfectly sufficient; without it, an additional 2–3 mm folding allowance on the dogtail would be needed to complete the fold. In mass production, however, this step would be impractical, and the pattern would require recalibration.

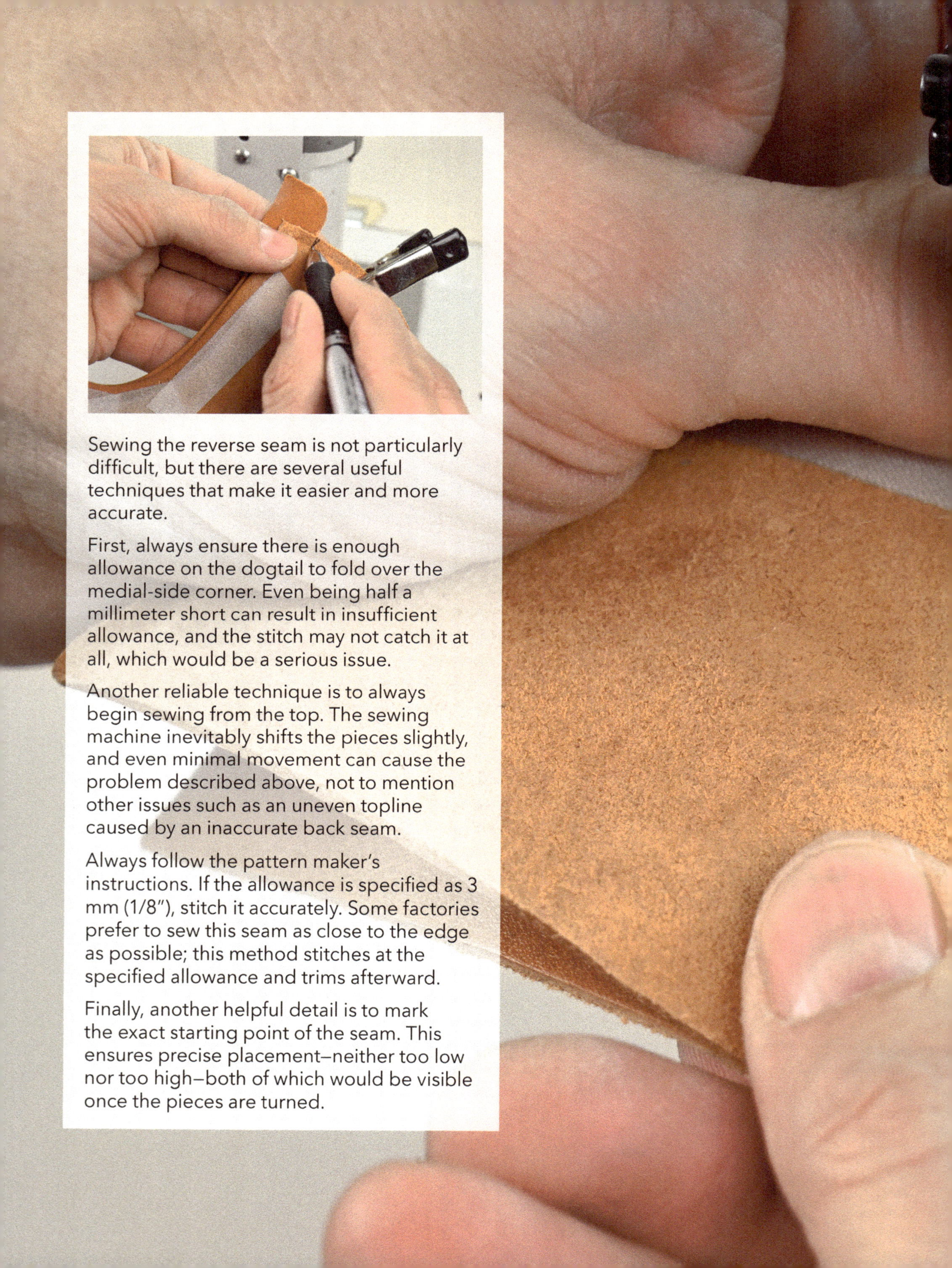

Sewing the reverse seam is not particularly difficult, but there are several useful techniques that make it easier and more accurate.

First, always ensure there is enough allowance on the dogtail to fold over the medial-side corner. Even being half a millimeter short can result in insufficient allowance, and the stitch may not catch it at all, which would be a serious issue.

Another reliable technique is to always begin sewing from the top. The sewing machine inevitably shifts the pieces slightly, and even minimal movement can cause the problem described above, not to mention other issues such as an uneven topline caused by an inaccurate back seam.

Always follow the pattern maker's instructions. If the allowance is specified as 3 mm (1/8"), stitch it accurately. Some factories prefer to sew this seam as close to the edge as possible; this method stitches at the specified allowance and trims afterward.

Finally, another helpful detail is to mark the exact starting point of the seam. This ensures precise placement—neither too low nor too high—both of which would be visible once the pieces are turned.

The reverse seam allowance is typically 3–5 mm, depending on the leather's thickness and firmness. This seam is a high-stress area and one of the most common points where stitches can deteriorate during lasting. For this reason, it is essential to use a generous amount of backstitching at the bottom, and even two to three triple stitches at the top.

There are two additional details that require close attention. First, the stitch length must not be too short. Very short stitches effectively perforate the leather, and even if the sewing is accurate, the seam can still tear because the needle holes form a weakened line. Second, when backstitching, it is critical to return to the exact same holes, on the same line. This avoids unnecessary perforation and prevents an unprofessional "double-row" appearance on the outside once the piece is turned.

Using a different needle point is a small but worthwhile adjustment. This is an ideal moment to use a flatbed machine with a rounded-point needle. Leather-point needles create small cuts rather than holes, and these cuts may tear later in the process. Always consider what the specific material can tolerate.

There are several approaches to achieving a precise reverse seam. One effective method is to trim the allowance after sewing. This technique requires confidence with the knife, but the results are consistently accurate. A useful rule of thumb is to leave an allowance equal to the combined thickness of the two skived layers. For example, if one layer is approximately 0.8 mm, an allowance of 1.5-1.6 mm is ideal. With this amount, the remaining leather is too short to shift left or right. When trimming, make sure to cut the leather in a single pass, as a second cut may show through on the finished side. If the allowance is left too short, however, the seam may tear.

Hammer the stitches from the flesh side with great care. A specialized seam-rubbing machine can also be used for this task, although it may not be worth the effort for a single pair. While this is a straightforward operation, there is still room for error, especially when using a regular hammer. Excessive force can force the layers apart, and the thread may cut into the skived leather. For better control, a folding hammer is recommended.

Some leathers are sensitive and prone to damage from hammering. When working with a new or unfamiliar material, always test it first.

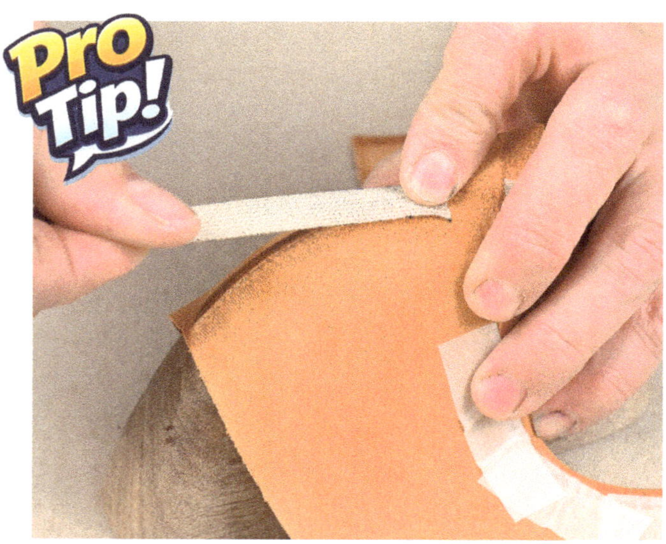

Reinforcing this seam may seem excessive at first, but it is entirely justified. This special type of tape helps keep the two layers perfectly centered, significantly increases the overall strength of the seam, and also serves as a stable base for the top-stitching.

To apply it, stretch the tape taut while adhering it. This, combined with the non-flexible nature of the backline tape, ensures that no pleats appear after turning. Without this technique, the orange-peel effect can cause bulky pleats to form—even with tape—which may show through the upper and interfere with stitching.

After applying the reinforcement tape, a light tapping from the grain side can be useful. As before, this must be done with great care, as excessive force can cause the thread to cut into the upper leather.

As shown, the small excess portion of the backline tape is folded upward at the bottom. This area is the most vulnerable part of the upper during lasting, so reinforcing it helps prevent potential issues. In more demanding cases, an additional small piece of reinforcement tape can be added here for extra protection.

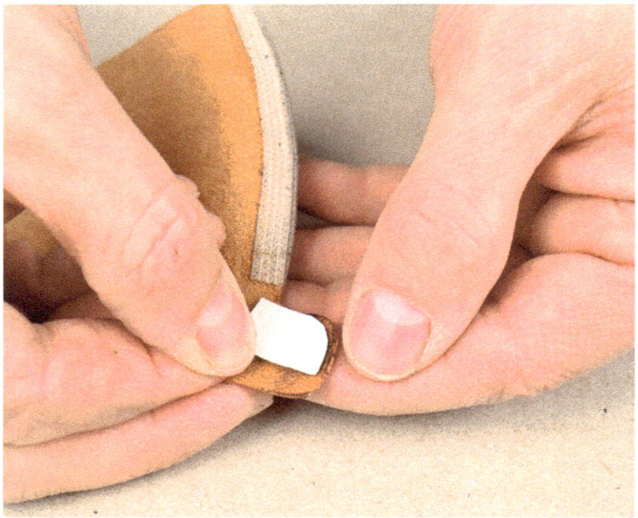

Even the most careful wearer is unlikely to use a shoehorn every time. Finishing the backline with a reverse seam in such a critical area would leave it highly vulnerable. The dogtail provides essential reinforcement for the topline at this high-stress point.

When finishing the topline stitching, a practical way to keep this area both clean and strong is to use a small piece of double-sided tape. Cut it slightly smaller than the dogtail shape. A dedicated pattern can be made for this, but for a single pair it is unnecessary.

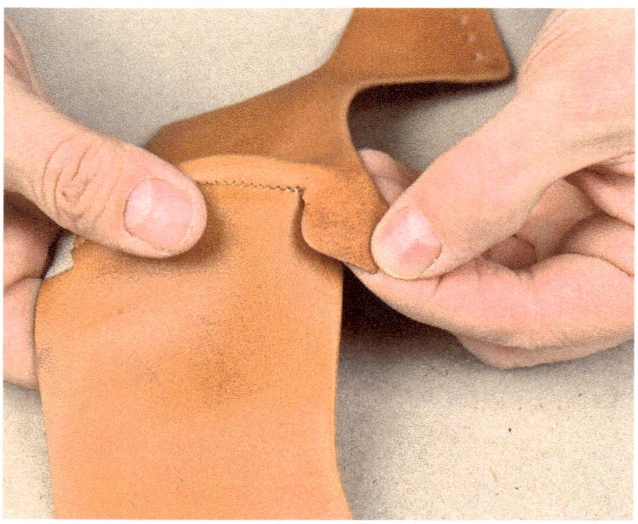

Another useful technique is to briefly pass the dogtail through the flame of a lighter. This causes the edge of the dogtail to curl slightly inward, allowing it to sit flush against the strongly curved surface.

After heating, bring the piece to the outside and gently press it into the position where it naturally settles, applying only light tension. Stretching it too much, or not enough, can both cause issues. Also ensure that the dogtail does not sit too low, as this would reduce the folding allowance and weaken the structure.

Now it is time to complete the folding. Apply cement to the edge and allow it to dry. Note that the fold on the underlying piece remains intact and unskived near the edge of the dogtail. This detail is important. This area is structurally the weakest part of the shoe, even with careful reinforcement. If the skiving were carried all the way to the edge, it would thin and weaken the material underneath, significantly increasing the risk of tearing.

After the adhesive has properly dried and is ready to bond, press the edge down. This is not a particularly difficult task, but a common mistake is leaving the edge loose and creating a small gap. The fold must be as snug as possible without collapsing the underlying layer. Beyond the aesthetic issue, an incorrect fold will also result in an uneven topline stitch. It is impossible to sew cleanly along an irregular edge. Whether the unevenness originates at the dogtail (on the left shoe) or on the opposite side (on the right shoe), the result will be a visible step in the topstitch.

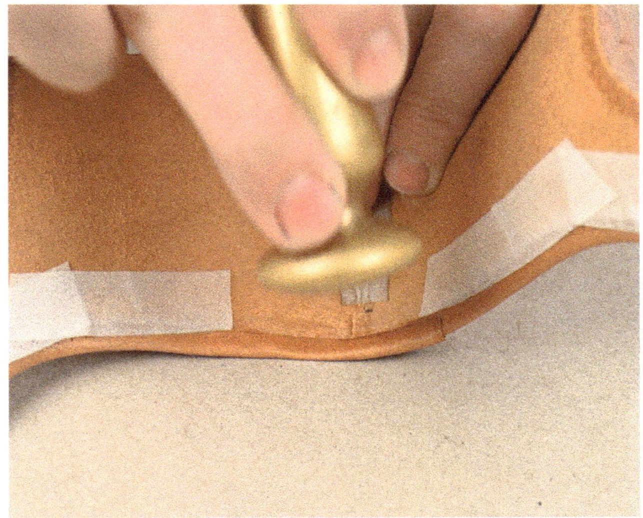

Here is a small additional technique to ensure long-term durability. Add a wide reinforcement tape beneath the dogtail. This tape does more than simply hold the piece in place; it is also stitched into the back seam, providing extra structural support for the dogtail. Because this section of the backline has a slight curve, proper adhesion requires a few small relief cuts along the bottom edge of the tape, as shown in the image. These cuts allow the tape to overlap slightly, enabling it to follow the curve while remaining firmly secured where it is most needed.

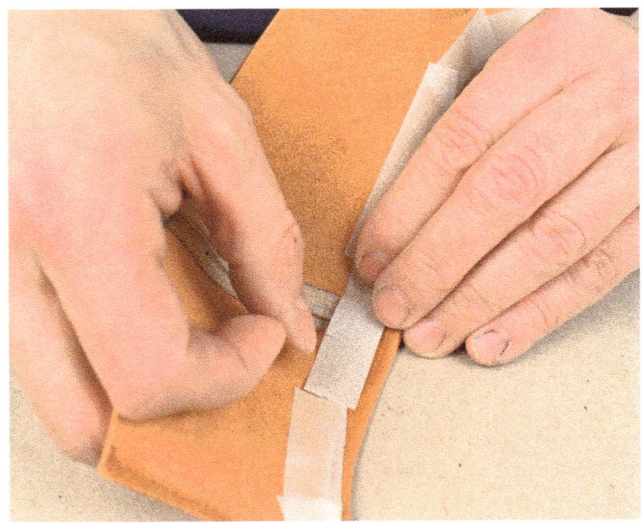

With the right technique, sewing the dogtail is not as difficult as it appears, despite the tight curve. A common approach is to "step-sew" with the machine, keeping the stitch line as close to the edge as possible. The sequence is as follows: (1) make the stitch; (2) do not let the needle complete the full cycle–when it starts to rise and the thread loosens slightly, lift the presser foot; (3) while the needle remains in the material and the presser foot is raised, pivot the material so the next stitch aligns precisely with the edge. Using this method, the stitch length remains consistent, and the line follows the curve cleanly and accurately.

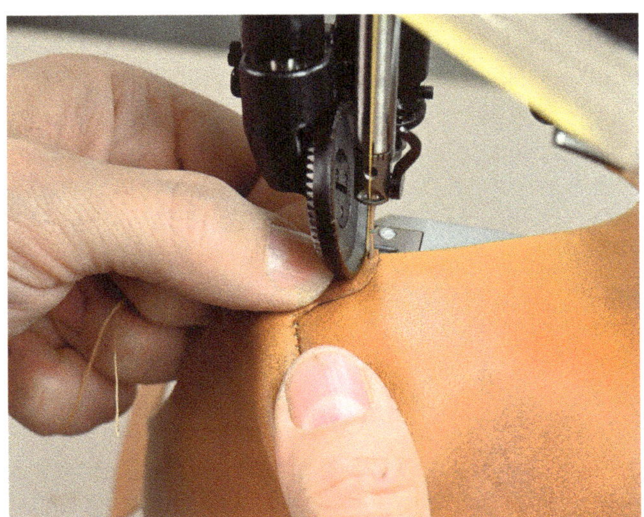

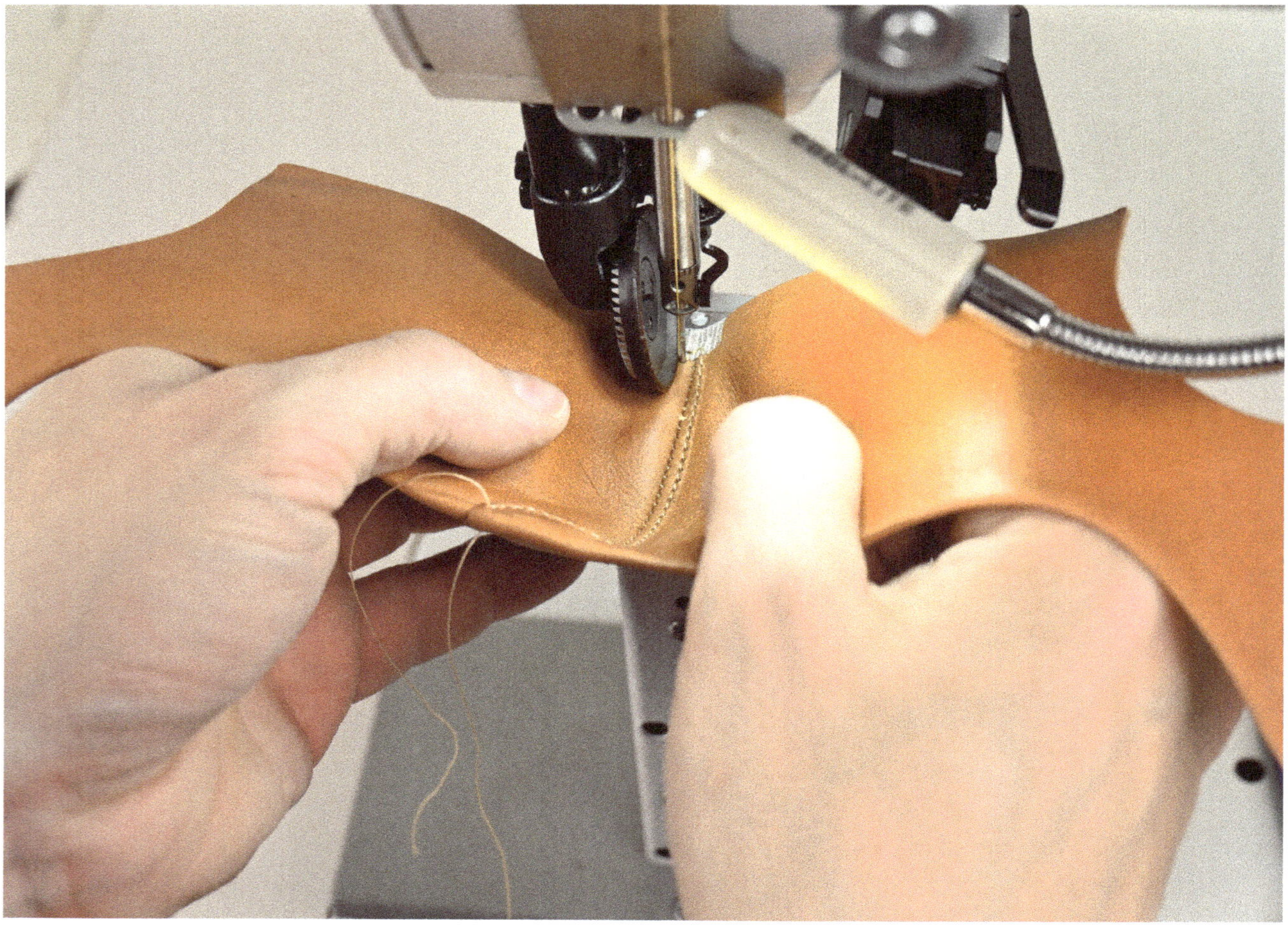

This step is optional rather than mandatory. In practice, this stitch can be challenging for beginners, so the following tips help achieve a clean stitch line.

In the case of a back seam, a "perfect" stitch line runs cleanly parallel to the reverse seam, at a consistent and not overly wide distance, and is placed on the flat area rather than on the curved section. Achieving this requires precise skiving, a clean reverse seam, and accurate trimming. Even when these are done correctly, the shape itself presents another challenge. The short flat section of a post-bed machine rarely aligns perfectly with the backline, making it difficult to judge the correct stitch distance. A useful technique is to turn the seam outward, as shown. This allows for slight stretching of the seam, focusing on the area directly under the needle and one or two stitches ahead, creating a clearer and straighter reference for stitch placement.

Another important tip–often applied instinctively by experienced makers–is to avoid overcorrection. Beginners tend to anticipate errors and make abrupt adjustments on a line that should remain smooth. This usually makes the flaw more noticeable than if the sewing continued steadily. During stitching, the roller foot presses down the reverse seam, and it may seem as though the distance to the edge suddenly disappears. This is the moment to resist steering away. Continuing straight, without sudden corrections, greatly increases the chance of achieving a clean, even, and professional-looking seam.

The finishing step of the back seam is straightforward. Stitch-in-the-ditch means sewing directly along the edge of the seam—in this case, the dogtail—as close as physically possible. In addition to being an attractive decorative detail, this stitch significantly reinforces the entire structure. The two seams running along the dogtail also secure the reinforcement tape underneath, creating a very strong bond.

Sewing this line correctly ensures that this vulnerable area of the shoe will not tear under stress.

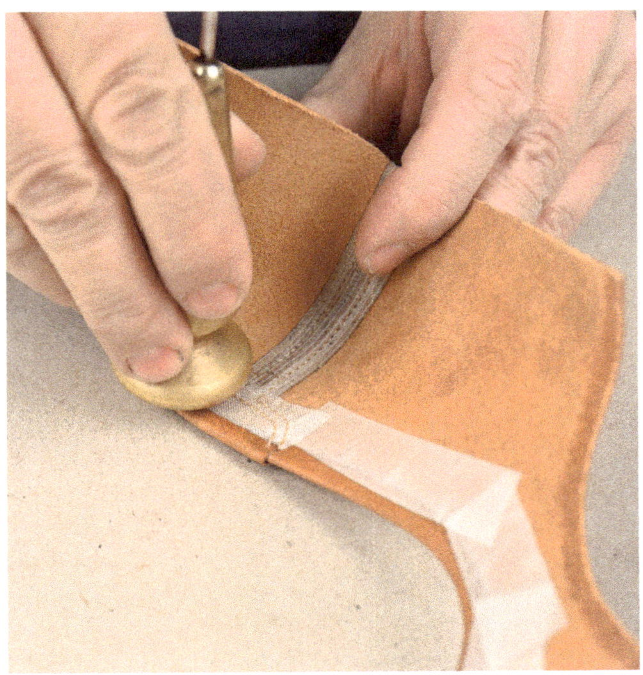

Here is the final back seam. For practical reasons, smaller stitches are preferred: they allow easier turns, reduce the chance of visible errors caused by uneven stitch length, and are stylistically the most appropriate for this design.

The start and end of the topstitch may require a single backstitch, or alternatively, the outer thread can be pulled to the flesh side and melted to prevent it from opening.

In general, it is a good practice to lightly hammer the back seam - or any seam, for that matter–to give it a more polished appearance after sewing. This half-circular wooden piece provides an excellent support for performing this operation.

As discussed earlier, the backline tape folds over at the bottom. It is important to note that the backline tape is not primarily used for its tear-resistant properties; however, this small detail greatly helps prevent the seam from starting to tear.

The Backseam Sequence

The sequence of the backseam is different between the left and the right shoe, because the roller foot must run on the stitched piece by default.

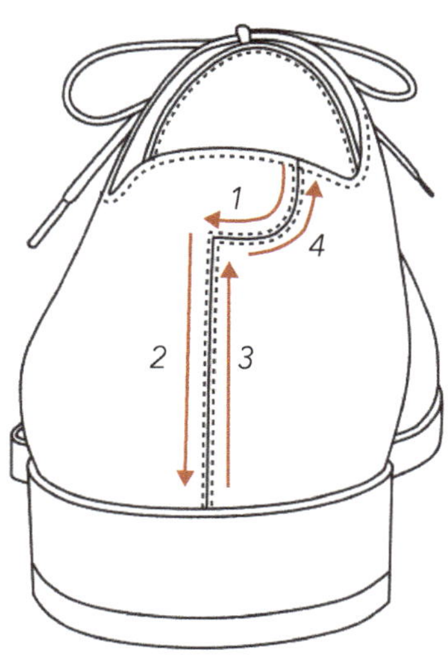

left

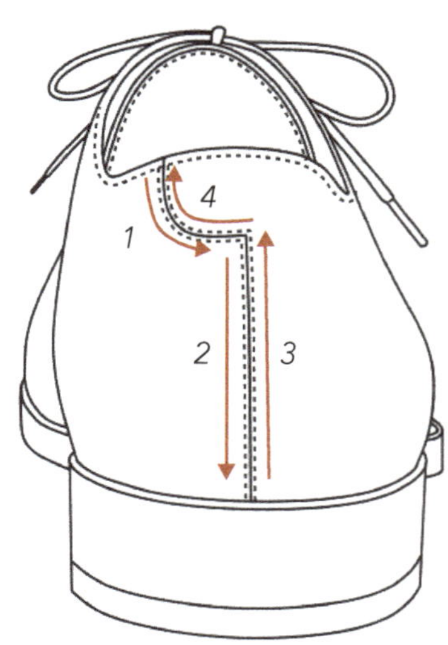

right

Preparing the quarter lining starts with the small reverse seam at the back.

After sewing it, trim the allowance, leaving only a narrow excess.

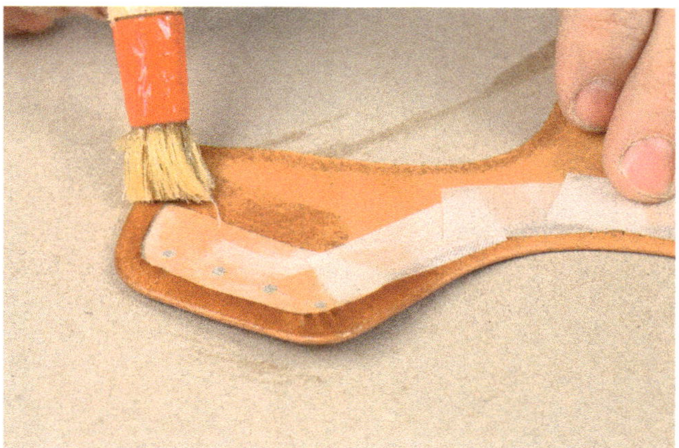

Apply a light layer of rubber cement to the topline and the lacing area, and repeat this on the lining.

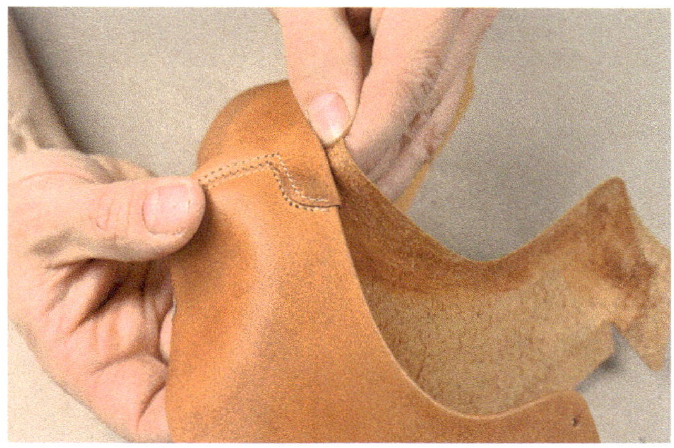

Start assembling the pieces from the back seam.

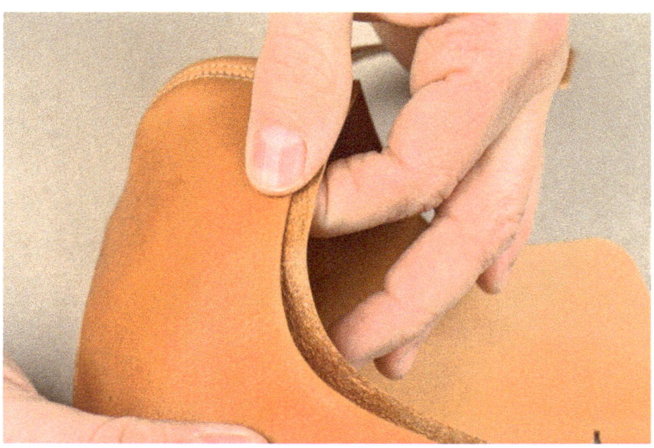

Work from the back toward the front, making sure to maintain the 5 mm allowance on the lining.

Finish the attachment at the lacing. The corners should be especially accurate due to the vamp lining overlap.

The overlap allowance is a good indicator of pattern accuracy. Ideally, there should be 5 mm all around, including at the front of the lacing, as specified in the design. Any deviation—either more or less—indicates that the back axis is not in the correct position and further adjustment is required.

It is very useful to mark the placement of the first and last stitches at the corners, for a simple reason. During sewing, the leather cannot be seen from a true top view. This topstitch begins by sewing away from the corner and finishes toward the corner. This angled perspective is misleading, and it is easy to place the first or last stitch incorrectly.

These reference marks ensure that the needle hits the exact intended point at both the start and the end, keeping the corner clean, accurate, and professional.

The topline stitch is sewn as one continuous seam from corner to corner. Backstitches can be added at both ends, but they will remain visible, however small. Since additional stitches will later be added for reinforcement, a cleaner visual result is usually preferred, and backstitching can be omitted.

When starting precisely at the marked point, the first stitch is very unlikely to come undone. To secure the seam, the top thread is pulled to the underside and fastened—either with a knot, by burning the thread ends, or by using both methods.

Make sure to keep an even distance from the edge. The width should be no less than the thickness of the folded edge, but not too wide either. Otherwise, there is a risk of not catching all layers, resulting in a very weak topline. At the corners, walk the machine if necessary, ensuring that the stitch length does not change through the turn.

At the end, when reaching the corner, there are two options: (1) finish the stitch at the marked hole and secure the thread; or (2) add a single backstitch, which helps prevent the seam from unraveling, a risk that is greater at this point.

Lining trimming by hand is usually challenging and takes practice, especially along a folded edge. When a skived edge is folded, the top layer is stretched considerably–on pull-up leather this is clearly visible as color fading. Even the lightest contact with the blade can cause a deep cut, yet the knife must pass very close to this vulnerable edge.

There are a few useful techniques to manage this. First and foremost, the sharper the knife, the better - even though a sharp blade can cause more damage if mishandled. The reason is simple: a sharp blade offers better control, requires less force, and cuts more cleanly, resulting in greater overall precision. Another technique is to avoid cutting directly toward the folded edge. Turning the blade a few degrees away from the edge can protect the upper while still producing a clean cut. The blade can also be angled slightly underneath the edge, which further refines the cut.

The wet-glue technique is a good choice here for attaching the vamp lining to the upper vamp. It is sufficient for temporary bonding, and its main advantage is that it can be applied to only one side, resulting in a very clean finish. Apply the cement to the lining, while avoiding the trimming allowance to maintain a clean result.

Attach the vamp immediately. The leather layers will absorb the small amount of solvent present. This is an effective technique for creating temporary bonds when the glue is not intended to serve as the primary means of attachment. Align the pieces along the bottom edges and around the tongue area. Even a slight misalignment can cause the overlap between the lining pieces to become inaccurate.

As the image shows, glue is applied only in the tongue area, with none along the edges. This is very important. This part of the lining must be easy to separate from the upper; otherwise, this assembly order becomes impossible. Here, the tongue is flat, but later chapters demonstrate how to pre-shape the tongue.

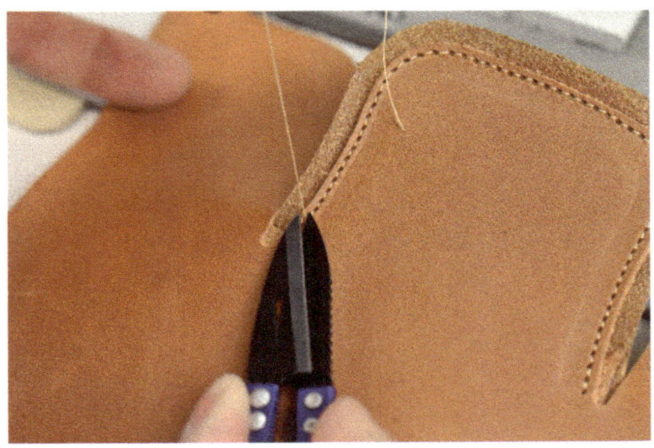

Sew the tongue from corner to corner. Here is a small but important detail: normally, loose thread ends are pulled through to the lining side. The tongue is different. This is the area where the quarter pieces overlap the vamp, so most of the tongue seam–and therefore the thread ends - will be securely hidden between the layers and locked in place by the reinforcement stitches.

Trimming the tongue lining is not particularly difficult, as this piece is flat – or very close to flat – depending on the process used.

This trimming can be done in several ways, depending on tradition, production volume, or available machinery. In some traditions, the lining pattern is cut a few millimeters smaller than the upper, which requires hand trimming with an edge trimmer or small curved scissors. In small-scale production, a simple clicking knife is a practical and effective tool.

With machinery, trimming can be done using a trimming attachment on a sewing machine. This method is fast and precise, though it carries some risk of error. There are also specialized edge-trimming machines widely used in the footwear industry. These are typically used after the seam is completed and operate with a combination of an oscillating blade and an edge guide, allowing them to follow intricate curves accurately at high speed.

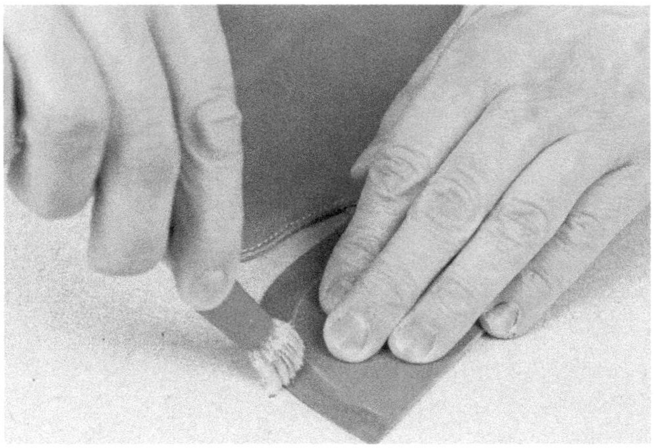

Add cement to the prepared seam allowance on the vamp and the quarter.

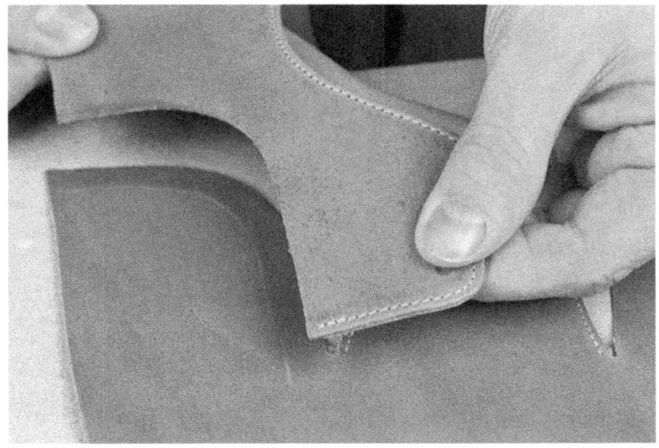

After drying, start attaching the pieces on the right side.

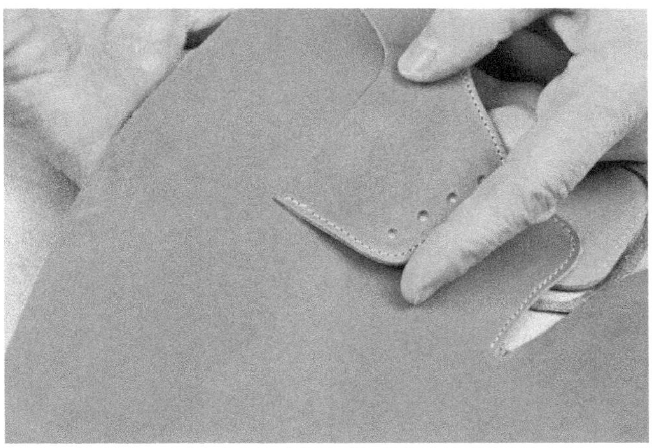

Starting on this side provides better access for sewing. Accuracy is critical; otherwise, in addition to visible imperfections, the vamp lining may not retain the correct allowance.

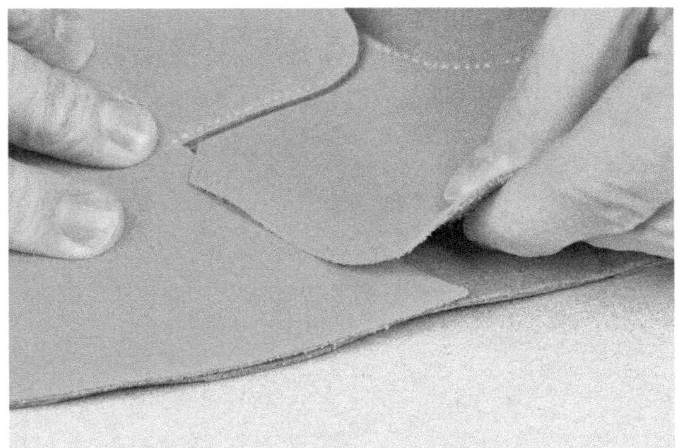

After the upper pieces are firmly attached and before sewing, check the lining overlap. The allowance is defined by the pattern, so no marking is necessary. Cement the pieces together.

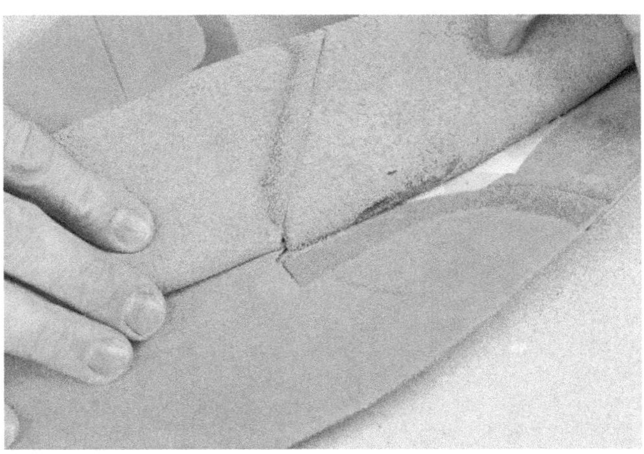

With this assembly order, the lining can be easily separated from the upper, providing good access to the vamp line where the vamp and the quarter are sewn together. There are methods where all four layers are sewn together, but these have drawbacks: a long heel counter is no longer possible, nor is a side stiffener; the seam becomes bulky as all edges meet; and precision within the layers is difficult to achieve. For bespoke-quality work, this option is not recommended.

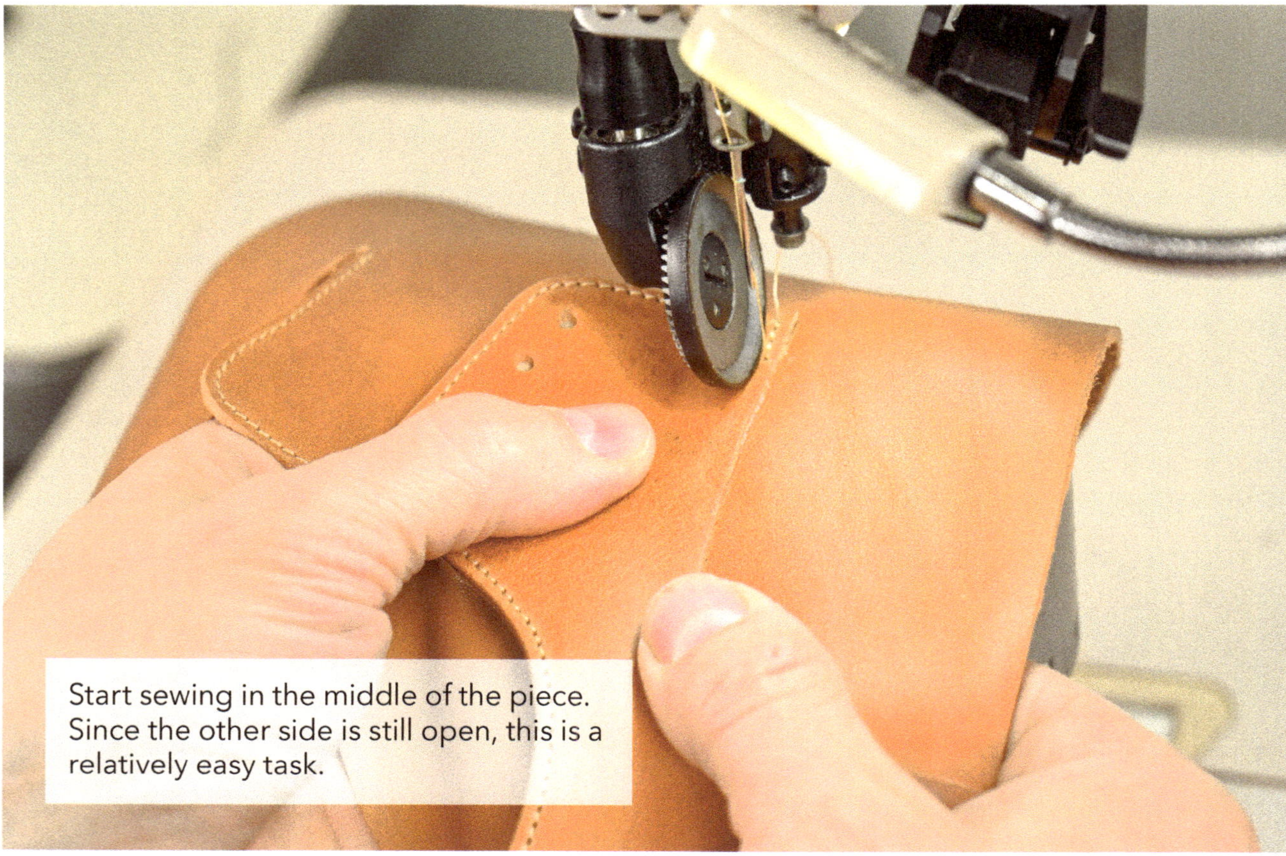

Start sewing in the middle of the piece. Since the other side is still open, this is a relatively easy task.

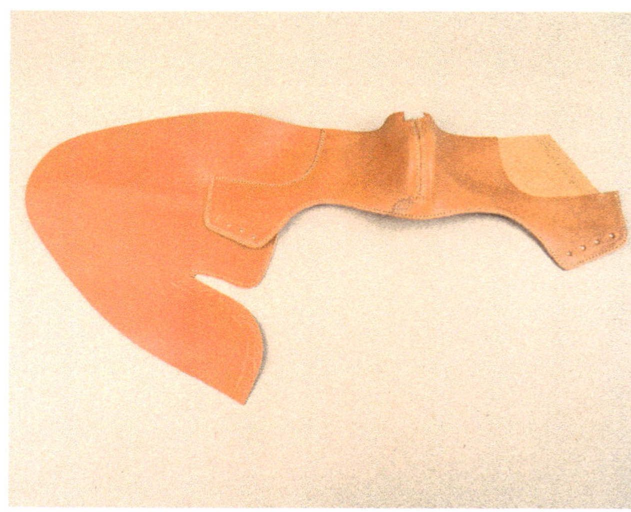

A generous amount of backstitching around the lasting allowance is recommended. A second, decorative stitch line running closely parallel is optional for a dress shoe such as a derby. Workboots and casual footwear may use three lines or more.

At this point, the upper is still open, which makes this seam very easy to access. This way, there are no pulls or tension from the other side.

Although it may be tempting to start the assembly on the left side, there is a good reason to begin on the right. Sewing on the left side starts at the lasting allowance, which is relatively easy regardless of the sequence. Sewing the right side, however - starting from the lacing–would be significantly more difficult if the other side were already fixed in place. By assembling the right side first, the sewing process does not present additional challenges. Using cement at this stage helps ensure accurate alignment.

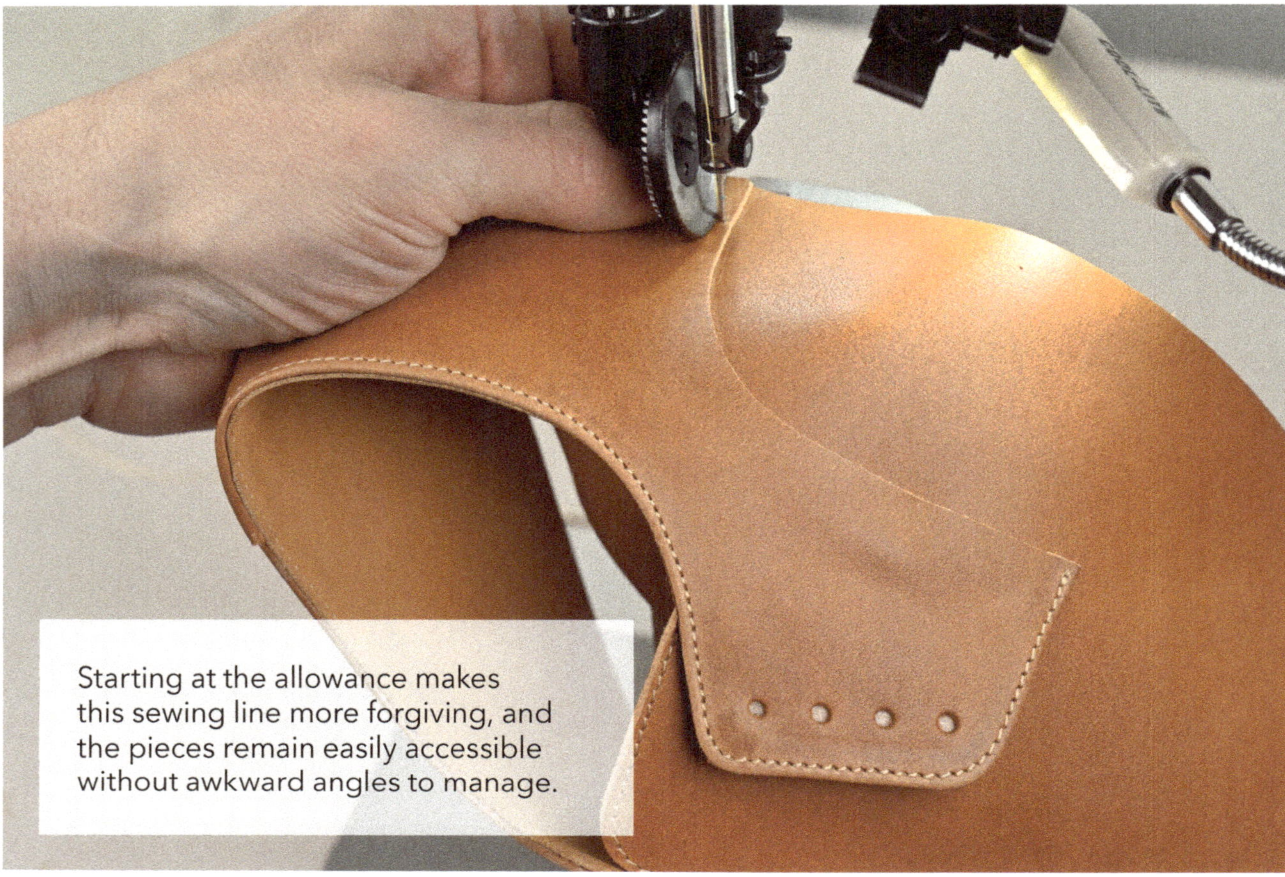

Starting at the allowance makes this sewing line more forgiving, and the pieces remain easily accessible without awkward angles to manage.

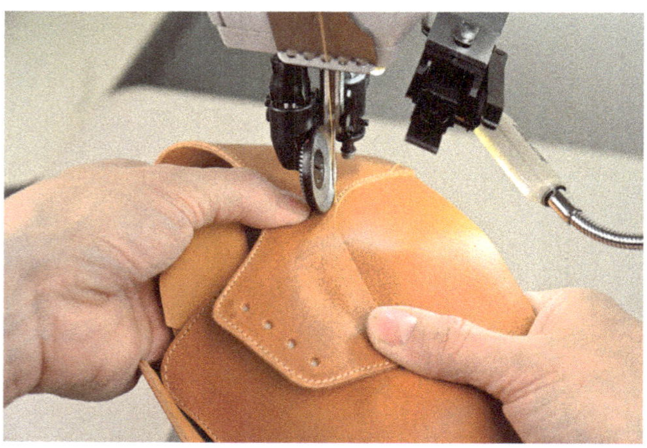

The lasting allowance also conceals the generous amount of backstitching required for the construction. A common and serious mistake at this stage is accidentally stitching through the lining. To prevent this, it is advisable to cement the lining overlap together so it can be held securely out of the way.

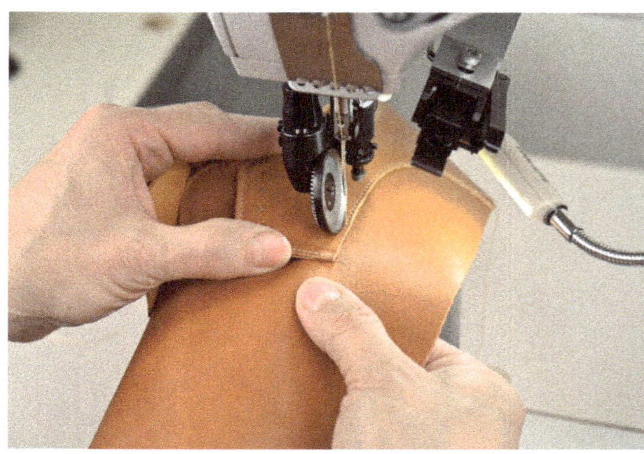

Even the best cementing job may not prevent the layers from sliding apart. This would result in the vamp line shifting, making the overlap too narrow or too wide, and potentially causing the corner to land in the wrong position. Continuously check the vamp line to ensure it corresponds with the marking.

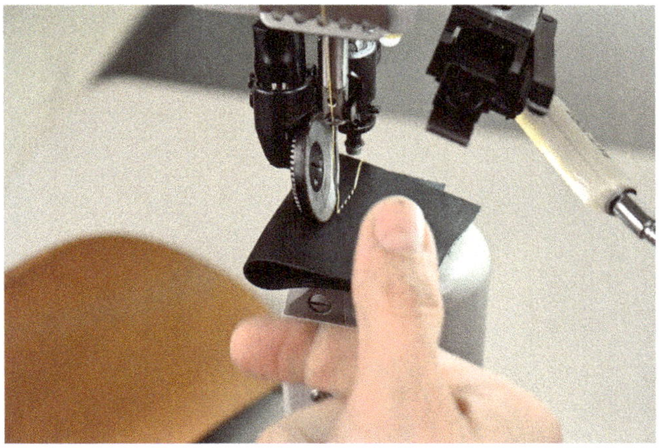

Here, a thread pig is used: a piece of scrap with a few stitches that holds the thread ends at the start of sewing. Without this, instead of cutting the thread, simply pull enough length and continue on the other side of the lap seam.

The thread pig makes it easy to keep the thread ends secured and taut when starting the next stitch line. Begin a few stitches in, add backstitches, and then complete the line.

The stitch line is finished. Stitching downward is one way to secure the corners of the quarter lining in this manner. This is not the only method, as other techniques will be demonstrated in later chapters.

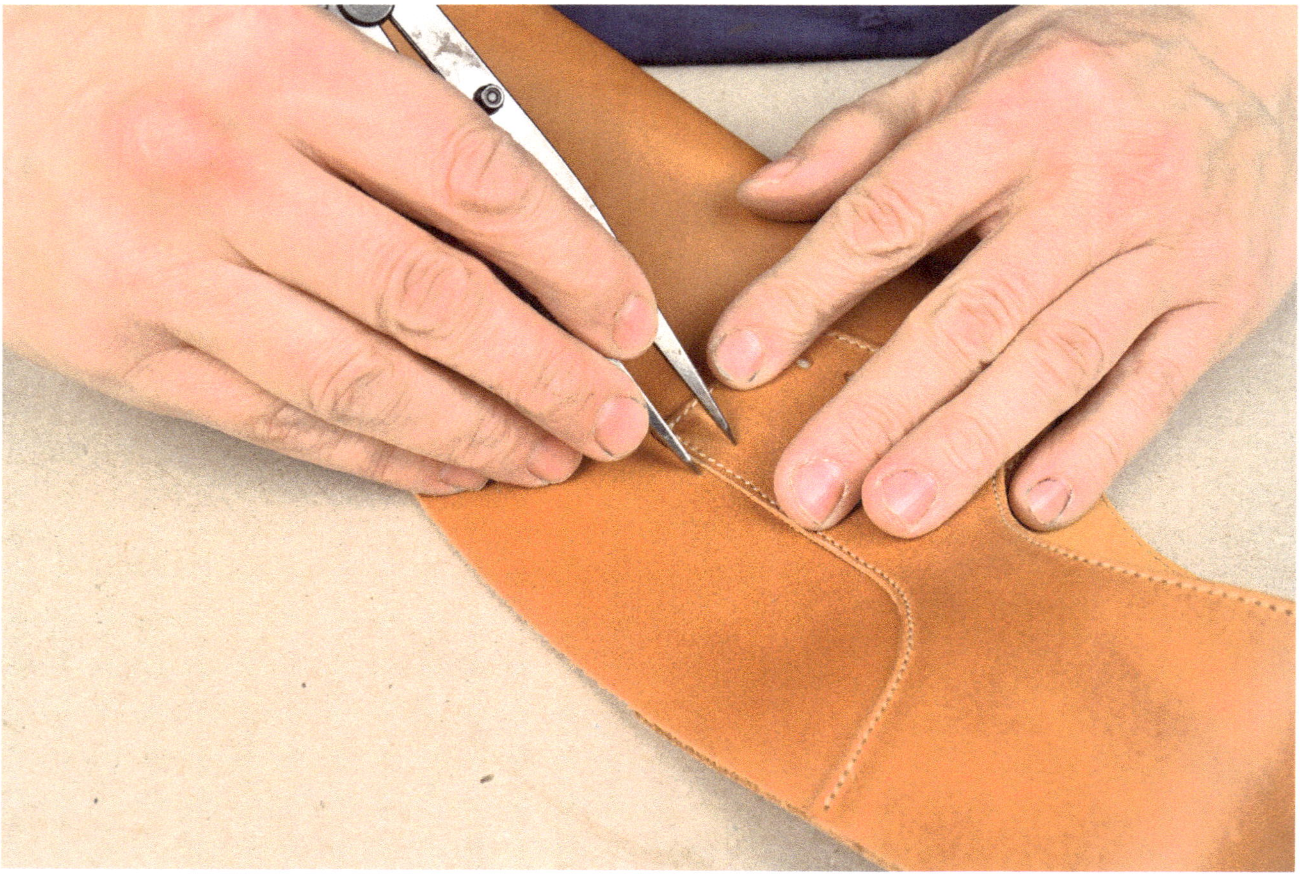

The reinforcement stitches can be marked on the pattern and then transferred, or marked directly on the upper. Either way, they should run parallel to the vamp line, using three, four, or five longer stitches of consistent length, especially around the corners.

Start by sewing the desired number of stitches, then reverse direction. Once the first hole at the edge is reached, return to the original direction. This method creates three threads and results in a very secure seam.

The stitch is clean and strong, holding the corner securely. This line can be sewn through all layers or only the top three, skipping the vamp lining. In all cases, the tongue must be avoided.

The process is now complete, with only a few final touches remaining before the upper is ready for lasting. Carefully hammering the seams is one of these steps, along with burning any loose thread ends. Some layers are intended to remain open, allowing stiffeners to be inserted between them, while the remaining layers can be glued down. A light application of glue may be added at the center of the vamp, approximately halfway toward the lace holes. This must be done carefully, as no eyelets have been installed and the layers should not be separated around the lacing area. If the lace holes have not yet been punched through all layers, this step remains straightforward to execute.

9 THE OXFORD SHOE

The oxford is widely regarded as one of the most refined and enduring styles in classic footwear. Its defining characteristic is the closed-lacing system, where the quarters sit beneath the vamp and meet neatly at the instep. This creates a streamlined silhouette that closely follows the contours of the foot, giving the oxford its elegant appearance. While the origins of the oxford can be traced back to 19th-century Britain – where the style became popular among university students and later in the city – it has since evolved into the most formal lace-up shoe in traditional menswear.

The oxford is not a single model, but a family of variations. Plain-toe, cap-toe, wingtip, wholecut, and Adelaide oxfords each interpret the style through different pattern constructions, decorative elements, or seam arrangements. Some of these designs emphasize minimalism, while others express ornamentation through brogueing, medallions, and stitching. All share the same closed-lacing principle that defines the category.

This book discusses three different uppers from the oxford family: a classic cap toe, a wholecut, and a Balmoral. These models differ not only in their pattern lines, but each chapter also presents a distinct assembly order, representing the three most commonly used construction sequences for this type of upper.

Creating the Basic Pattern Lines

Like any other footwear pattern, the style lines of an oxford can be interpreted in many different ways. The upper may feature a clean one-piece vamp, a straight toe cap, wingtips, or numerous other variations. Each section of the pattern offers design opportunities, from the toe to the quarters and beyond. This chapter focuses on the fundamental structure, while recognizing that exact style lines and components depend on the designer's intent. Common decorative and functional details that can be added to the upper are also covered.

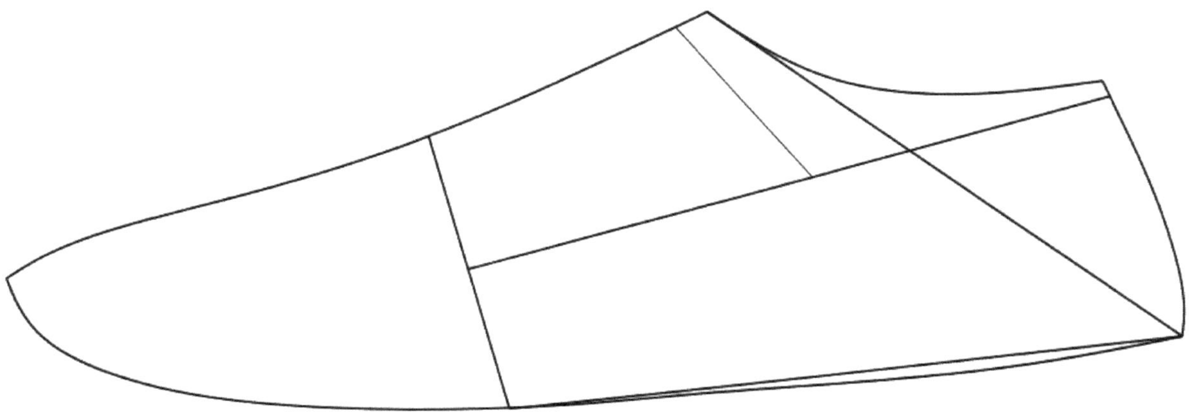

Since the oxford upper cannot open in the same way or to the same extent as the derby discussed earlier, the topline plays a crucial role in both fit and comfort. Wider feet or lasts may require a longer–and therefore deeper–topline to allow adequate entry and to accommodate slipping the foot in and out of the shoe.

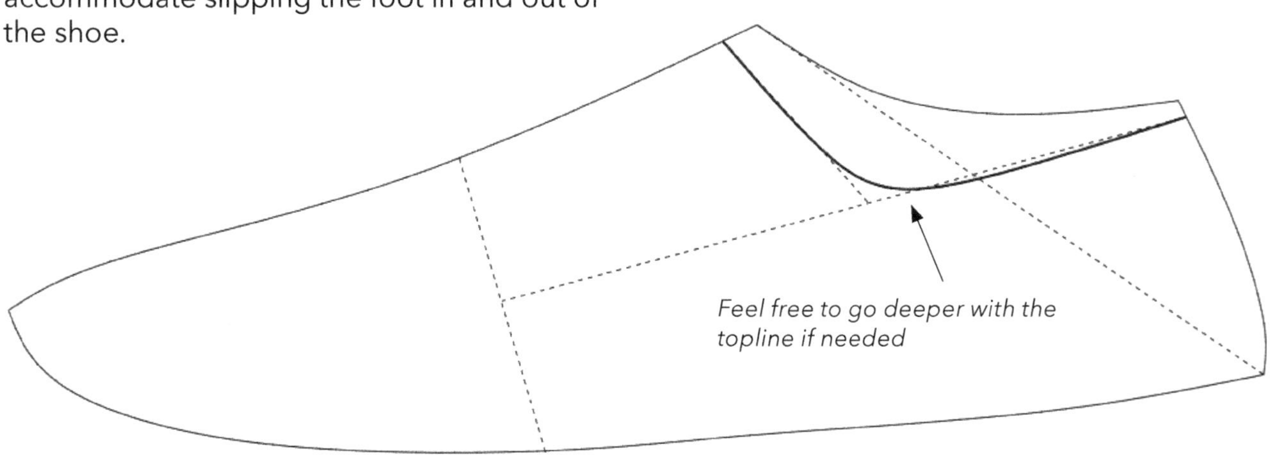

Creating the Vamp and the Toe Line

The same basic method for establishing the vamp line is used as described in the derby chapter. This is primarily a design decision, with functionality playing only a minor role. The placement of the vamp line is largely determined by the designer, but there are several practical considerations when selecting a suitable position. The vamp should harmonize with the overall last shape and the topline. An elongated toe often benefits from a slightly longer vamp line. The shape of the topline and any decorative elements along the edges should also be taken into account. If a counter is present, a shorter vamp may be more appropriate. If the line is intended to meet the heel breast, a longer vamp is required.

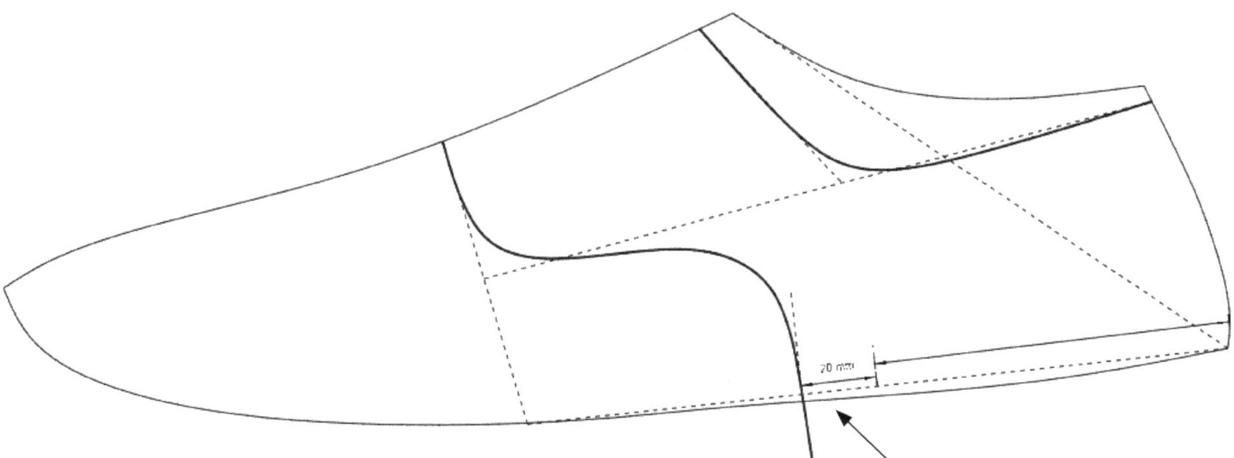

Measure 20 mm toward the toe from the center point.

The position of the straight toe cap is also a matter of fashion and proportion. A more elongated toe may require a correspondingly elongated cap to maintain visual balance. There is another important point: a "straight" toe cap is never truly straight—only visually so. This is easy to verify on the last. A toe cap line that appears straight on the three-dimensional last will look curved once the last is flattened onto paper.

The explanation lies in the geometry of the last. The line begins with a slight curve around the ridge—where the top surface of the toe transitions into the side wall—and then approaches the feather edge at a near-perpendicular angle. When viewed on the last, this shape reads as a straight line.

The toe cap line turns at the ridge and meets the edge perpendicularly.

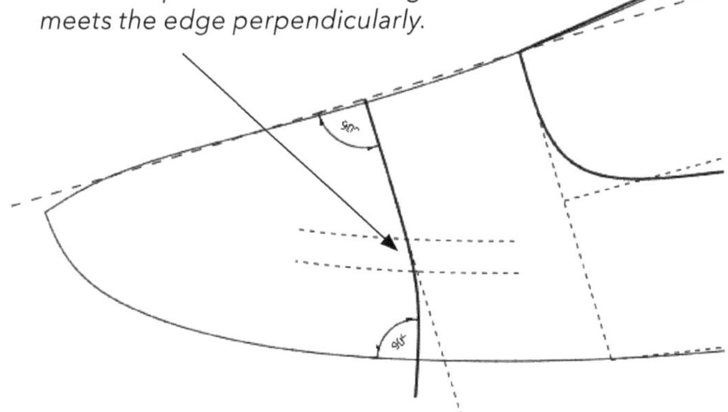

Aligning the Pattern Lines to the Heel Breast

This section addresses a very common mistake in pattern making. The upper pattern is not an isolated element–the outsole and heel are equally part of the design. All components must work in visual and structural harmony.

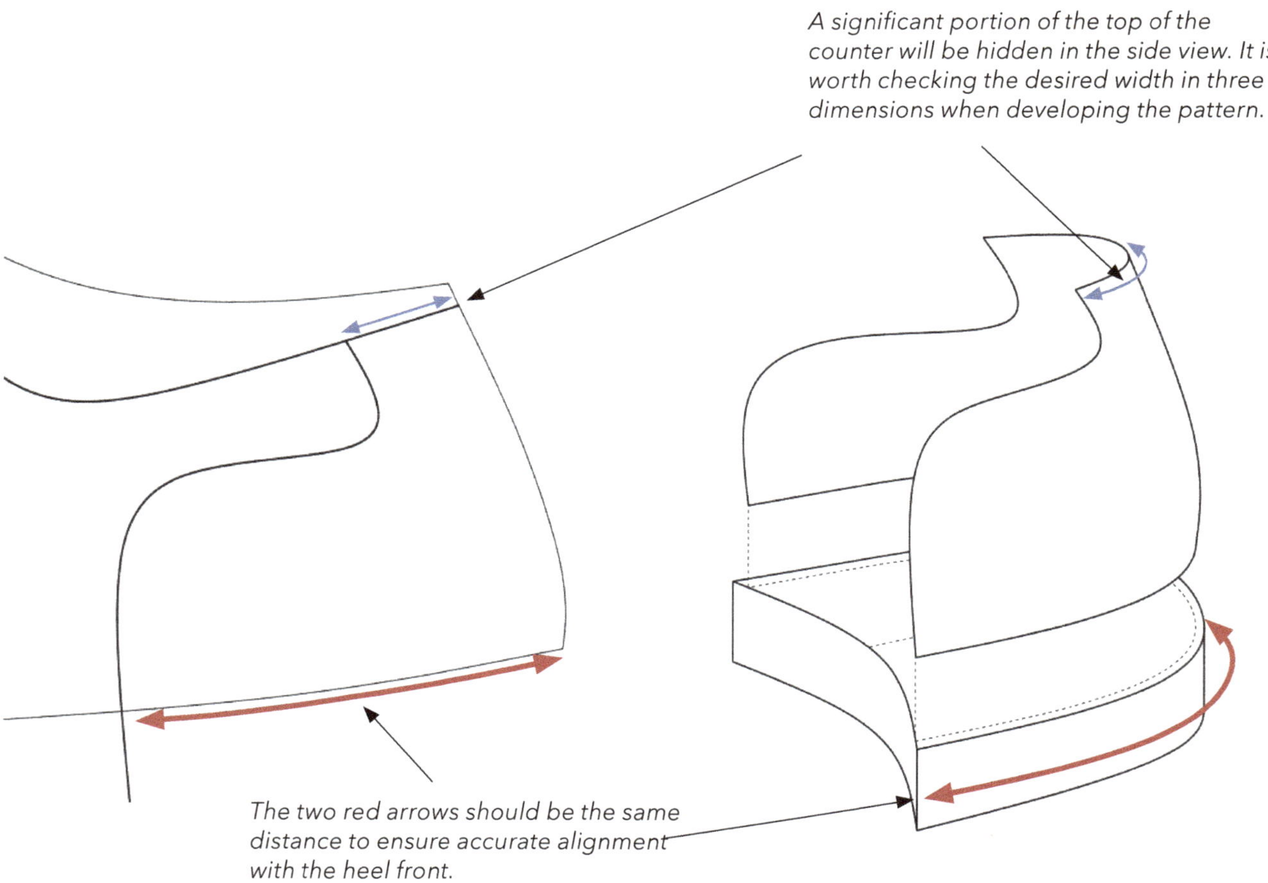

One of the most frequent examples of misalignment occurs when an upper style line does not properly meet the heel front. Even a minor deviation is distracting to a trained eye. The visual cortex is wired for pattern recognition as a survival mechanism. Continuous patterns are perceived as harmonious, while visual breaks create what neuroscience describes as a prediction error, which the brain interprets as irregular or potentially problematic. This explains why a poorly aligned pattern can be genuinely unsettling for a design-conscious shoemaker or an experienced pattern cutter.

A useful rule of thumb for the counter is to remember that the pattern is a flattened representation of a curved form on the last. Distances on the flat layout always appear disproportionately longer when they follow a curve on the last. This is often why beginners miss the heel front. A reliable technique is to measure the heel from the midpoint of the lasting margin to determine the correct counter length.

The Backline

The construction method for the backline is very similar to that of the derby, but there is a major difference: instead of the curved line normally used for a full backseam, an axis with a very small reverse seam is used, which is more demanding.

As a first step, create the backline in the same way as in the derby.

A slight adjustment on the topline to make sure it hits the back axis perpendicular

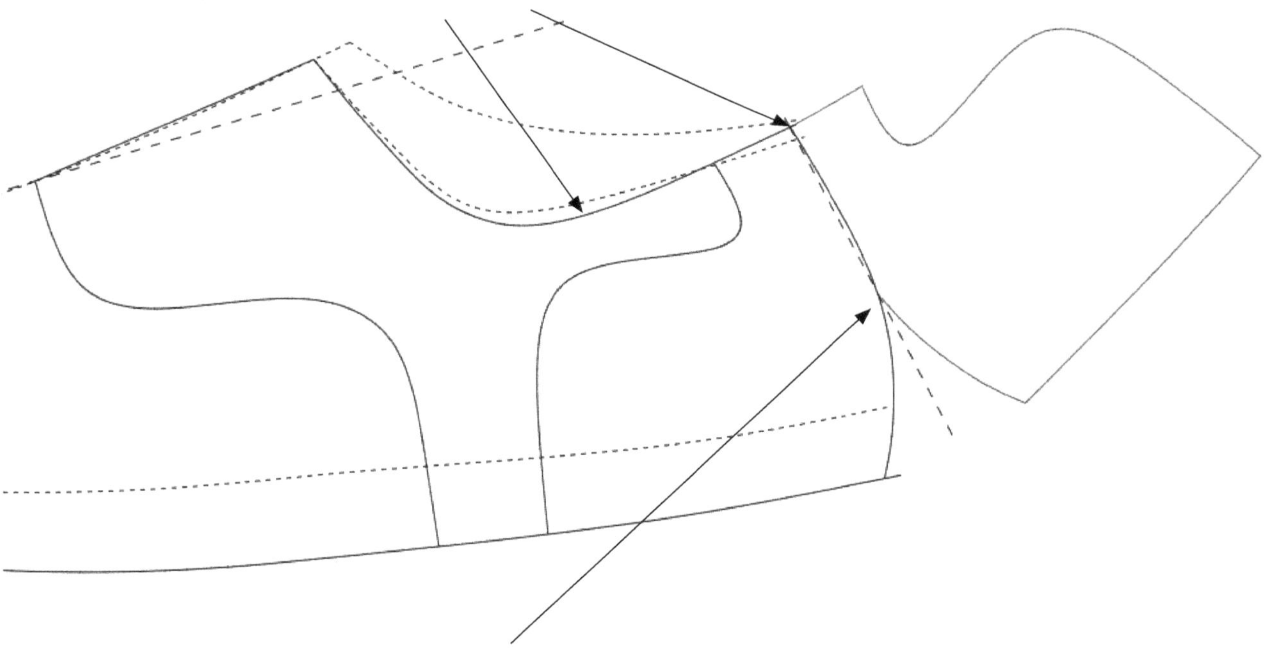

The axis starts at the corrected counter point and meets the adjusted backline curve at approximately one third of its exposed height above the feather edge. Ideally, the axis should remove only a minimal portion of the curve, although stretchier leathers allow the removal of slightly more material. The small reverse seam must remain visible after lasting; if it disappears or becomes barely noticeable, it was unnecessary from the outset. A reverse seam that does not remain visible after lasting indicates a technical patterning error.

The Eyelets

Placing eyelets has technical requirements, but it is also a stylistic decision. Two different approaches are examined here. First and foremost, eyelets are functional elements.

The bow-shaped lacing features holes placed along a curve. This distance can be doubled to establish a decorative stitch line, or the same logic can be applied in reverse by placing the holes along the centerline.

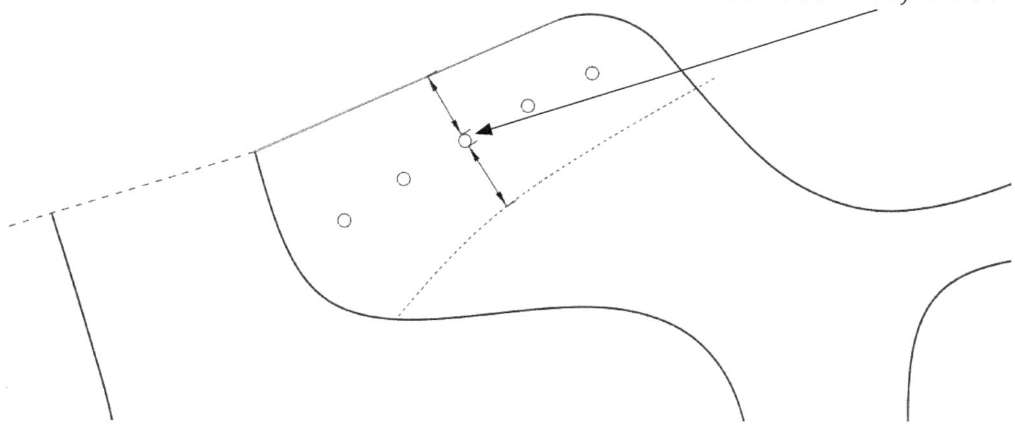

They must hold the lace securely and withstand years of tension; therefore, a minimum distance between the eyelet and the edge must be maintained. (In the footwear industry, this is a standard test used to evaluate how firmly the hardware is installed and how much force the laces can withstand.)

A practical minimum is 11-12 mm, usually supported by a lacing tape. For an oxford model, exposed hardware may be undesirable, so hidden eyelets are an excellent option.

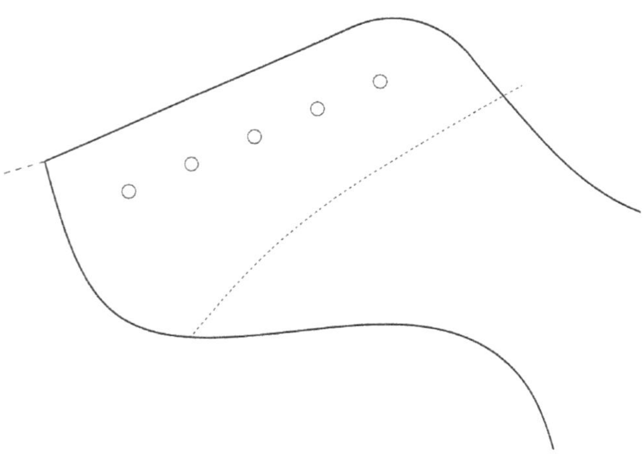

It can also be kept simple and classic by running a straight line to the edge. Even though this line does not interact with the decorative stitching, it does not read as a mistake. The two elements are independent of each other, so their alignment is not critical and does not appear as a misplacement.

The Tongue

The tongue on an oxford functions very differently from that of a derby. While the latter may shift during lasting, the oxford tongue is permanently anchored at the bottom. Once positioned by the pattern maker, it remains fixed in place.

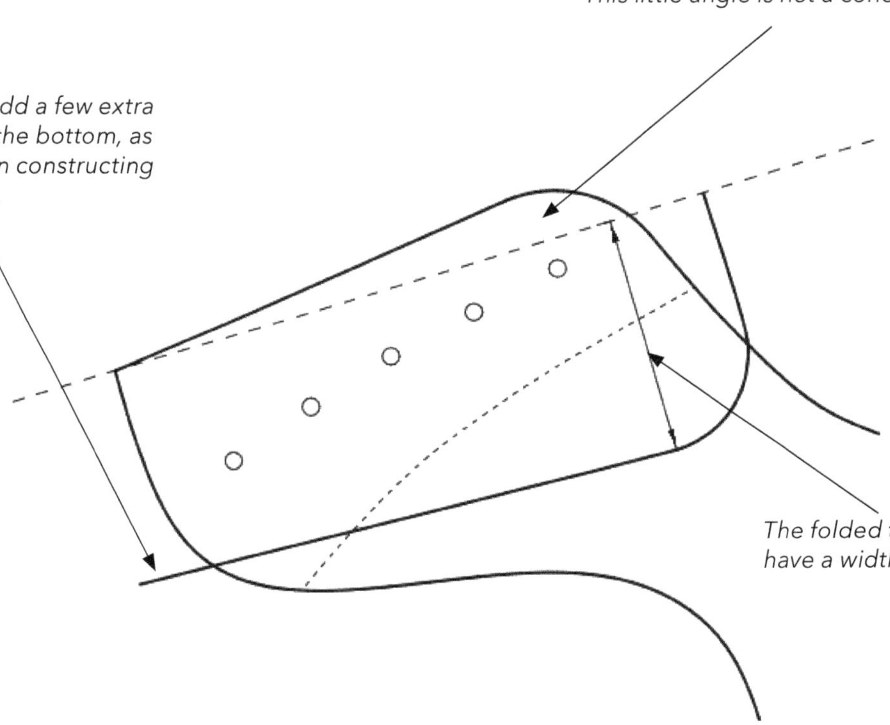

As with the derby, the tongue axis (or vamp axis) runs beneath the lacing. This is generally acceptable as long as the axis remains within approximately 15°. This value also depends on the material–if the angle exceeds this, the vamp lining can be shaped to compensate.

For a standard model size, the tongue width is typically around 70 mm (35 mm on the master pattern, since it is folded). The tongue is primarily a stylistic component, provided that it fulfills its main function of separating the laces from the foot.

Some pattern makers treat the tongue as an independent piece and may even omit it from the pattern set, using a generic form and attaching it only after the upper is assembled. While this approach offers convenience, it also introduces drawbacks, such as an unnecessary seam facing the foot.

Ultimately, these differences may appear subtle, but many refinements in shoemaking fall into this category. They are applied not because they are strictly required, but because precision and craftsmanship take priority over convenience.

Upper pieces

The Vamp

The toe cap and vamp pieces are straightforward: leave a lap seam allowance between them—typically 10-15 mm—and then unfold the components into their final pattern shape.

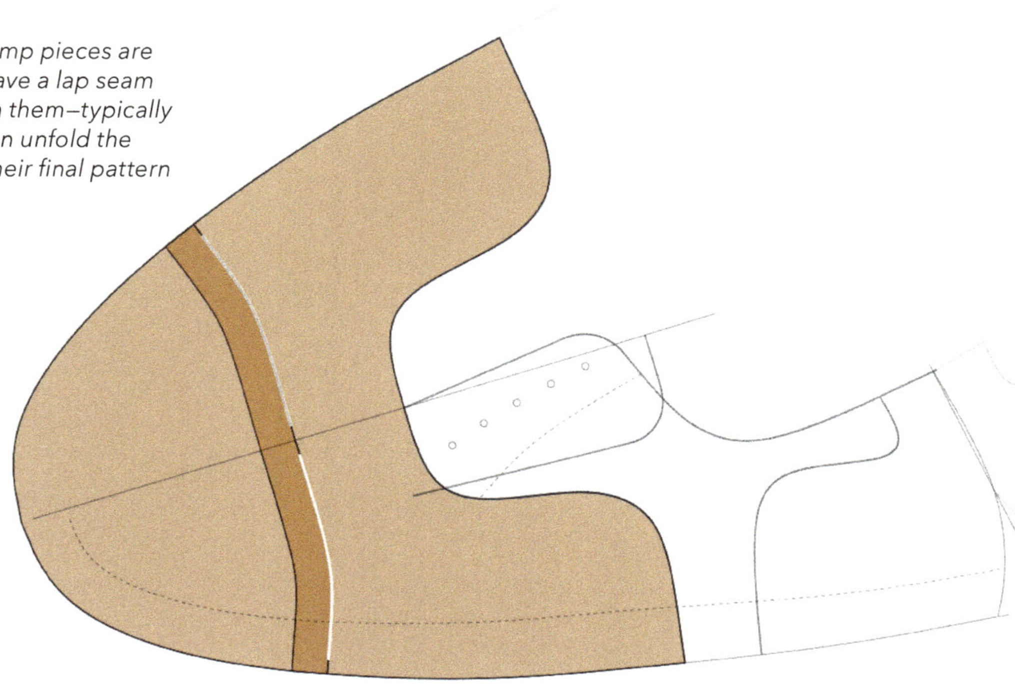

Variations for the Quarter

The simplest solution is to add allowance to the pattern piece for a lap seam. This approach is examined in detail in the making section. However, this method also has several limitations. It creates a weak point where the quarter meets the counter, and this area is highly stressed. It is common for wearers to slip into their shoes without a shoehorn, and this is where the topline is most likely to fail and tear.

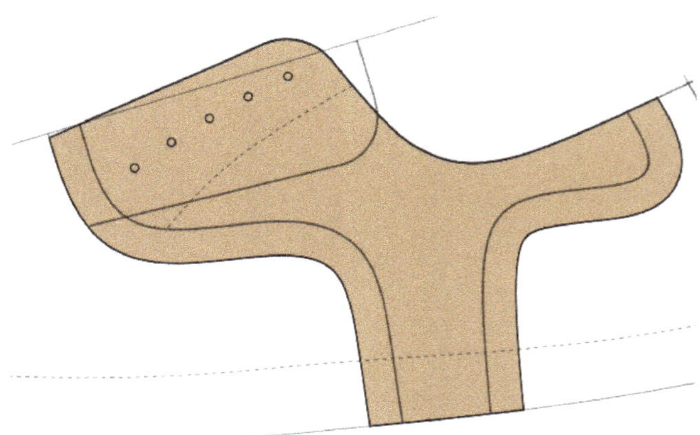

Another potential issue occurs at the front. Since the quarters do not include an overlapping extension, there is an increased risk that the pieces will separate during the vamp-line stitching process.

The One Piece Solution

The quarters can also be integrated into a single continuous piece, connected as illustrated. This eliminates the need for skiving at the quarters and produces a much stronger seam, capable of withstanding significantly more stretching.

The drawback of this method is increased material consumption. Nesting a component of this size and shape is highly inefficient and results in considerable waste. For this reason, this approach is generally reserved for bespoke work or the highest grade of footwear.

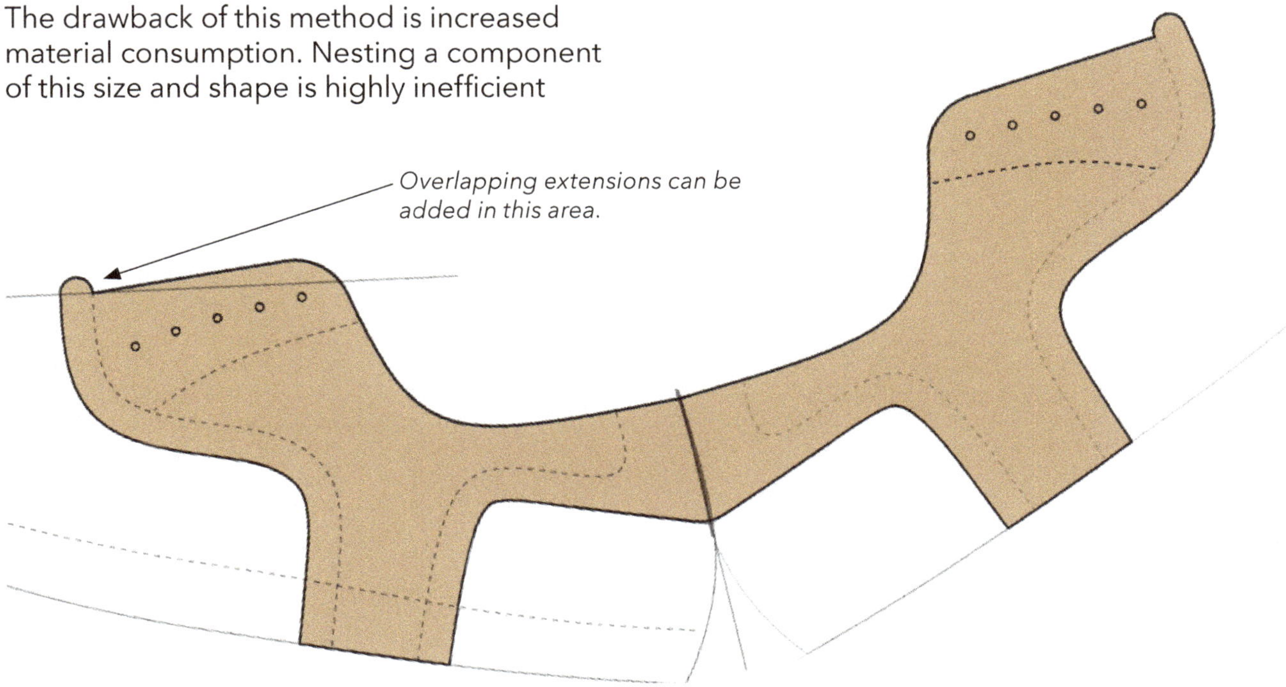

The Zigzag Seam Solution

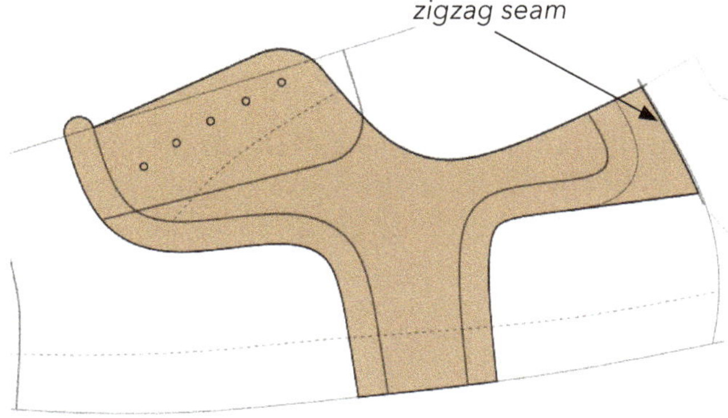

This approach represents a hybrid solution between the two previously described methods. The quarter receives an extended allowance at the back seam (not on the axis), and the two sides are joined with a zigzag stitch. This seam remains flat and is therefore invisible once it is covered by another component, in this case the counter.

The Quarter Correction

There is a small but important refinement to consider when creating the quarter pieces. This applies to any of the construction methods discussed above. The issue to address is the tendency of the quarters to spread when flattened during stitching.

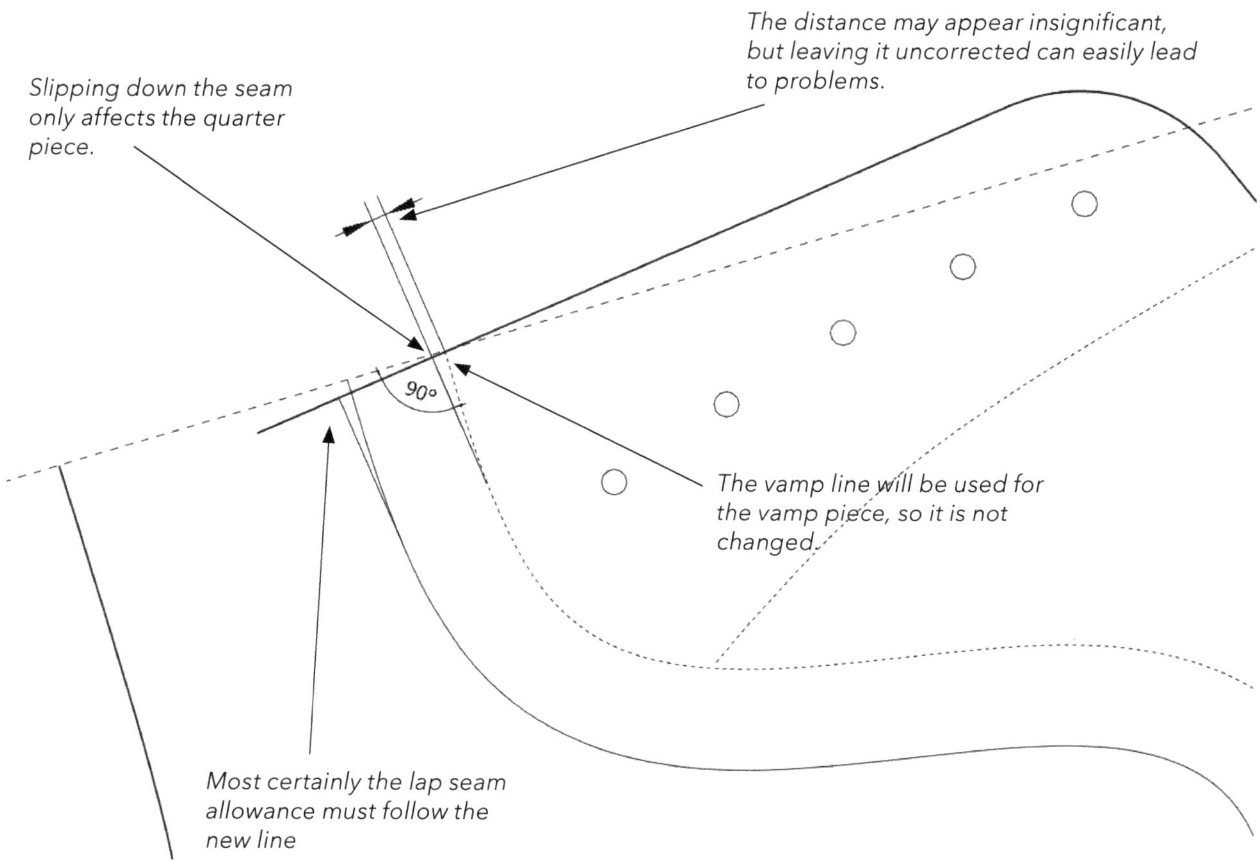

As a result, the lower edge of the quarter may extend slightly beyond the intended line, revealing overlapping extensions, skived or sanded edges, or even seams.

To prevent this, the quarter can be extended slightly and set to a true 90° angle. This ensures that even when the pieces are pulled apart, the vamp line does not reveal more than intended. Adjustments of this kind are very common in higher-level pattern making, where small corrections contribute significantly to clean construction.

Finalizing the Master Pattern

Only a few additional lines are required to complete the lining pattern. These elements were discussed in detail in the derby chapter. The lining requires an additional axis at the back, and the lap seam between the quarter lining and the vamp lining must also be added.

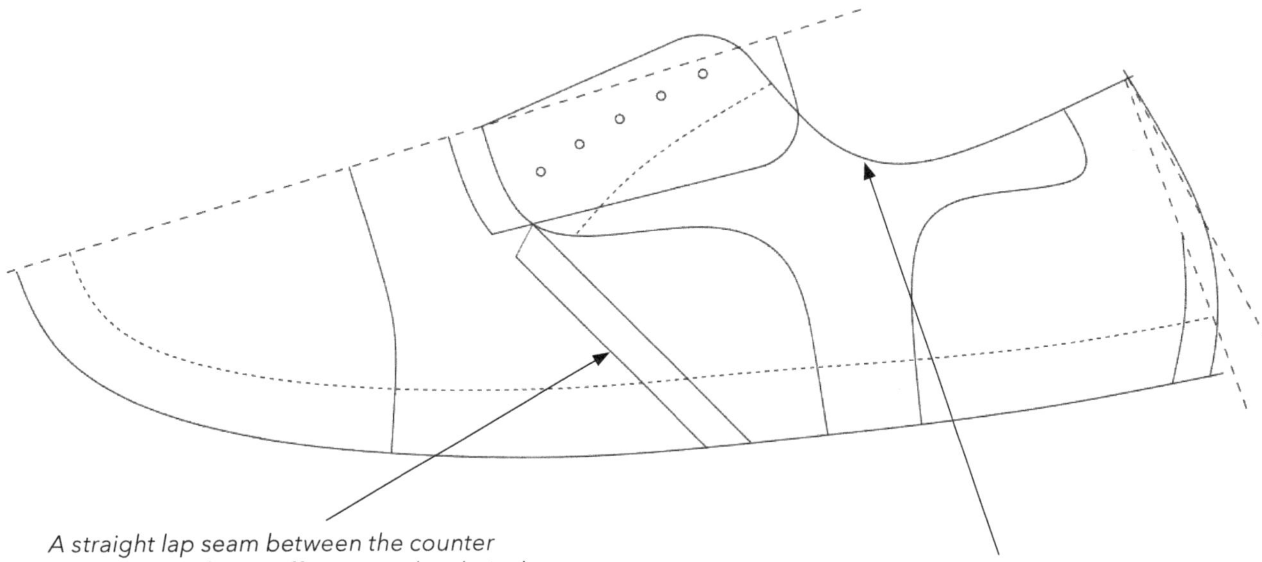

A straight lap seam between the counter and the vamp lining offers several technical advantages. It allows the use of double-sided tape for temporary bonding, and the linear shape makes the sewing process considerably easier.

The edge treatment allowance, if required, does not need to be part of the master pattern.

Lap seams and other allowances may be incorporated directly into the pattern pieces or indicated as specifications. Adding these specifications can be helpful, although the master pattern itself does not need to be overly detailed, as it may later serve as the basis for another design with different edge treatments.

With these final lines in place, the master pattern is complete, and the appropriate outlines can be transferred to create the individual pattern pieces.

This book features a symmetrical oxford pattern, which is very common. Some makers, however, prefer an asymmetrical topline, where the medial side runs slightly higher.

The Completed Upper Pieces

Unfolding the upper components (except for the quarter in this solution) is relatively straightforward. Ensure that sufficient allowances are included for the lasting seams and the topline edge treatment. In this example, a folded topline is used, which requires approximately 5 mm.

The counter allowance is handled somewhat differently—regardless of the chosen quarter construction method—because it must cover an already folded edge. The standard 5 mm is rarely sufficient, so an additional 2 mm at the topline is recommended.

Overlap allowances must be added to the vamp toward the toe, to the quarter toward the vamp, and to the counter. These allowances depend not only on leather thickness but also on decorative edge treatments. Brogueing, for example, requires a larger overlap to ensure that the perforations do not expose the raw edge.

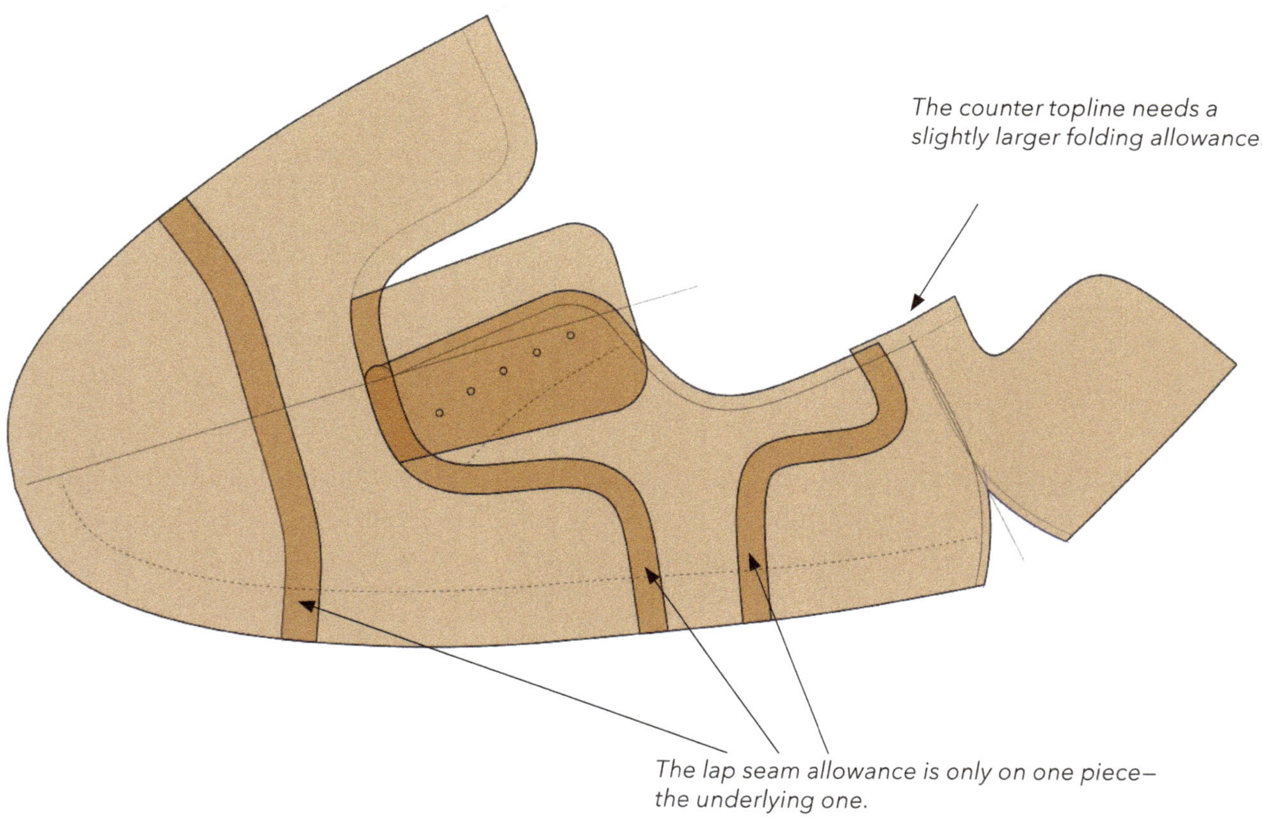

Lining Pieces

The lining consists of only two components: the vamp and the quarter. The quarter may be divided into separate pieces if material economy is a primary consideration. This solution is common in higher-end and bespoke shoes.

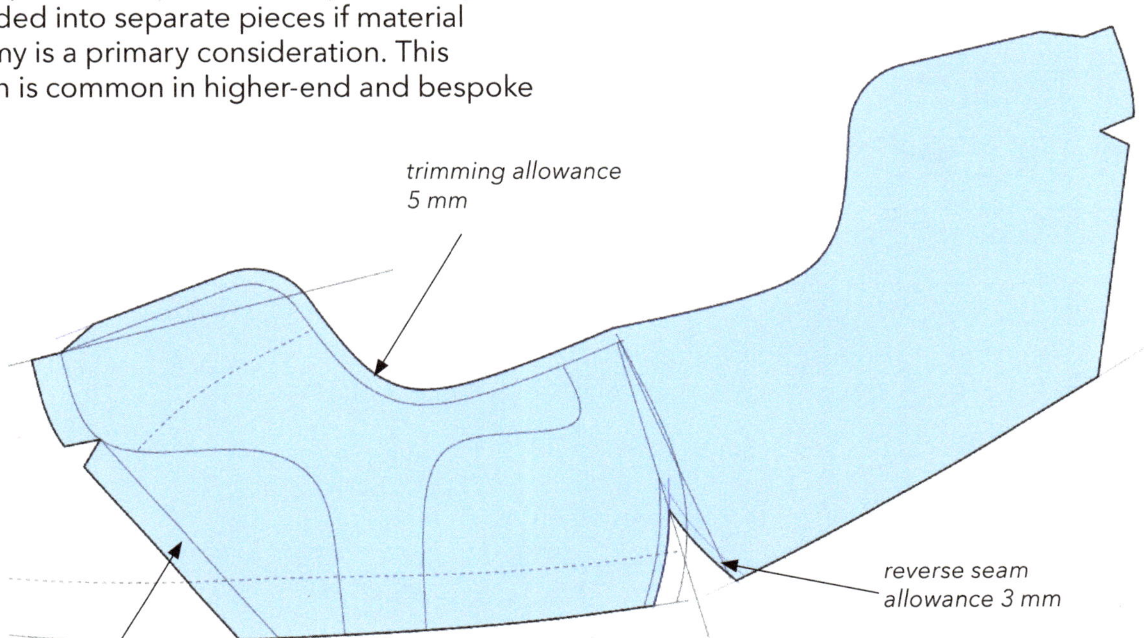

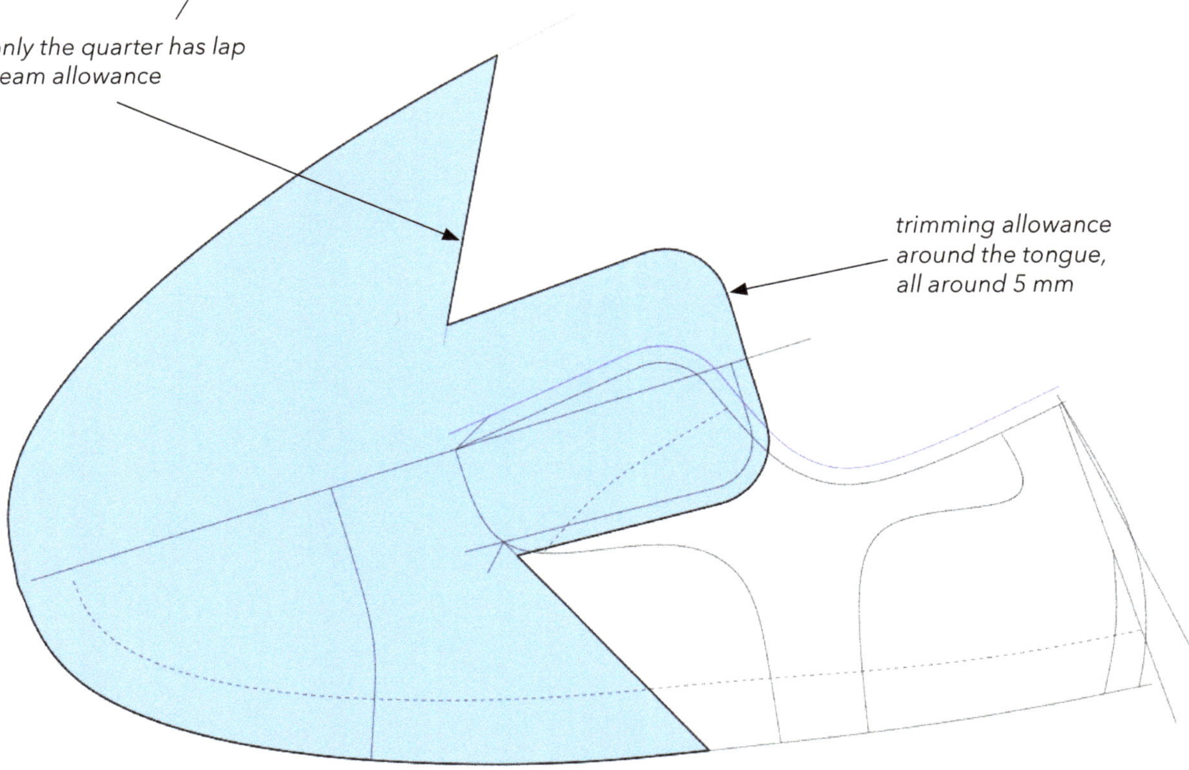

The Complete Pattern

1	NO SKIVING
2	SHORT SKIVING HALF THICKNESS
3	MEDIUM LONG SKIVING HALF THICKNESS
4	LONG SKIVING HALF THICKNESS
5	LONG SKIVING NO EDGE THICKNESS
6	SPLITTING

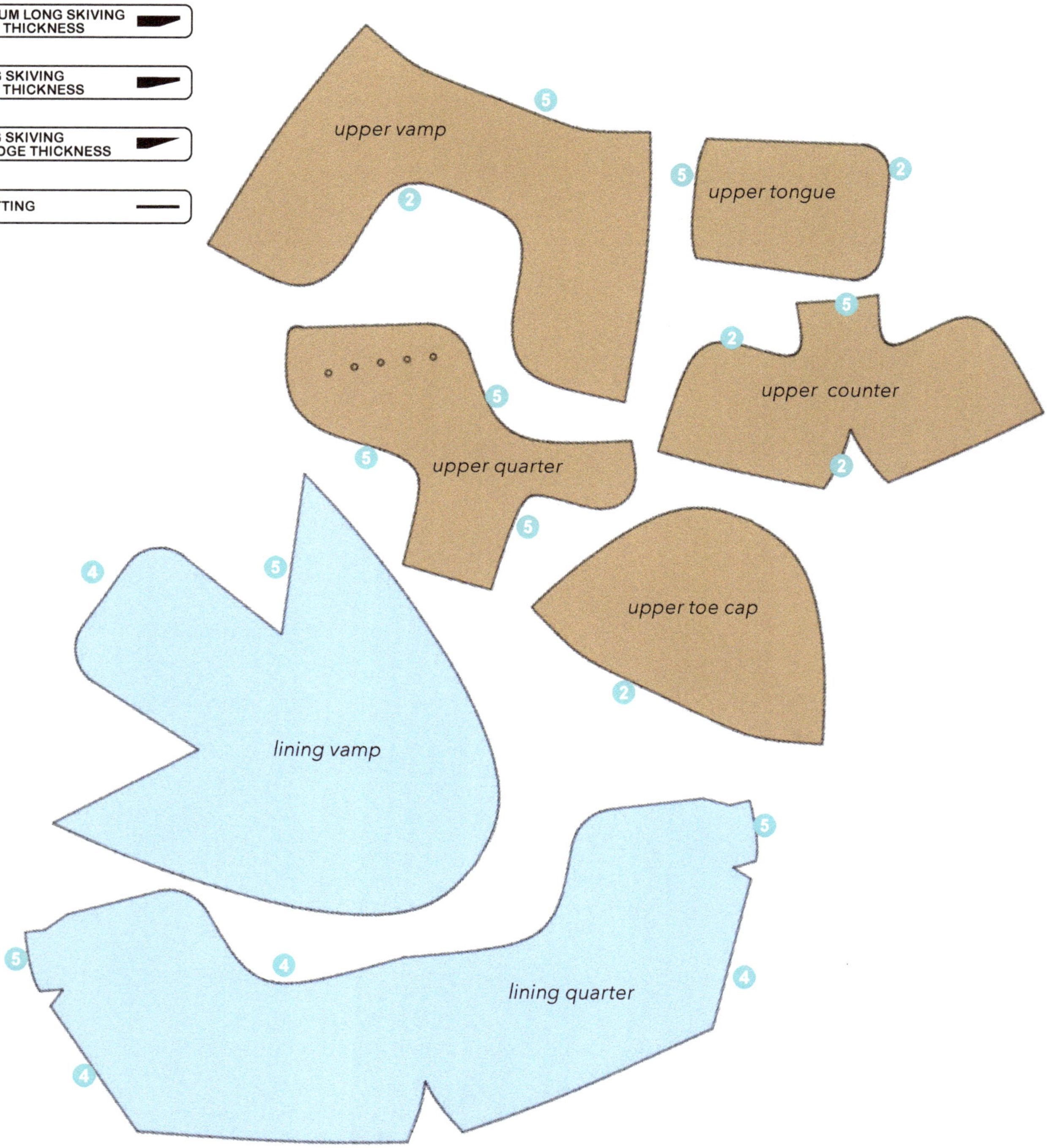

Variations

Although the oxford is a traditional and highly classic style, it exists in numerous variations. The following are several technically distinct examples. There are variations considered to be oxfords – a good example is the wholecut model – but since there is a dedicated chapter for it, it is not listed here. The shape of the toe cap is also not a distinctive factor; the major differences are seam types and changes that may affect the assembly order.

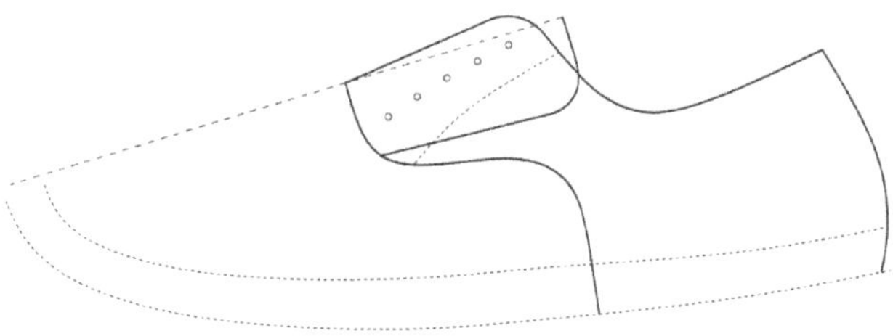

A plain oxford is considered one of the most elegant variations of the style. The backline is curved, allowing either a dogtail or an axis to be used. In the latter case, a small reverse seam may be added at the base, or the shape may be refined through hand-shaping methods or by using a counter-shaping machine.

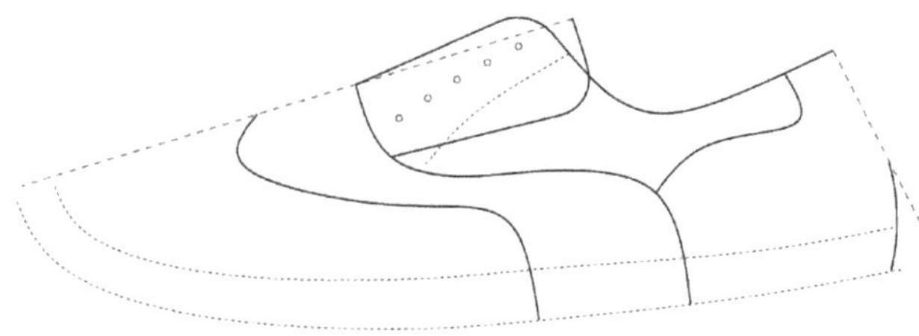

The vamp may also overlap the counter. This is a very traditional construction and, although somewhat old-fashioned (in fact, it appears in many older pattern books), it remains a refined solution. The assembly sequence is straightforward: the rear section of the upper is completed together with the counter, and the vamp is then sewn over the top of these components.

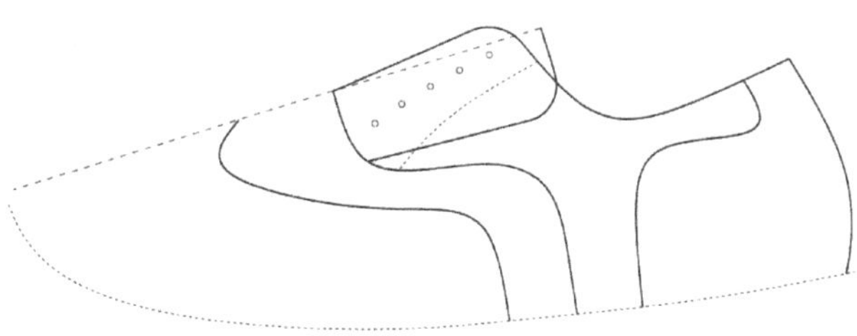

A wingtip with a counter is another classic variation. This model works well on its own, but when brogueing and gimping are added along the edges, it becomes what is known as a full brogue. A full-brogue oxford features decorative perforations and gimping along all major seams, making it the most ornamented and visually rich version of the classic dress shoe.

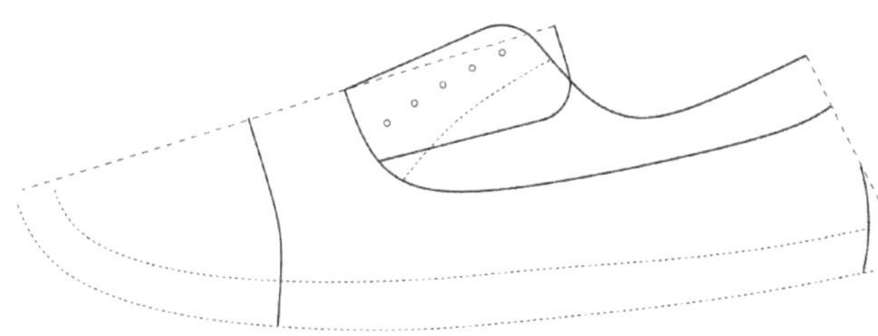

The U-shaped vamp is closely related to the Adelaide model, but the vamp extends all the way to the back seam. Because the design requires an opening, there are two options: a back seam or a small medial lap seam–the latter generally being the structurally stronger solution.

Style and Technical Errors

There are a few typical style and technical issues worth mentioning and discussing. The style-related ones are purely subjective and can simply be ignored; the technical errors, on the other hand, may affect manufacturability, comfort, and durability. This is, of course, not a complete list of possible errors, and to be fair, trying anything new will generate mistakes. Errors are therefore not necessarily bad – only when they occur repeatedly.

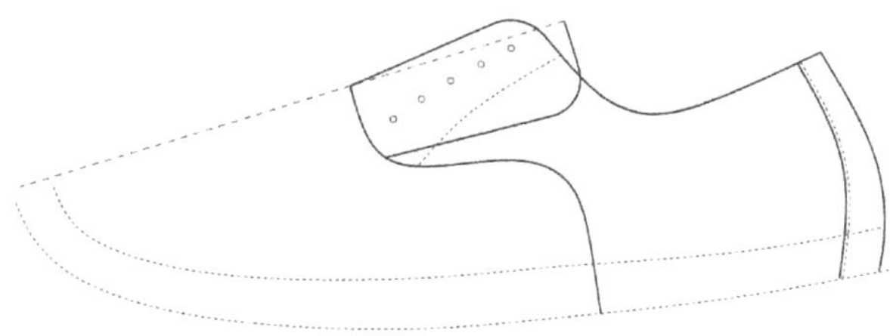

Using a backstrap. As one highly respected former colleague once remarked, "backstraps are for farmers," a lighthearted way of stating that backstraps are generally considered more casual and utilitarian, while the oxford is meant to represent elegance and fashion; therefore, combining the two is often not considered appropriate. On the other hand, fashion is about breaking rules, so oxfords can be found with backstraps – although mostly in factory production, which either reflects a deliberate rule break or, more likely, a lack of awareness of these traditions.

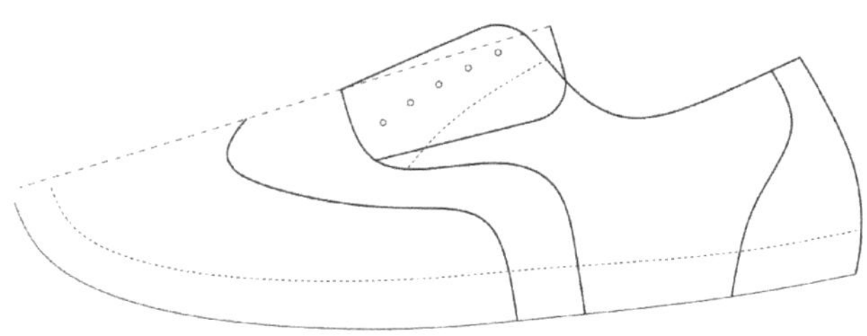

A short counter is purely a style issue. It works well on short-toe models such as court shoes or ballerinas, but it is generally not the best choice for elegant dress footwear.

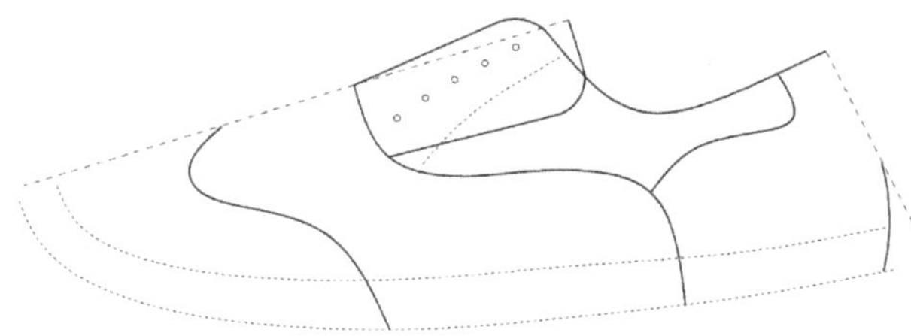

Short wingtips. This is a very subjective topic and may explain why some people are drawn to a style while others are not. It is another example of reinterpreting a design in a less-than-classic way. Classic shoes are generally expected to follow established proportions, lines, and shapes that align with what is commonly perceived as "correct," even though no absolute standard exists. In the case of a wingtip, this means the piece should be proportional to the toe shape and length, and in harmony with the overall style. A short wingtip – unless used as a deliberate statement (for example: clown shoes usually feature this pattern piece from a colorful leather, mocking the elegance) – is more commonly associated with children's footwear or novelty shoes than with an elegant, elongated toe. While this is not a structural error, it is generally advisable to avoid it.

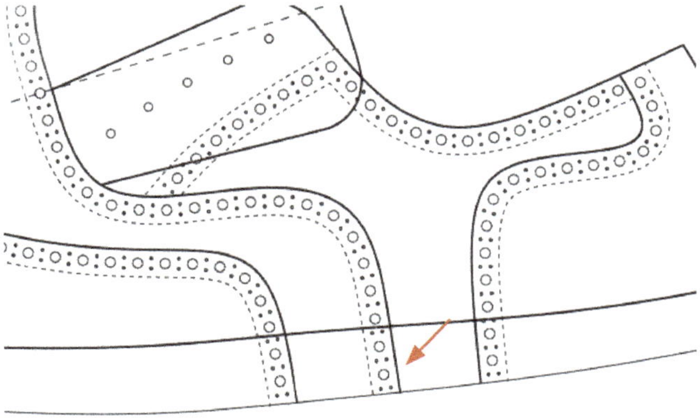

Brogueing decorations in the lasting allowance. Brogueing requires significant effort; when done by hand, it can take hours to punch the perforations accurately, ensuring correct placement at corners, center points, turns, and even spacing throughout. Adding brogueing to areas that will not remain visible is a minor issue, as it mainly results in wasted time. The more serious problem is that brogueing always weakens the material, and placing it in the lasting allowance can create substantial risk. In areas where the upper must be stretched excessively, often involving pivoting – such as along the vamp-quarter edge – the perforated leather may tear. The shoe can still be completed and the defect concealed, but durability will be compromised. A responsible shoemaker should avoid selling such work, knowing the issue is likely to surface later. It is therefore advisable to avoid unnecessary decoration in the lasting allowance, where structural strength is critical and decoration has no functional value.

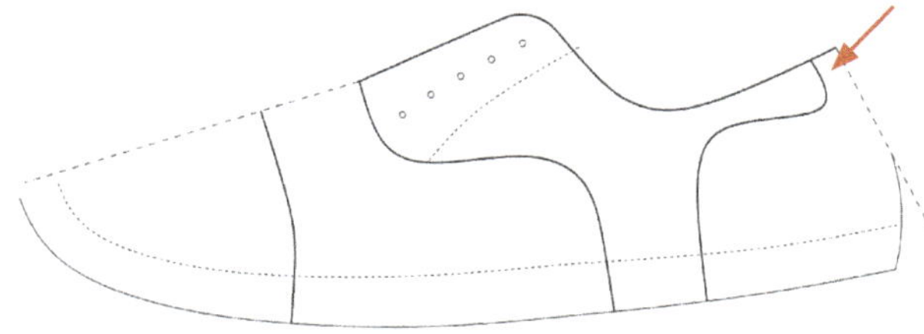

Narrow counter top. A last copy is not a side view. Making the top too narrow can result in this area not being visible from the side. If the intention is to make it appear narrow from the side, drawing it onto the last copy can help determine the correct width.

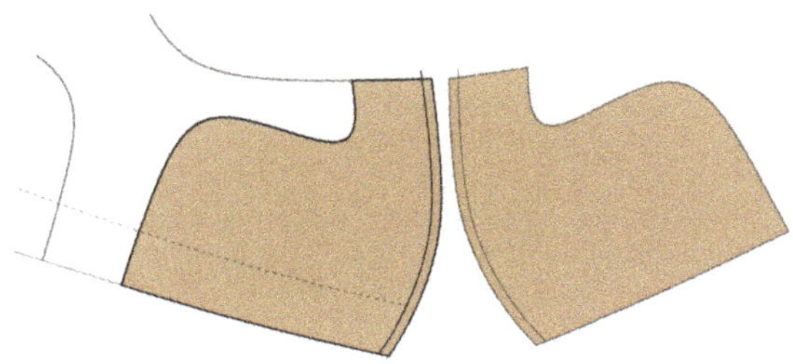

Using a counter with a reverse seam. Usually one or the other, not both. There are two possible reasons to divide a counter this way: (1) it makes upper assembly easier, as stitching automats can be used efficiently; or (2) "we always did it this way," which is probably the worst sentence a craftsman can say, as it cuts off the opportunity to improve. While this is not a fatal error, it ruins the classic look of the counter for the wrong reasons.

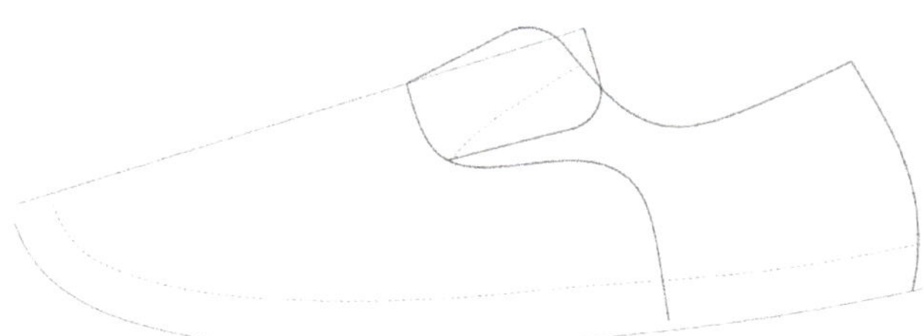

Too high vamp line. This is not only unattractive and disproportional but also negatively affects wearability. It makes slipping the foot in more difficult, places unnecessary stress on the front of the quarters, and is likely to fail and tear in a relatively short time.

Decorative elements

Brogueing

Using brogue decorations on all major seam lines makes a wingtip oxford a full brogue. For structural reasons, brogueing is not added along the lacing edge, and there is little justification for applying it across the entire tongue, except possibly on the visible section.

Brogueing is always a challenging task for both the upper maker and the pattern maker. Ideally, the hole positions–at least the largest of the three–are placed at specific locations such as the centers of pieces and turning points. Grading these patterns requires additional care to ensure the holes remain correctly positioned.

The perforations significantly weaken the edges, so additional reinforcement is required. They also expose the underlying layer, which means an extra strip of thin upper leather must be added beneath the brogueing.

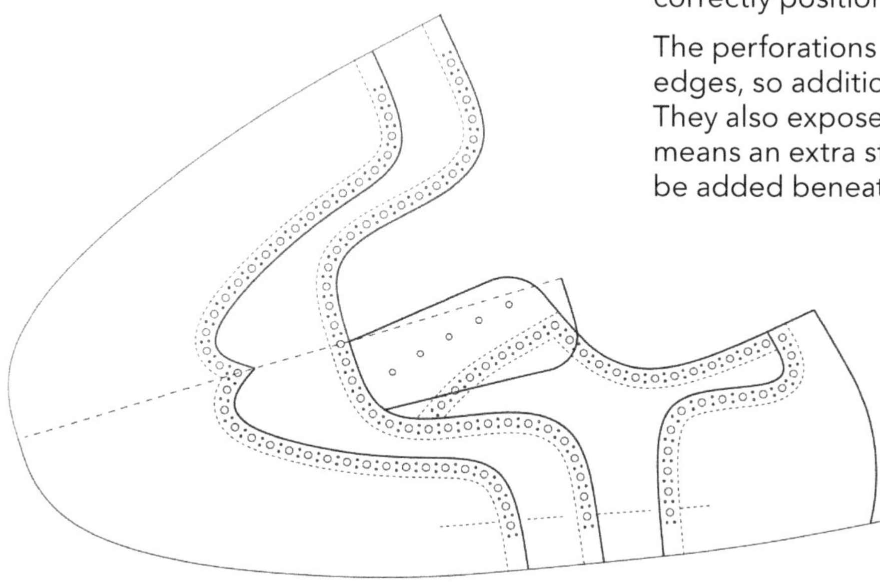

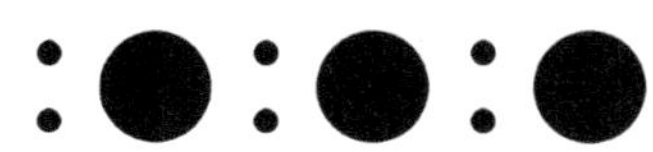

The traditional brogue pattern is determined by these three holes and the distances between them

Gimping

Gimping is an edge decoration consisting of a zigzag-shaped cut. It can be produced by careful hand cutting, pinking shears, a dedicated gimping-brogueing machine, die cutting, or digital cutting machines. For most edges, these methods make no functional difference, but the topline treatment is strongly affected. Gimping results in a practically raw edge which, like an unfinished edge, cannot be left exposed, so piping is usually applied underneath.

Medallions

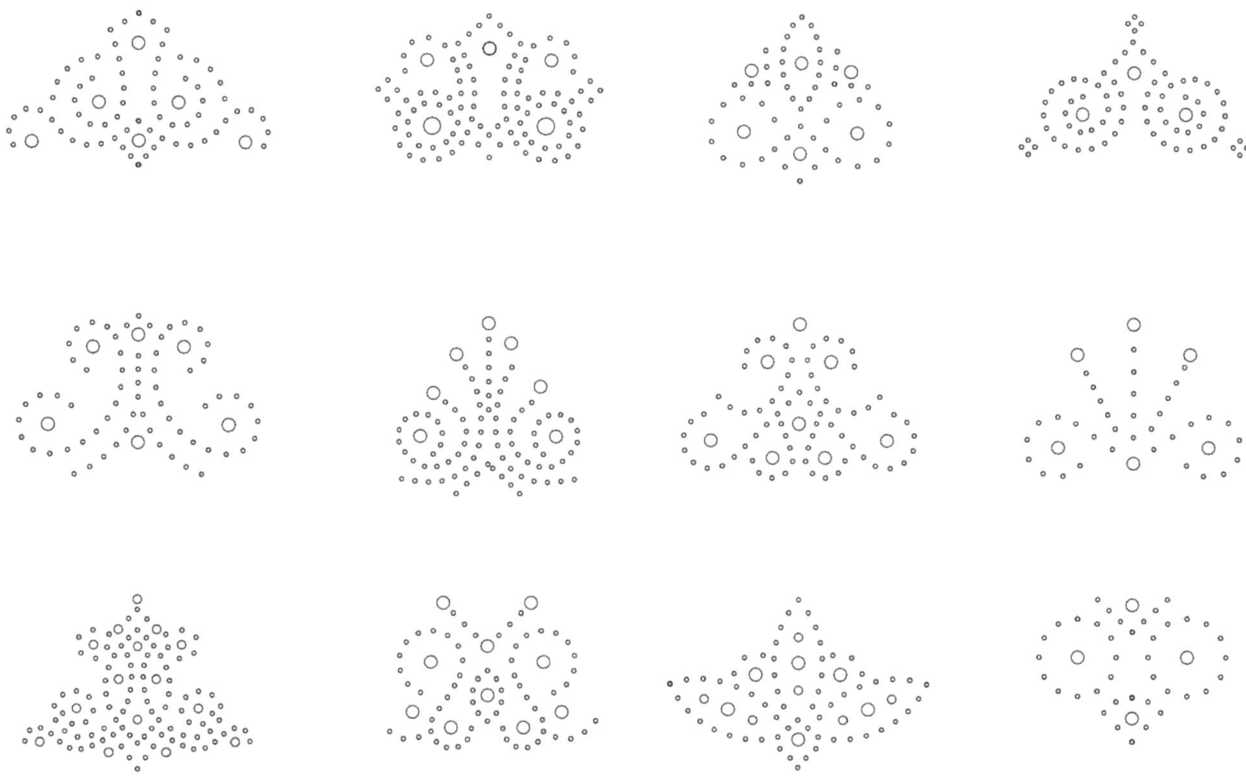

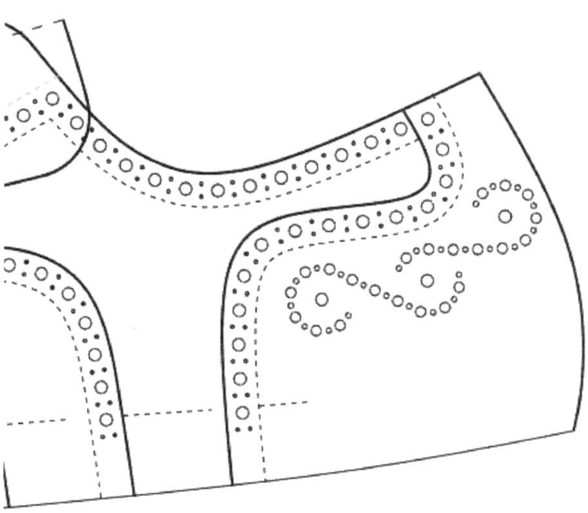

A medallion is a decorative perforation pattern applied to the shoe upper, typically seen on brogues and wingtips. It consists of a combination of larger and smaller punched holes arranged in a symmetrical motif, which can range from simple geometric groupings to highly elaborate ornamental designs. A medallion should also harmonize with the toe shape – for example, a wider, rounded toe often benefits from a larger pattern, while a narrow chisel toe will not suit the same layout.

Medallions serve no structural function, yet they remain a defining stylistic element in classic footwear. They may also be applied to areas beyond the toe, depending on the desired visual effect.

Actual Distance vs. Visual Distance

When it comes to edge decorations or seams, perceived spacing can easily become misleading. Brogues offer a typical example: although the measured distance between edges may be the same, the eye does not read it that way.

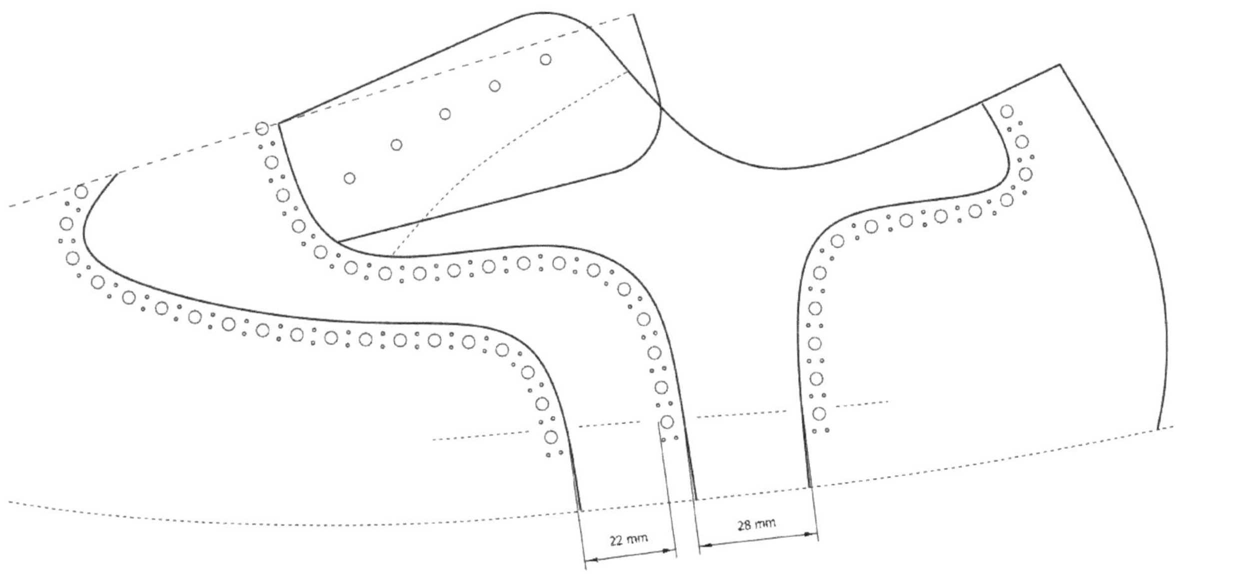

The clean, undecorated area is perceived as the true distance. As a result, the spacing between the wingtip and the vamp appears narrower than the spacing between the vamp and the counter, even if they are technically equal.

Leaving the pattern unadjusted can make the shoe look subtly "off." It may be difficult to articulate why, but the visual harmony is disturbed. To correct this, the edges must be modified so that the spacing appears equal, even if it is not mathematically so. Shifting the vamp line back by a few millimeters creates the visual impression of balanced and evenly distributed edges.

Moving the vamp line slightly backward equalizes the clean area and produces a more visually balanced result. If the brogue edge receives a double seam, this is an even stronger reason to shift the vamp line further back to maintain the same visual proportion.

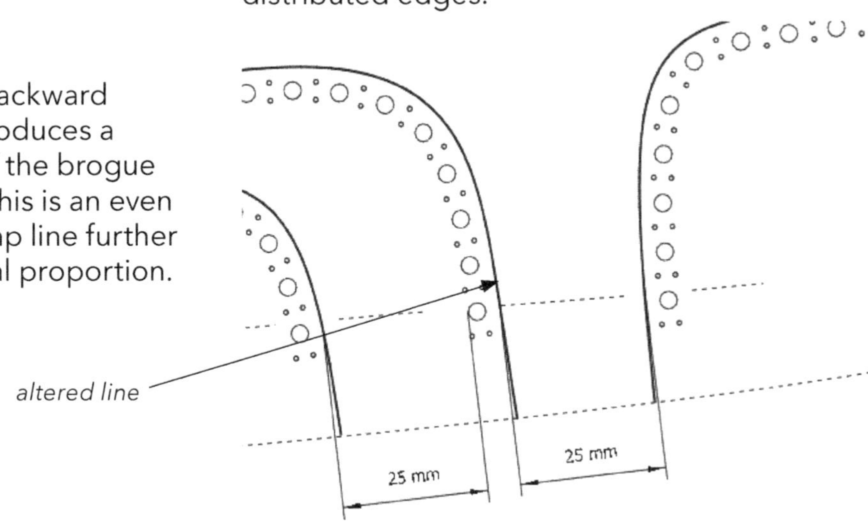

Alternative Lining Structure

The oxford pattern – like any other – offers numerous lining variations, and understanding these options is an important step toward becoming a more advanced pattern maker.

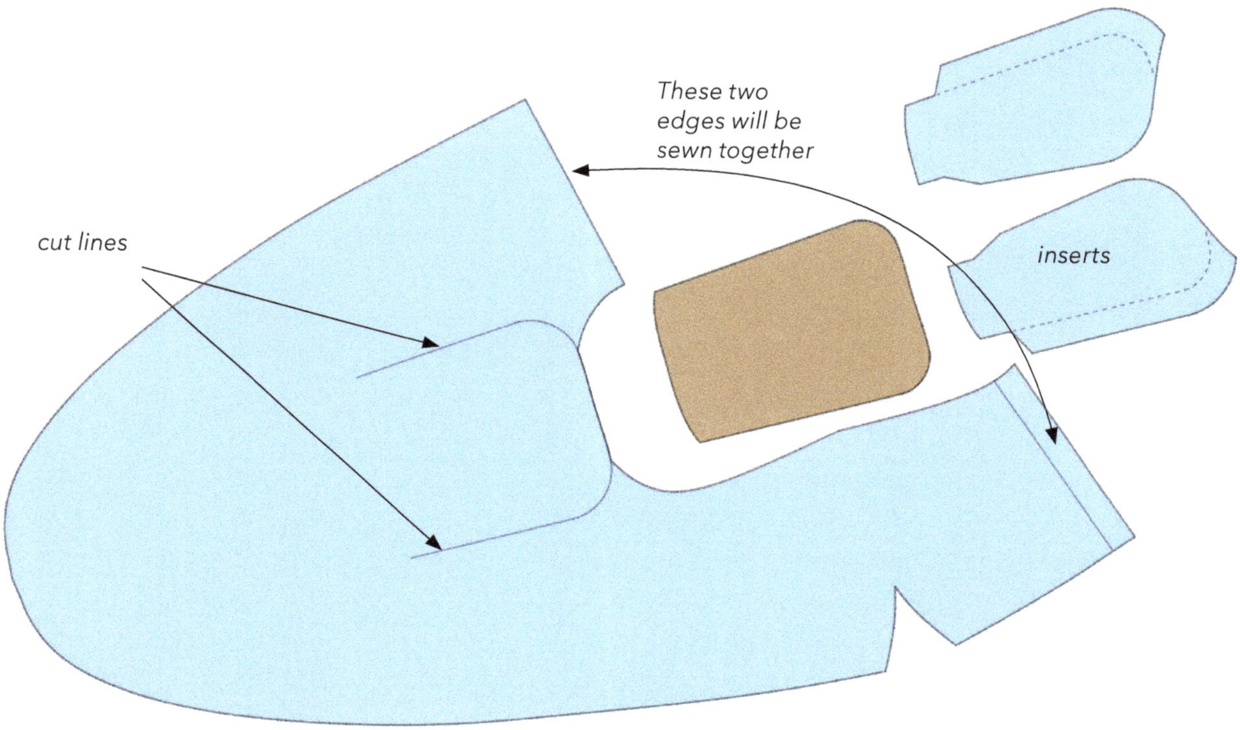

This particular solution is very simple, as it eliminates the lap seam between the vamp and the quarter and replaces it with a single seam. A heel counter lining may also be added if material economy is a concern.

In this construction, the two inserts are placed beneath the vamp lining along the marked line, and at the bottom they pass under the tongue, as in other models. The assembly sequence is slightly more complex, as the entire lining unit must be assembled beforehand. A detailed explanation of this order is presented in the following chapter on the wholecut.

Assembly Order

1. Clicking, skiving, marking if necessary
2. Cement the straight toe cap to the vamp, then sew them together (this can be done at any time)
3. Reinforce the topline and the lacing
4. Fold the edge of the quarters
5. Cement the counter to the quarter pieces
6. Fold the edge of the counter over
7. Sew the lacing decoration seam on the quarters
8. Attach the tongue to the vamp lining
9. Sew the counter edges
10. Sew the counter reverse seam, rub, and reinforce
11. Sew the reverse seam on the lining quarter
12. Attach the lining quarter to the upper quarters
13. Sew the topline
14. Trim the quarter lining
15. Tack the quarter fronts together
16. Attach the vamp upper to the quarters
17. Sew them together
18. Cement and sew the lining lap seam

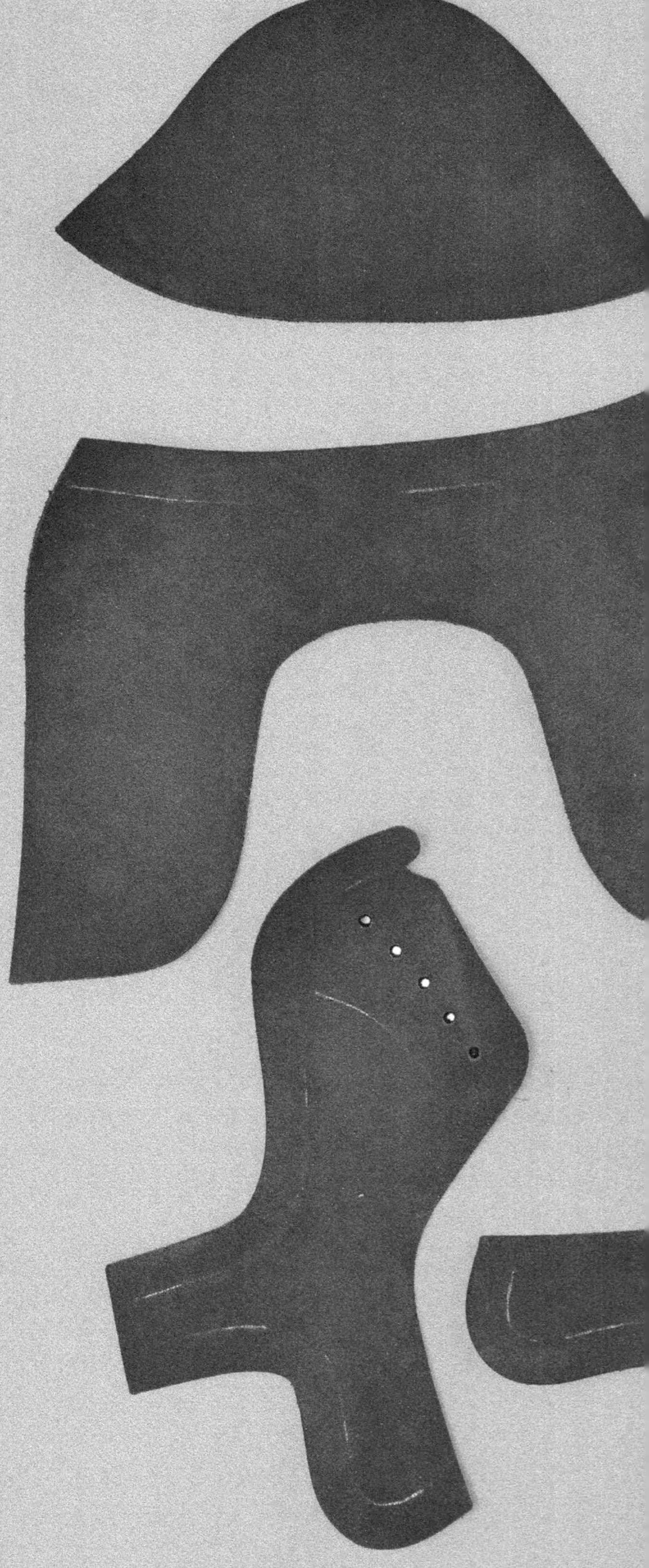

Hand Skiving

Even if all components are precisely skived on a machine, hand skiving still has a place and purpose. There are edges that are difficult or impossible to reach with a standard skiving machine, and in some cases the edge must be taken all the way down to zero. A good example is the small reverse seam at the back. This can be done with a special skiver foot, but a curved knife often achieves the result faster and cleaner. Using a flat-surface tempered glass with a curved edge (as shown in the image) - two quick cuts will produce a clean, precise edge.

The same result can be achieved with a push knife - the choice comes down to the maker's preference. The goal is to reduce the edge to about half its thickness and leave a 3.5-4 mm skived area, although this varies with the leather. Another important detail for a smooth, clean-looking seam is the shape of the skive. Ideally, it should form an inverted "V," with the point centered and both sides of equal width. This becomes important during stitching, as the needle should not land on a thicker area, which would affect the appearance of the seam.

Another important area for hand skiving is the overlap allowance on the vamp, and in fact any overlap allowance. The toe, however, is the most critical. Unlike the counter - where the stiff heel counter can hide some excess material—or the side of the vamp, where it is less visible, the toe is always exposed. A second visible line of thickness here does not look professional. Especially in fine or bespoke work, this extra bulk must be removed to achieve a clean, refined result.

At this point, gluing two pieces together is nothing new, but a few small techniques can improve the result. First, do not apply cement all the way to the edge. Leaving one or two millimeters clean will produce a crisp edge and eliminate the need for later cleanup. The second technique is more subtle.

When burning off the fibers, do not stop at removing loose leather fibers; wait until the edge begins to curl inward. Different leathers react at different temperatures and speeds. This ensures that the toe cap tightly follows the vamp and prevents it from curling toward the top side.

Add the lacing reinforcement to strengthen the lace holes and to assist the folding process. If a single layer does not seem sufficient, a second layer is justified, or consider a thicker material in the future. The topline tape continues where the lacing tape ends. This preparation for folding is no different from what has already been shown. Make small cuts along the curved area to help the leather fold without distorting the edge.

The corners are not too difficult to fold if the skiving is done properly – meaning neither too thin nor too thick. Adjust the pleats gradually, splitting larger pleats into smaller ones for as long as possible. When this is no longer feasible, use a hammer to flatten the area.

A truly precise skive is only complete after it has been flattened from the grain side. This is how the smoothness of the curve can be checked, although there is little opportunity for correction if it does not meet quality expectations, other than redoing it.

Use the same technique as before by lightly flaming the edge. Any curved shape with a lap seam can make the edge more vulnerable. This small step not only curls the edge slightly but also makes it a bit firmer.

Apply cement to the bottom layer and place the counter on top immediately. Accurate placement is critical here, as with other components, so if the placement marks are not clean or clearly visible, redraw them using a pencil, scratch awl, or silver pen – whichever works best.

The two quarters and the counter are cemented together. There are a few additional points to note. The counter extends slightly beyond the edge, more than the standard five-millimeter folding allowance, and for good reason: this allows the excess to fold over the other, already folded edges.

Mark the desired folding edge. This is not absolutely necessary, but it helps in the next step. Skiving the corners of the quarters is also advisable to reduce bulk in this area.

Add a piece of reinforcement tape directly under the intended folding line. This aids the folding process and, more importantly, reinforces this area.

Adding reinforcement tape under the sewing lines serves two important purposes. First, the seam becomes more resistant to stretching, which is not a theoretical issue–it will occur during lasting. If the leather is soft and stretchy, this additional reinforcement helps preserve the pattern as it was designed.

The second, equally important reason is the appearance of the stitching. Adding a firm support layer beneath a soft material consistently results in a cleaner, more defined stitch.

When the cement is sufficiently dry, the counter can be folded over the quarters. Aim for a tight fold without distorting the quarter edges. Leaving a gap here would be highly problematic and would be clearly visible in the finished shoe. Ensure this does not occur and pull the edge down as firmly as possible, without distorting the top edge.

The upper pieces are only cemented together at this stage. There are several ways to finish the counter from here. This chapter presents a clean, simple approach, but a few alternative solutions are worth noting.

One option is to sew the counter from corner to corner with one or two continuous lines. The challenge with this method is stitching the second line on the top of the counter before reaching the first line. The second line can also be placed slightly farther from the topline stitch, which creates a refined detail.

Another option is to sew the counter in place before folding and run the seam all the way to the edge. The disadvantage of this method is that the exposed thread at the edge may wear faster.

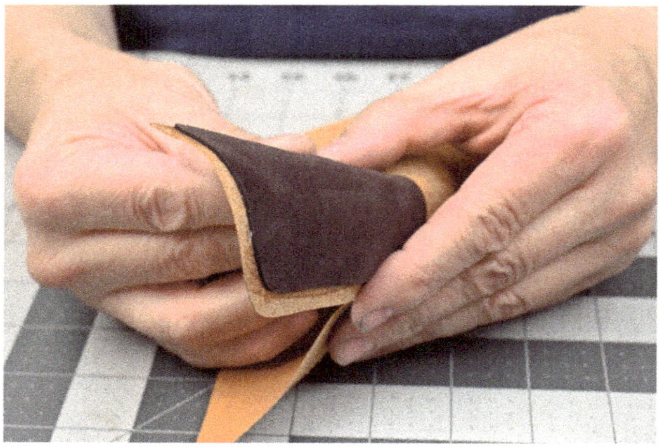

Cement the tongue together with the vamp lining. Many makers add a standard tongue at the final stage of the process, which is certainly simpler and more straightforward. This method requires precise pattern making, but the result is cleaner, more elegant, and seamless. The two layers can be joined with a slight curve.

Sew the two layers together along the edge. The tongue edge can be folded, and some makers choose to do so, but a raw edge – especially in nubuck or suede – also works well. Very little of this edge remains visible, so the additional work required for folding may not be justified in this case. Raw tongue edges can be found even in the finest bespoke shoes.

Before proceeding further, the decorative stitches can be added. These stitches could serve a functional role by holding the lining and the quarter together, but in that case this assembly order would be at least problematic, though not impossible. With the modern adhesives used today, these stitches are clearly decorative rather than structural, so adding them at this stage is entirely appropriate.

This line is traditionally sewn as a double line, although a single line also works. Start on the lap seam allowance at the bottom, sew upward to the edge – stopping so the topline seam will pass over or just beneath it – then turn with a single stitch and sew back down. This avoids a loose thread end at the top, which may leave residue regardless of how carefully it is cut or burned. Following this sequence results in a clean sewing line.

The counter has two curves, which can present a slight challenge, especially the smaller one. As discussed earlier, short curves can affect stitch length, so the same method should be used to avoid this issue.

In most cases, the two layers are the same color, making the leather edge difficult to see from above due to the lack of contrast. A useful technique is to use an additional LED light positioned at a very low angle, directed toward the edge. This highlights the edge and helps guide the stitch line.

Hammer the stitch line to close the holes and achieve a cleaner-looking seam.

Sew the reverse seam and use a generous amount of backstitch at both ends.

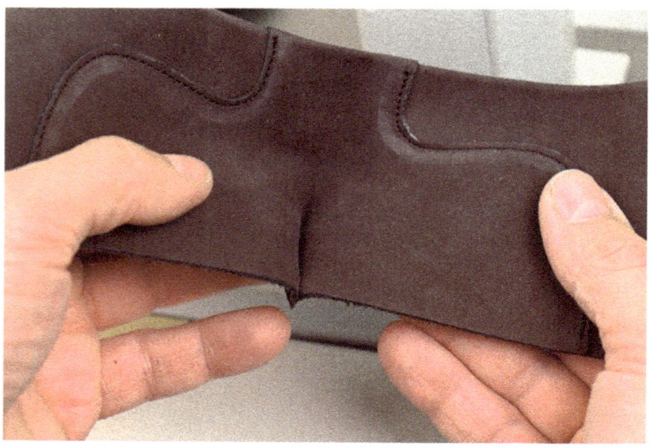

The seam should run above the cut, but not exceed one third of the counter height.

Trim the excess, leaving just enough to flatten. This must be done in a single cut.

Flatten the seam carefully. Excessive hammering can damage it.

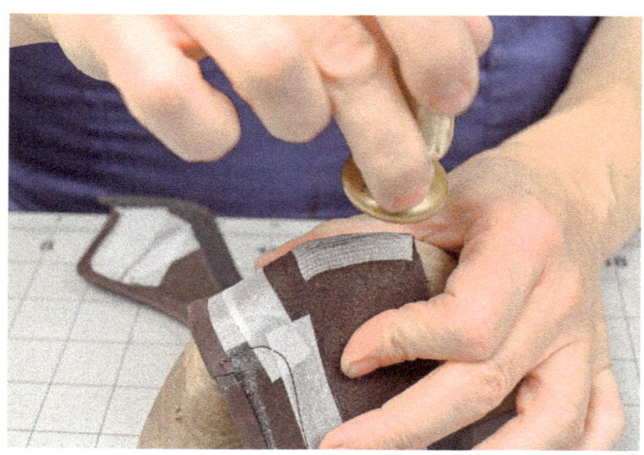

Add a backline tape along the top and repeat the hammering process.

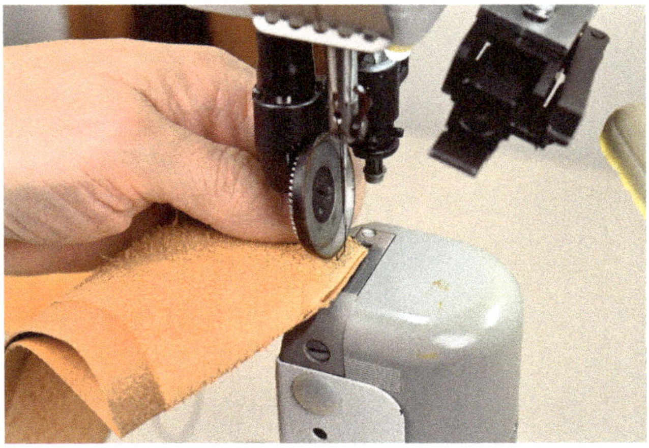

Sew the reverse seam on the lining, from the flesh side.

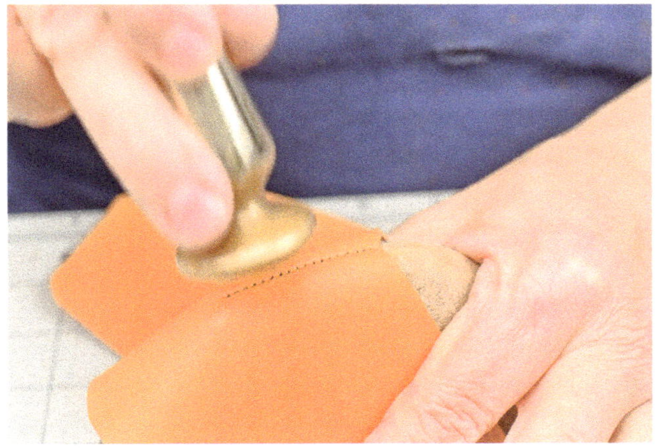

Trim the excess and gently hammer it flat from both sides.

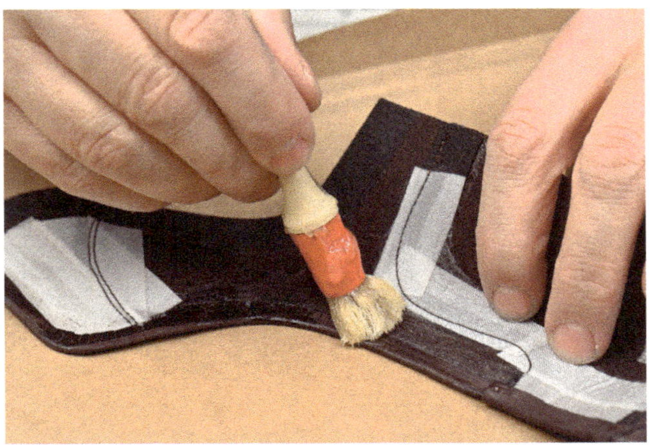

Apply glue along the top edge, but not too wide - aim for about 10 mm.

Add cement to the lining, but leave the trimming allowance clean.

Start assembling from the center. This can be marked on the pattern or by folding.

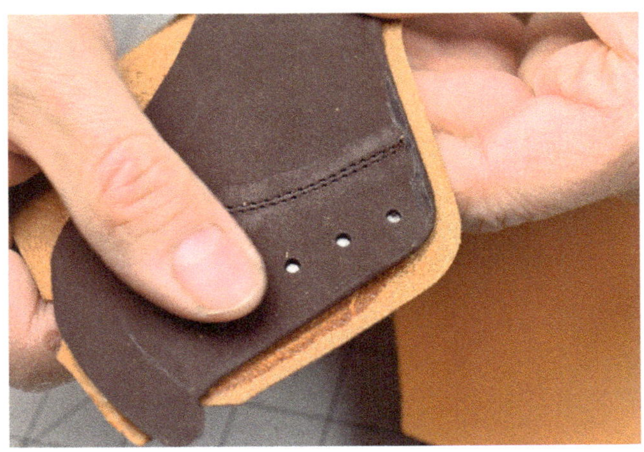

It should be an even 5 mm trimming allowance all the way around.

The quarter lining and the quarter panel (the quarter pieces and the counter together) are attached at this stage. This is where the three assembly sequences differ significantly. The main difference compared to the wholecut method – where all upper pieces are sewn together before any lining is added – is the vamp stitch. The scissor cut must be deep enough to allow the vamp line to be sewn without catching the lining before the lacing. This is the key requirement of this assembly order and is only possible with a sufficiently deep scissor cut.

The advantage of this method is easy access to sew the entire topline from beginning to end, followed by very straightforward trimming.

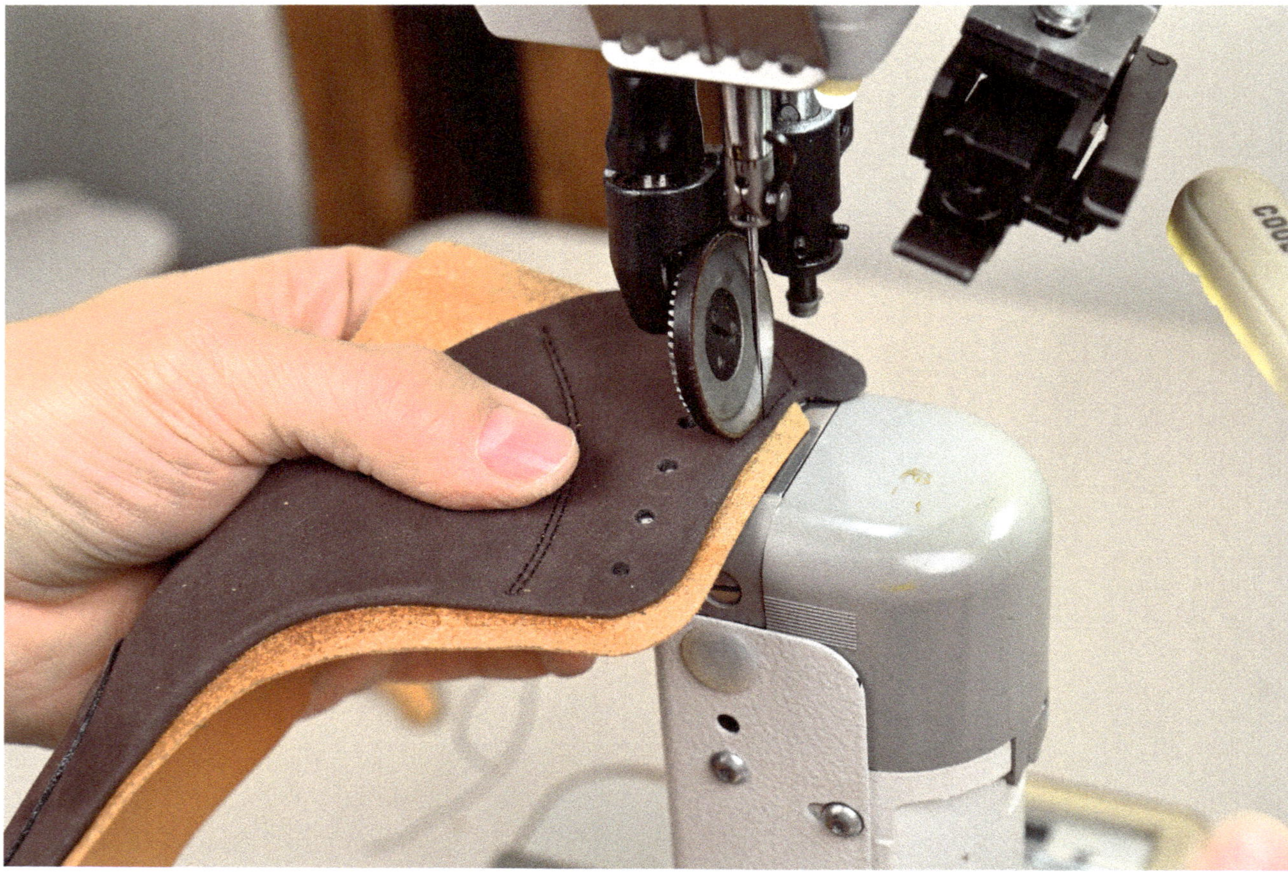

The topline stitching is probably one of the more challenging operations. Since this is a folded edge, sewing too close to the edge would be a mistake. Sewing too far in, on the other hand, looks poor and may expose too much of the opposite side of the fold as the layers separate over time. Aim to sew evenly at 1–1.5 mm (the thickness of the folded edge).

Trim the excess lining with extreme caution. The trimming can be done with a clicking knife or with small curved scissors.

It is a good idea to slightly turn the knife under the edge, as this is safer and reduces the chance of cutting into the upper.

Reinforce the toe cup from the flesh side with reinforcement tape or a full panel of stretch-resistant backing material. If the leather is too stretchy, lasting may otherwise introduce an unwanted curve along the line.

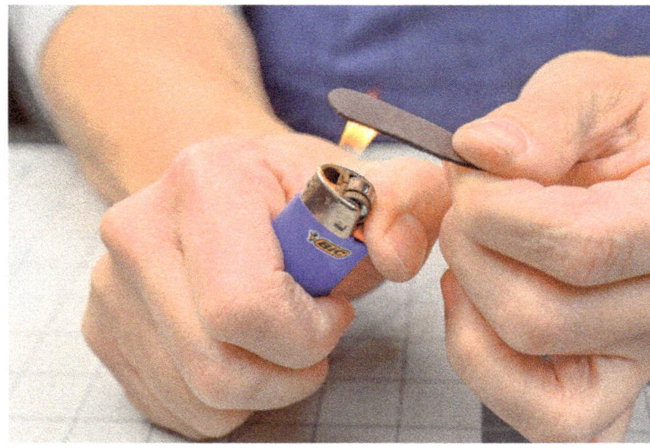

Cut a small tongue-like piece (a 12–15 mm wide strip with a rounded end) to cover the front of the lacing. This is a very traditional method and a good alternative to hand-stitched reinforcement. Curl it with flame.

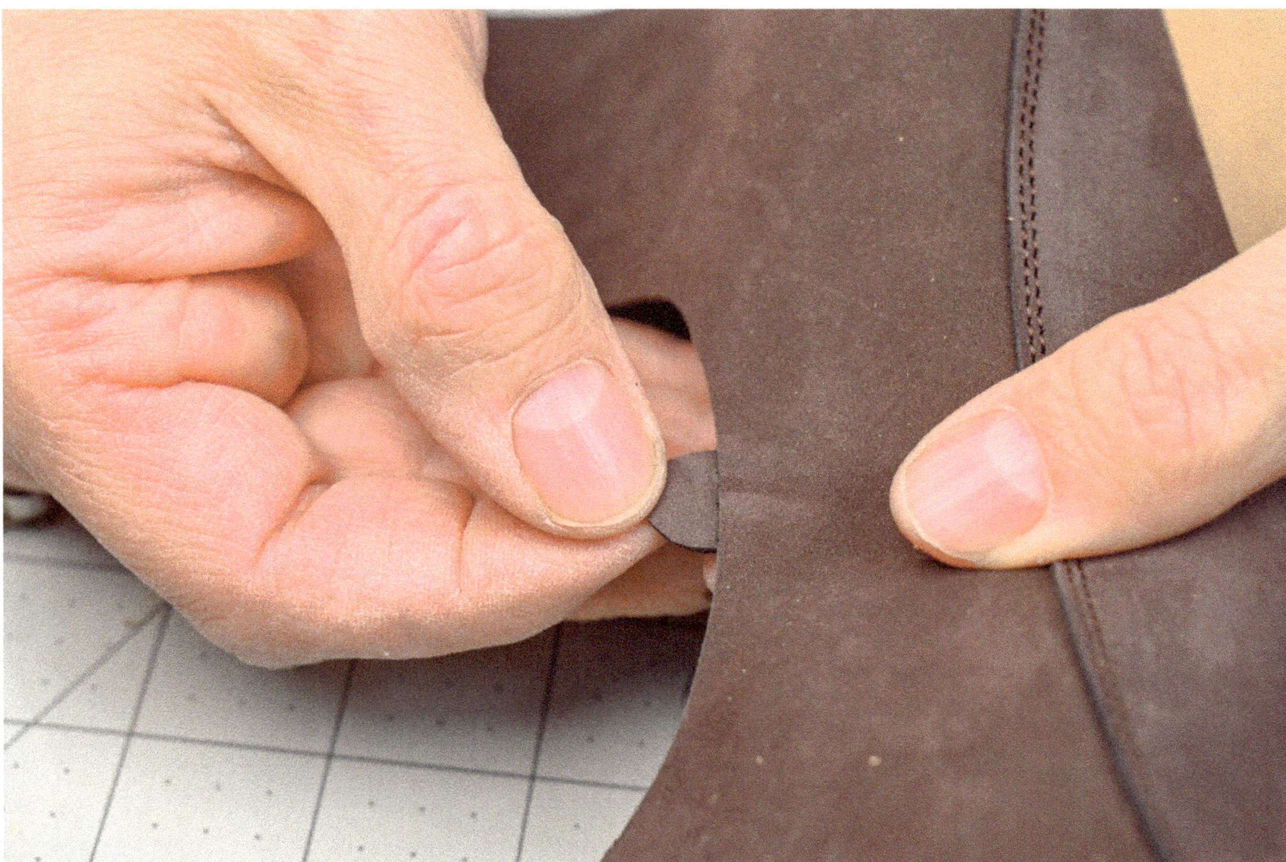

Add this small piece from underneath, accurately centered. Only a half circle should be visible from the top. Ensure the grain side faces up. This piece should be slightly thinner than the upper, which can be achieved with a skiving machine, as the piece is very narrow.

Closing the lacing

Pro Tip!

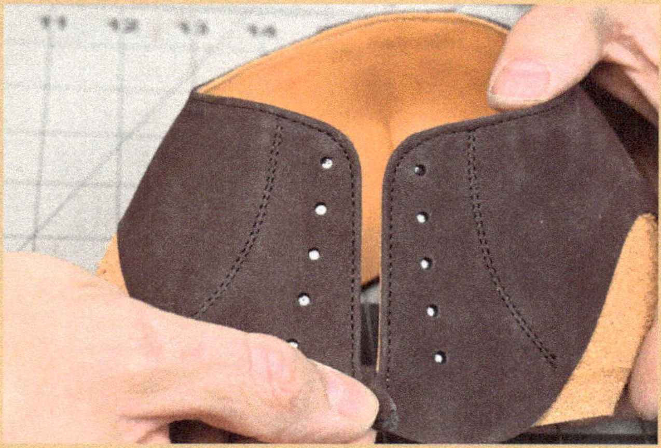

Leaving a gap between the laces is a typical beginner error. Check the position before continuing.

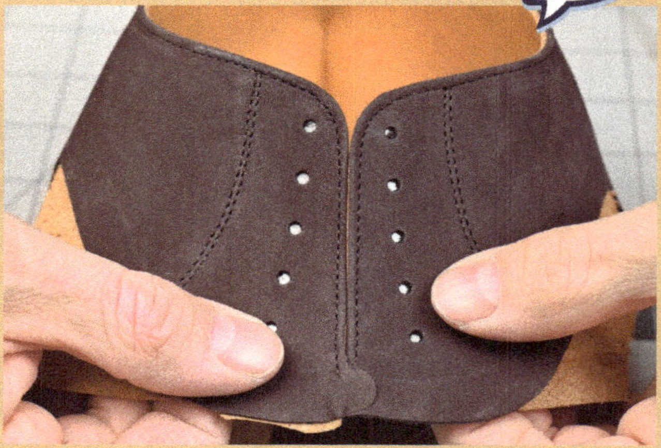

When the laces are assembled, they should line up, closing evenly against each other.

If the laces are simply placed together, the bottom will not overlap at the correct angle. If the process starts at the bottom, there is a good chance the lacing will slip out of position. A more effective method is to hold the bottom at the correct angle – so the top overlaps, as shown in the image above. In other words, a slight overlap at the bottom is necessary to ensure that when the lacing is opened, it appears properly closed.

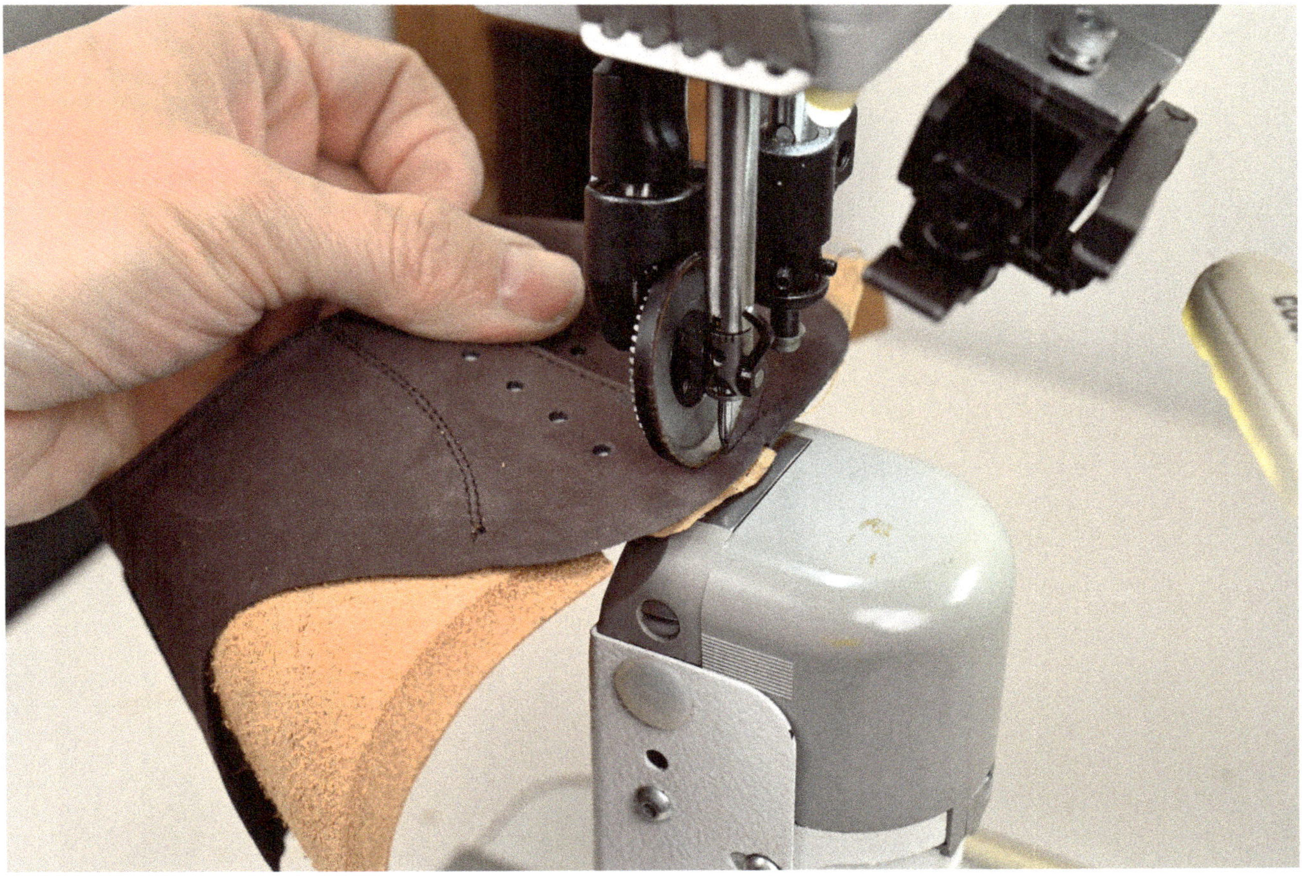

Make a few stitches across the overlapping section at the front of the lacing. These will be well hidden and will hold the two sides firmly together during vamp stitching. Even if the cement bond is strong, the sewing machine may push the sides apart, at which point it is too late to correct. This method has proven to be the most effective.

Make sure this stitch line is positioned well away from the marked vamp edge, but also not too close to the skived edge; otherwise, it will not achieve its intended purpose.

This is the best time to punch these holes, even though access was better earlier, because at this stage the holes can be matched perfectly.

This is how the other side should look. The scissor cuts are deep enough to allow the quarter lining to be pulled back, giving access to the vamp stitching without catching any unwanted portion of the lining.

Do not be concerned by the rough appearance of the lacing area; this will be covered by the tongue and closed with an additional stitch line. The key point is the added strength provided by these extra stitches.

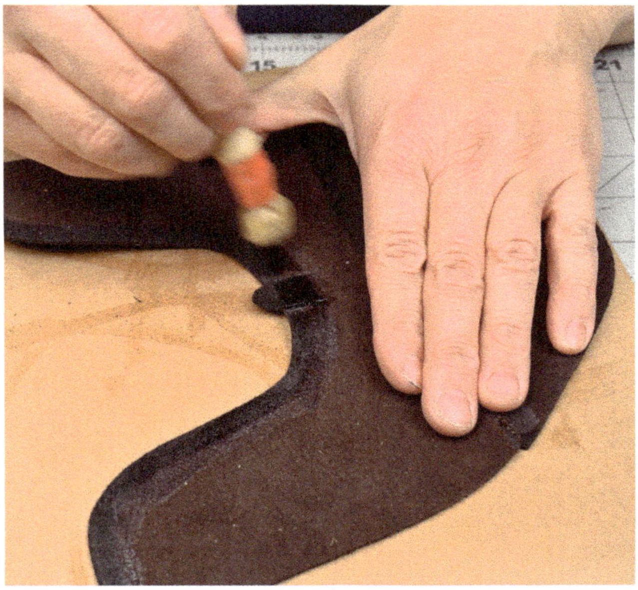

Add cement to the edge. This is a very three-dimensional seam, so a strong bond between the layers is essential to achieve a clean-looking seam. If adhesion is insufficient, the pressure of the roller foot can force the vamp out of position.

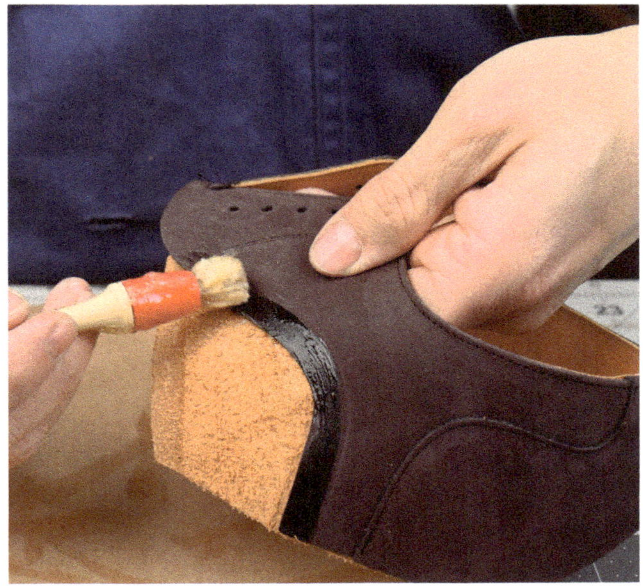

Also apply cement to the other side. Since this is a nubuck material, the edge must be kept clean, leaving the last one or two millimeters untouched, as removing cement is almost impossible. After this, allow the adhesive to dry properly.

Start the assembly from the center, especially if this is a new, untested pattern. Otherwise, you can start from the edge, which is easier. The issue is that even a small misplacement can lead to a larger problem.

Keep the cemented edges apart like this until it is time to press them together. Focus on one side, press the layers firmly together, then move to the other side and do the same.

It is time for the seam – single or double, as specified by the designer (in the technical specifications documents). This seam has a few challenges. First, keep the overlapping section of the lining away from the stitching. If the pattern supports it, this is not particularly difficult. If not, there is a small but effective workaround, aside from correcting the pattern and redoing the process: simply extend the scissor cut slightly deeper. If the required cut becomes excessive, the result can be considered a successful but not final prototype. If only a small increase is needed, the prototype is successful and production can proceed.

Either way, any change made at this stage must be reflected in the pattern. Keep in mind that materials have thickness, firmness, stretch, and other properties, and all of these factors can affect assembly, for better or worse.

Another issue is the shape itself. In the center area there are more than two layers to sew through. These layers can shift, and the feed mechanism (especially while turning) can create a pleat. Also ensure that the vamp remains flat under the machine; even a slight angle can deflect the needle, causing it to hit a previous needle hole and break.

In short, execute this step with great care and slow down as much as needed.

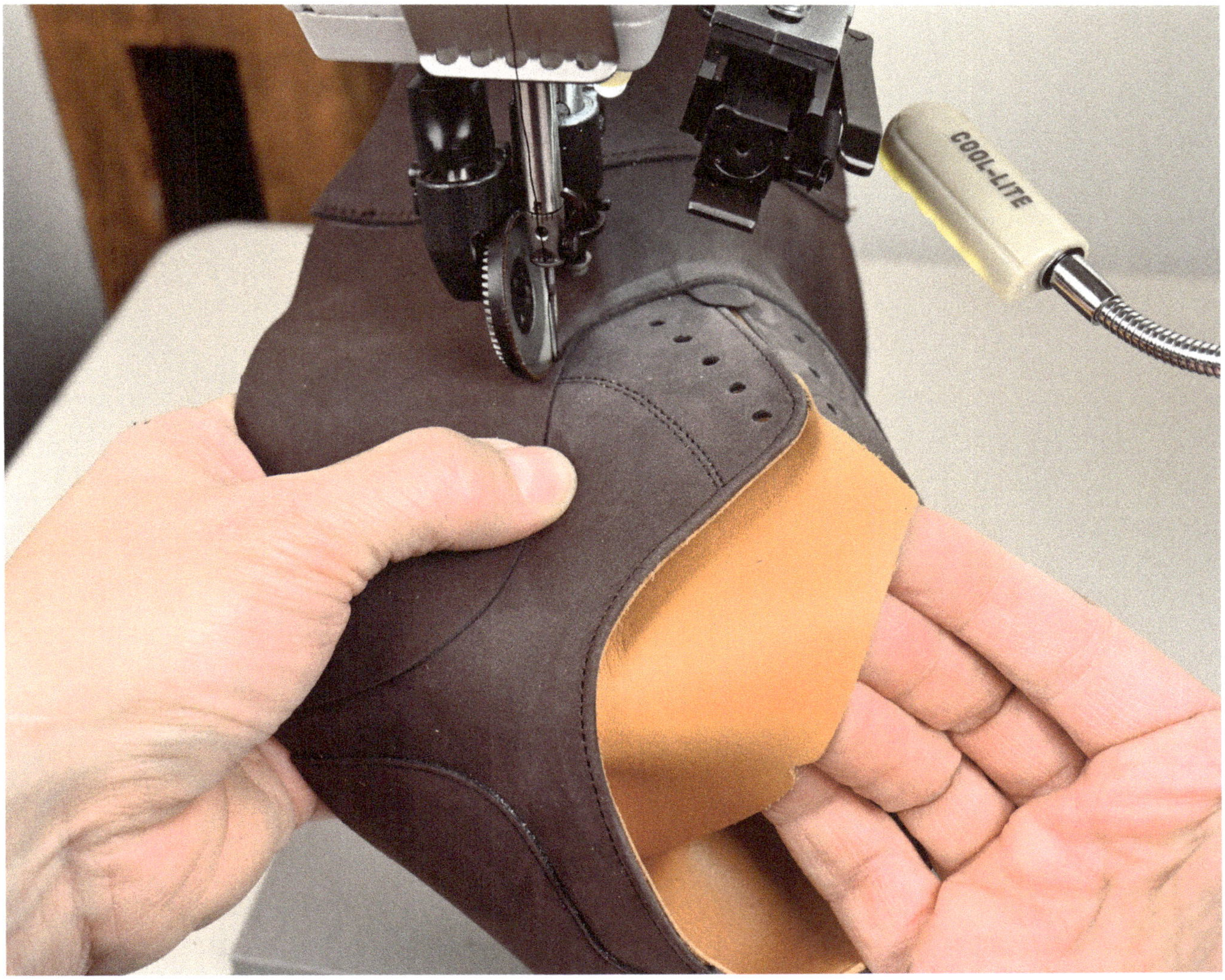

This is the point where the seam reaches the other side of the quarter lining. Make sure it does not slip under the stitching, as there is no way to correct this afterward. Keep the seam close to the edge, as a wide stitch line would be inconsistent with the style.

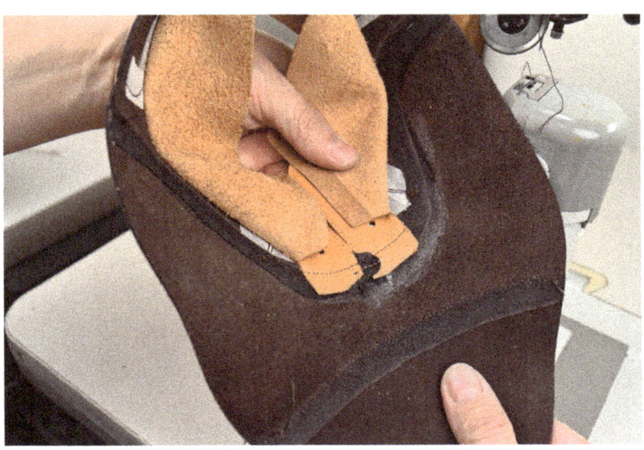

Here is the result from the flesh side. It clearly shows that the seam includes the lacing section but leaves the bottom area (below the scissor cut) loose. Another small but important detail is the decorative seam on the quarters. If the lining had been incorporated into this seam - which is not necessarily a bad approach, but would require a different assembly order - pulling the lining up in this way would be impossible, and the vamp seam would also pass through the bottom section, which would be undesirable.

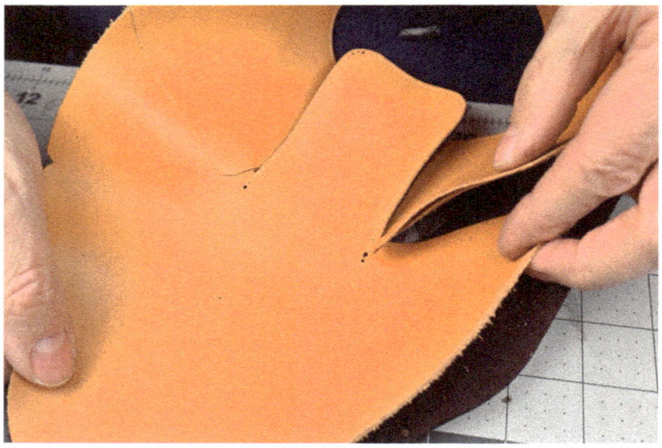

Drop the vamp lining into place. If everything is correct, the overlap will be accurate.

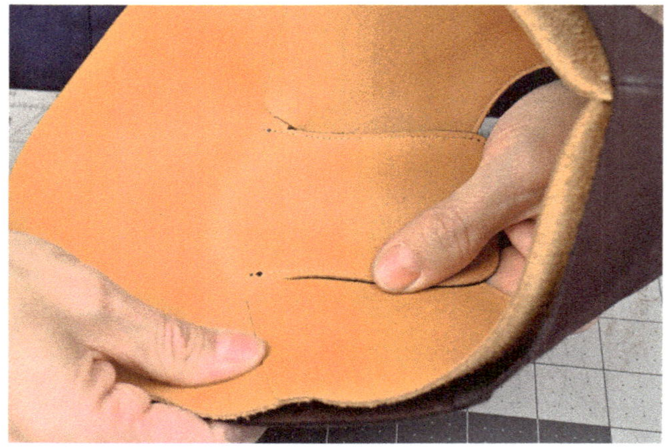

Check the amount of leather on both sides. For a standard lining, 8–10 mm is ideal.

Cement the layers together. As an alternative, double-sided tape can also be used.

Make sure there is enough clearance to run the post-bed machine along this seam.

Sew both sides close to the edge, then cement the middle section.

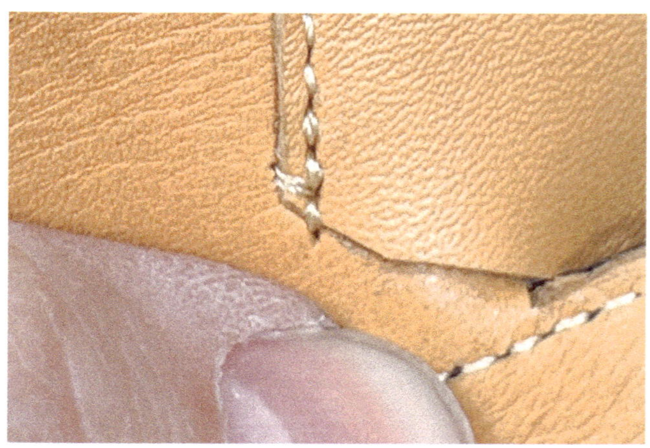

At the corner, this neat little "A" stitch can be used to hold the corner down.

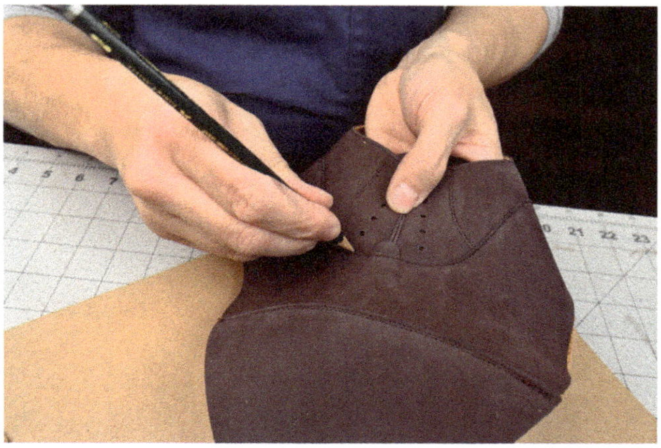

Find the tongue position at the bottom and mark a small dot just before the sewing machine would run off the edge.

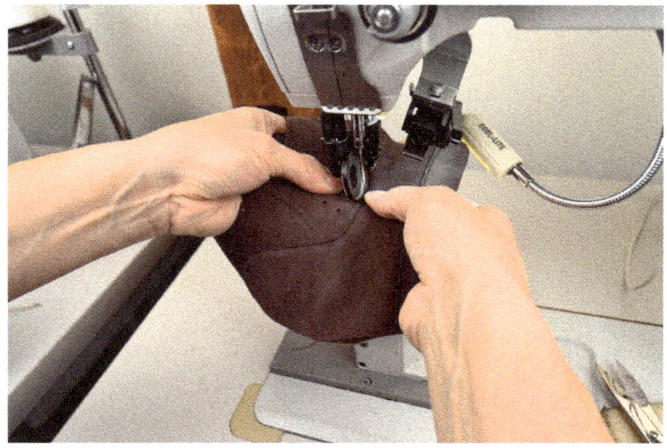

Sew this line. The small half-moon piece can go either inside or outside the stitch; here it is sewn underneath.

The upper is complete. It still requires a thorough cleaning to remove any glue residue and to burn off loose thread ends. Another step is to add a small amount of cement between the upper and the lining around the lacing. Adding hidden eyelets is optional.

10 THE WHOLECUT SHOE

The wholecut (design represents one of the purest and most elegant forms in classic men's footwear. Its defining feature is that the upper is made from a single piece of leather, with no seams except for the reverse seam at the back. (A completely seamless wholecut is possible, but it falls outside what is considered "classic.") The design has minimal stitch lines - the only visible ones run along the topline and around the backline. The appeal of this model lies in its sleek simplicity. Traditionally, it is constructed with five pairs of eyelets, although six pairs are also common. In terms of comfort, this type of footwear offers the same instep allowance as other oxford-style shoes.

Wholecut construction does not lend itself easily to mass production, which may have contributed to its reduced popularity after its introduction in the 19th century. However, interest in this model has increased over the past few decades. The pattern is very similar to that of an oxford but uses only a single axis, which leads some makers to apply slight crimping to the upper. This is often unnecessary, as a well-designed pattern can be lasted successfully without pre-shaping. The classic edge finish for the topline is piping, although it can be combined with folding, except between the laces, where there is insufficient excess. An example of this approach follows.

Pattern Lines

The wholecut is fundamentally defined by the last shape; the pattern lines themselves are almost non-existent. The only line that is visibly present is the topline and its continuation into the closed-lacing section.

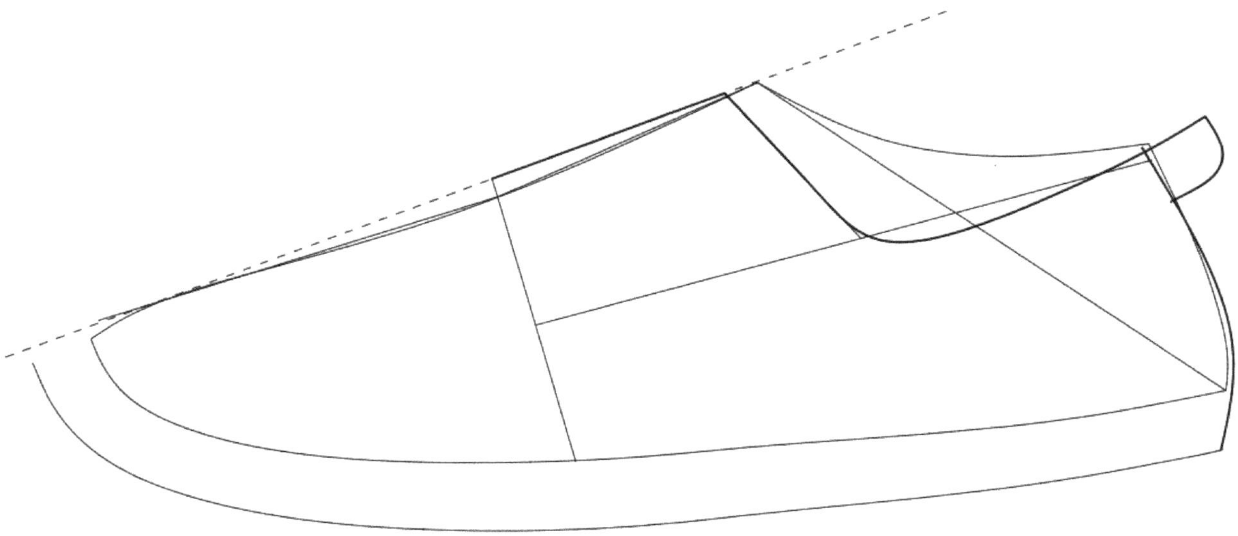

Functionally, a wholecut behaves like an oxford – the difference lies only in the number of pattern pieces. For this reason, a simplified oxford pattern is used, following the same structural lines. The only minor deviation is the addition of a dogtail to reinforce the backline, as introducing a separate pattern piece would defeat the very concept of a wholecut.

Lasting allowance is added using values appropriate for the chosen construction method. For standard hand lasting with a regular, not overly firm upper leather, the 16–18–22–16 mm combination from toe to heel is used.

The wholecut is generally not the ideal choice for someone with a very wide foot when making custom shoes. However, this does not mean the pattern maker cannot make adjustments to improve comfort at the instep. Two practical options are available: extending the topline by lowering it below the ankle bones, and lengthening the lacing section. Both modifications increase the topline girth, making it easier for the foot to slip in.

This is not only a comfort consideration but also a durability one - easier entry reduces stress on the seams and the counter, helping the shoe retain its shape and integrity over time.

Eyelets

The wholecut functions essentially the same way as a closed-laced oxford, so the same number of lace holes can be used – in these examples, five per side. Other counts work equally well, as long as they maintain proper closure of the lacing, as intended.

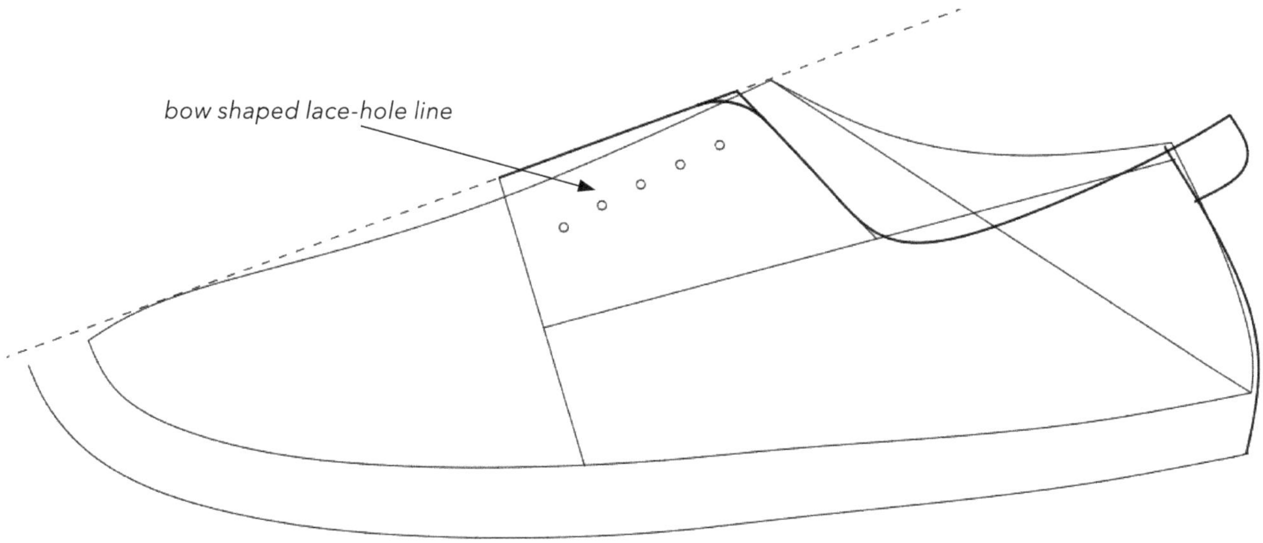

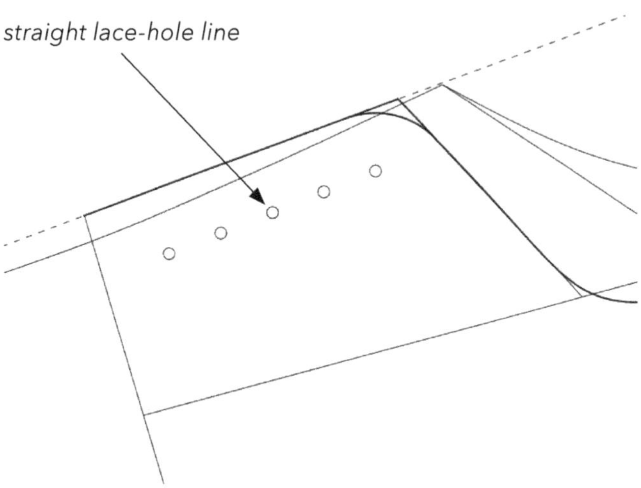

There is room for creativity in the lacing design. A subtle bow-shaped eyelet line can add elegance and a bespoke character to the shoe. This arrangement also exposes the lower section of the laces slightly more, which can increase strength. A straight eyelet line is, of course, always an option as well.

If a bow shape is chosen, make sure the curve is clearly visible. The only real limitation in placing the eyelets is that they must sit directly over the tongue.

Lining Pattern

The lining is made in two pieces, as is usual for higher-end shoes, although the four-piece construction shown in the oxford chapter is also possible. However, because a wholecut has no seams or decorative elements to conceal structural transitions, a counter lap seam may show through on the finished upper. For this reason, in this case a one-piece quarter lining is the better choice.

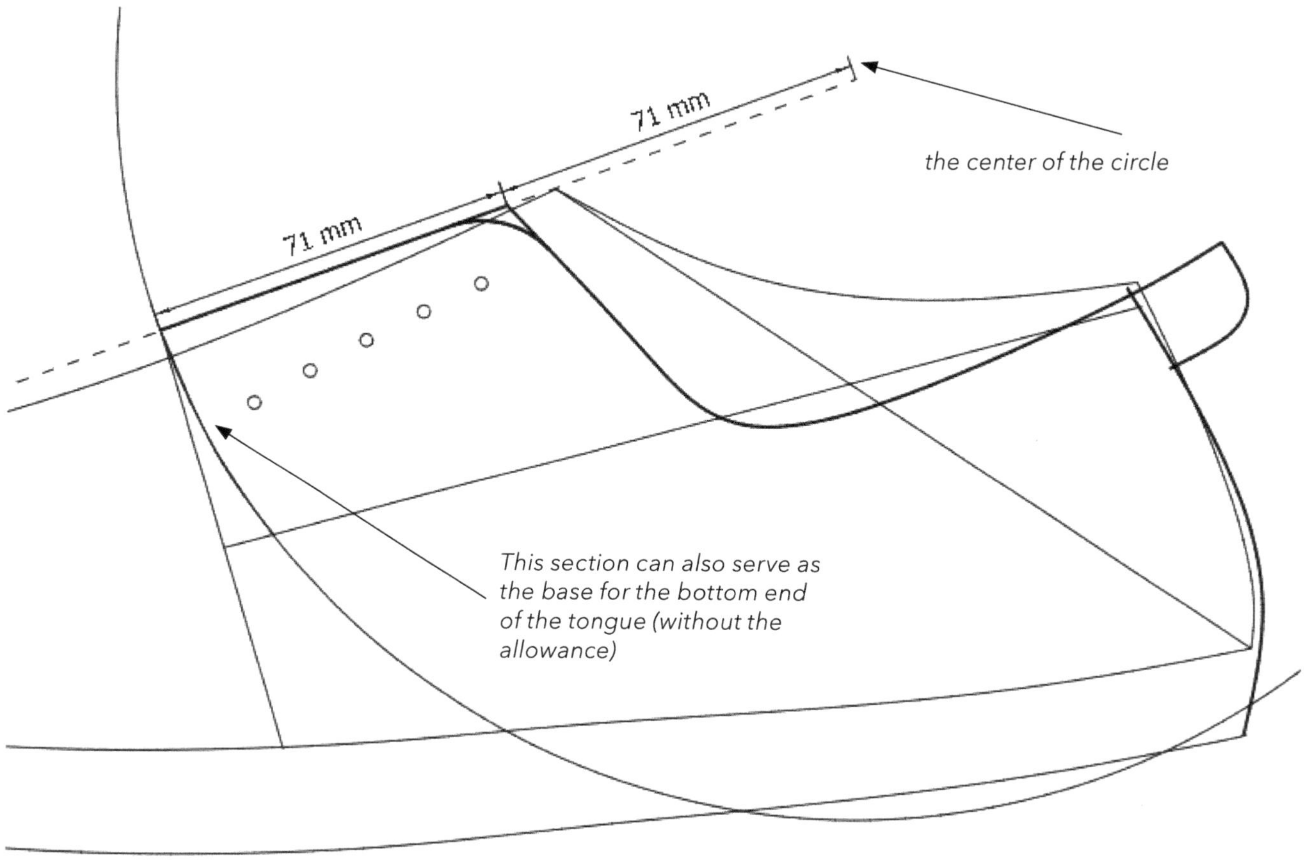

The front shape of the quarter lining - and the bottom edge of the tongue - is primarily a functional design decision, but it carries an important consequence: it determines where the lap seam between the quarter and the vamp begins. Both a slight curve and a straight line can work well. If a curved solution is chosen, it may follow a natural contour or be constructed more formally: extend the main folding axis, measure upward the desired lacing distance, and draw a circle from that center point. Keeping this center on the axis is essential so that the curve intersects the axis at a perpendicular angle.

Upper Pieces

The upper pieces are unfolded. The vamp has a reverse seam allowance at the backseam and a dogtail added to the lateral side. If the last geometry does not require it, there is no need to add further asymmetrical details.

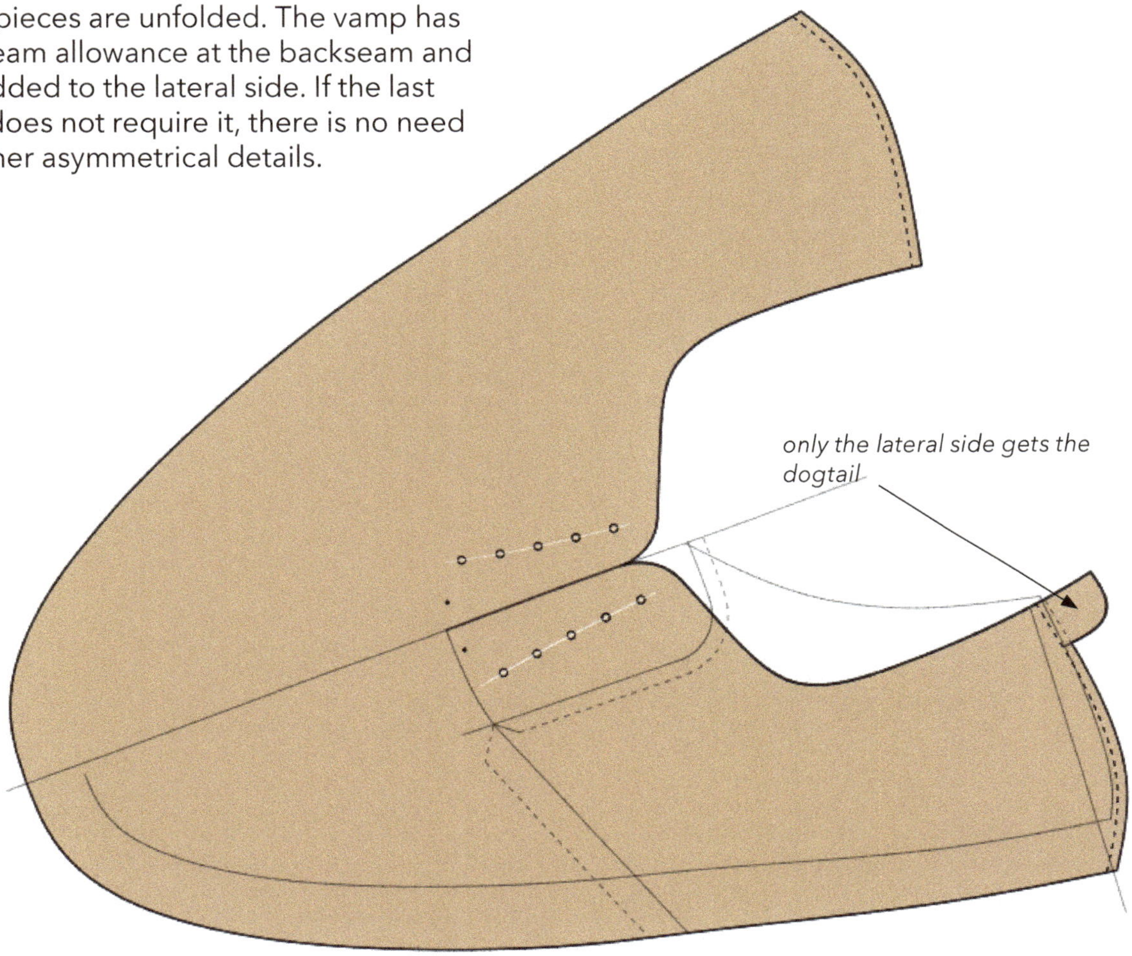

The tongue is a very simple piece. No excess is added to the edge, since piping does not require any. There is, however, a useful adjustment that can be applied if needed. The edge of the tongue is skived, just like the quarter edge at the same location, and this can create a problem. When two or more skived edges fall directly on top of one another, even well-executed skiving can result in unwanted bulk. To avoid this, the edges can be offset. In this case, adding a few extra millimeters of allowance to the tongue ensures that the skived edges do not coincide, preventing buildup.

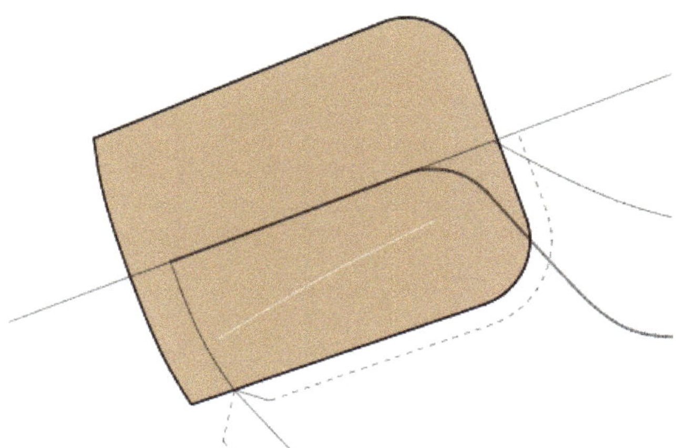

Lining Pieces

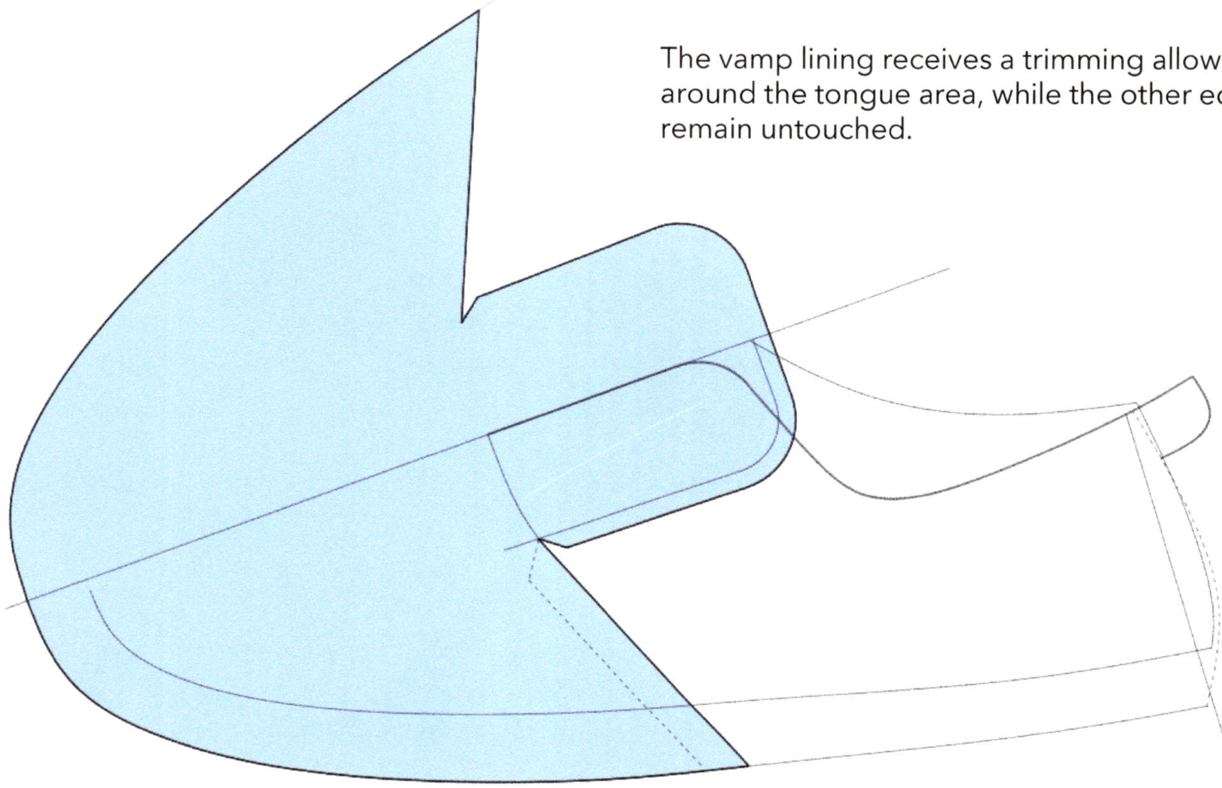

The vamp lining receives a trimming allowance around the tongue area, while the other edges remain untouched.

The quarter pattern is unfolded along the lining axis, and the lap seam allowance is added to its front edge. Instead of using the reverse seam method applied elsewhere in this chapter, another elegant solution is shown here: simply slicing into one side of the designated reverse-seam design line.

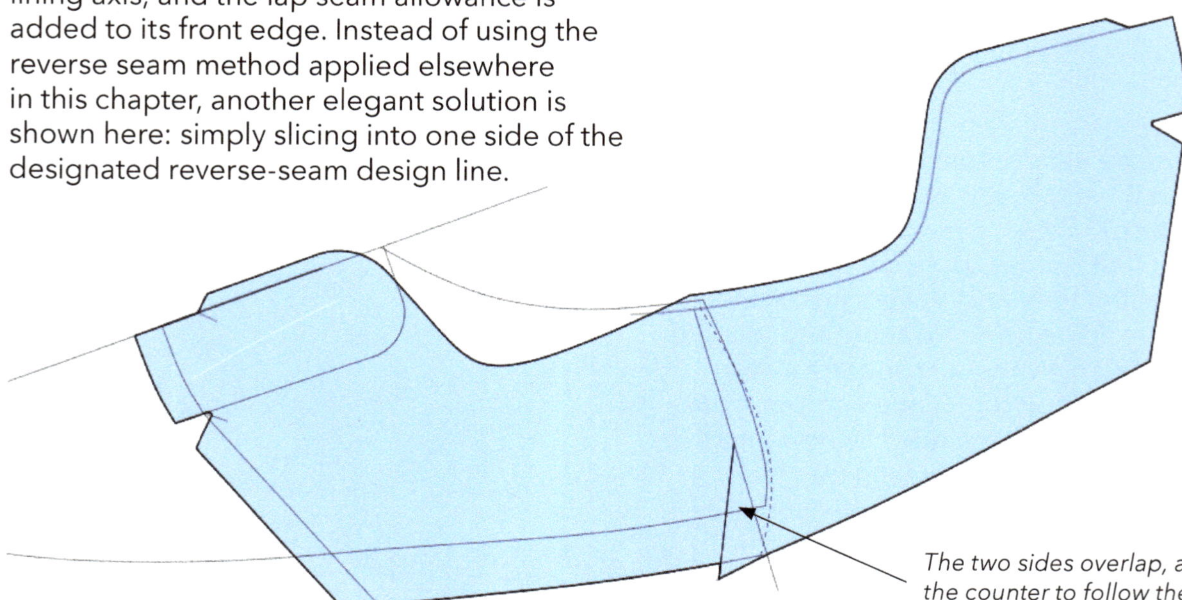

The two sides overlap, allowing the counter to follow the natural curve of the heel curve.

The Complete Pattern

1. NO SKIVING
2. SHORT SKIVING HALF THICKNESS
3. MEDIUM LONG SKIVING HALF THICKNESS
4. LONG SKIVING HALF THICKNESS
5. LONG SKIVING NO EDGE THICKNESS
6. SPLITTING

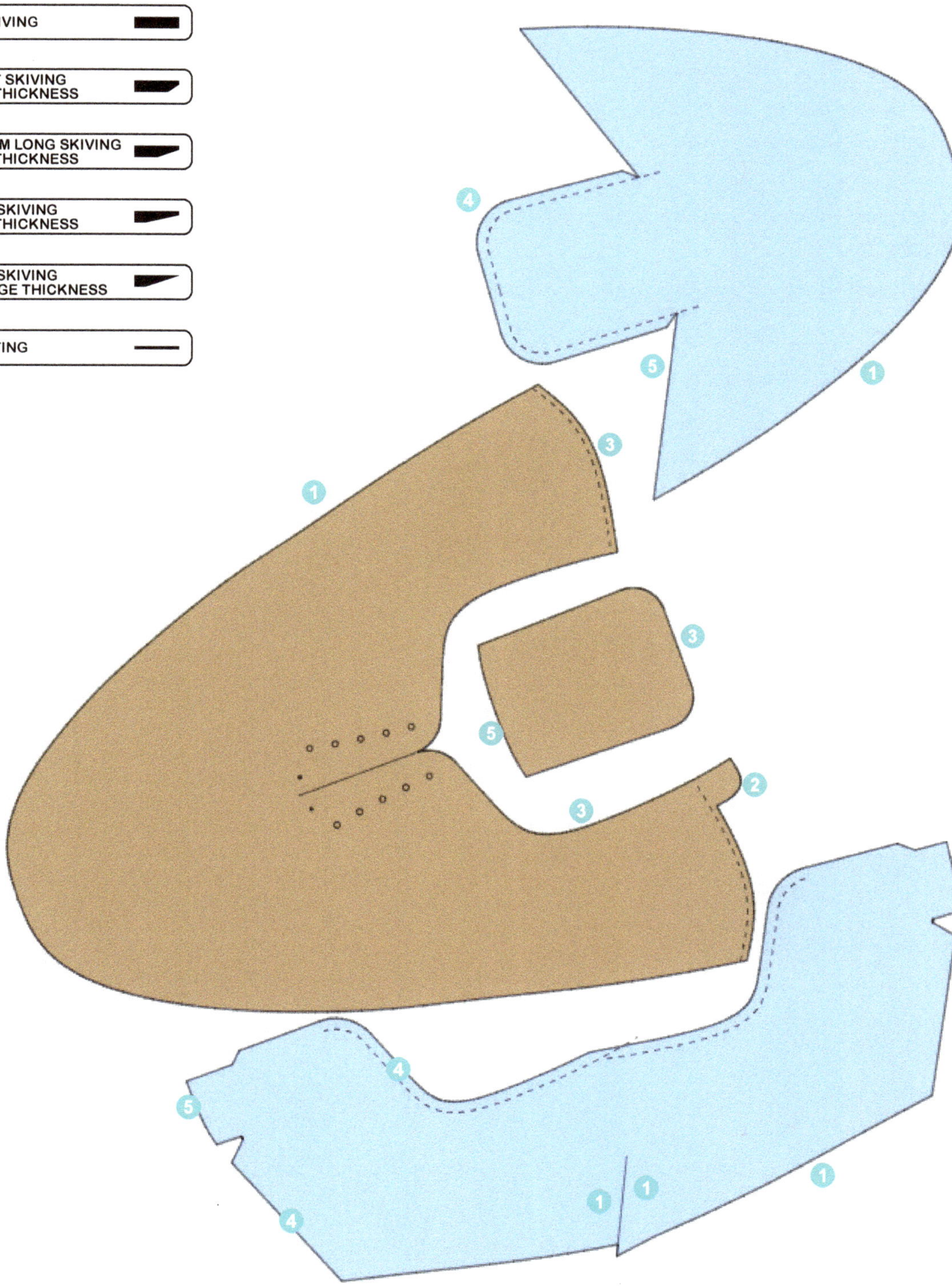

Assembly order

1. Clicking, skiving, marking if necessary
2. Burn the edges
3. Prepare the piping strips
4. Cement the piping to the tongue, attach the vamp lining, and sew them together
5. Reinforce the upper under the laces and the topline; add double-sided tape under the dogtail
6. Sew the backline, then trim and rub
7. Attach the dogtail, reinforce it, and sew
8. Add the piping to the topline of the upper
9. Add double-sided tape to the quarter lining and attach it to the vamp lining
10. Sew the lap seam between the lining pieces
11. Sew the reverse seam, then finish with trimming and hammering
12. Attach the complete lining structure to the upper
13. Sew the topline, then trim the excess lining
14. Punch the holes
15. Sew the reinforcement seam
16. Glue the upper and lining where needed

Piping is the classic edge treatment for the wholecut. To achieve a clean result, one especially useful technique is to lightly burn the leather edge from the flesh side. In addition to removing loose fibers, the heat curls the corners back so the leather hugs the piping more tightly. In the picture, the left side is untreated, while the right side has been flamed, and the difference is clear. Since a folded edge cannot be used on a wholecut, exposed raw edges are unavoidable, and this method makes them more presentable.

Flaming can be used all around the tongue, even if piping is not chosen as the edge treatment.

The piping strips should be thin, and the folding should be done off-center.

This picture clearly shows the asymmetric folding. Around the top of the lacing, a few relief notches help the piping strip follow the curve without distorting the edge or bunching up excess material. Make sure these cuts go roughly halfway to the edge.

Make additional cuts along the strip where the other side of the lacing begins. Without these cuts, the piping strip may push the topline outward, and since it has a skived edge, this can cause a real problem.

Measure enough strip for the tongue. The piping does not need to run from corner to corner, as the ends will not be visible. Skive the ends of the piping. Although the corners here are not very tight, relief notches are still helpful.

A piped edge should look as if the leather has been folded. This means the piping should not be too visible, only a thin strip all around. This results in a clean, elegant look, and the trimmed lining will appear consistent and complementary to the piping.

After the piping is finished, attach the tongue upper to the vamp/tongue lining and sew all the way around, as covered in previous chapters. Trim the lining edge. As usual, pre-shaping the tongue not only helps with the lasting process but also improves comfort.

Adding a piece of double-sided tape is a useful technique for achieving a neat and secure attachment, and it serves as a clean alternative to glue. Applying it while the piece is still flat keeps the process simple. Place the tape with the backing in place and hammer it down to ensure proper adhesion.

As usual, add reinforcements for the lacing. One important detail to keep in mind is that the piping already provides reinforcement along the topline, so the material should not be placed as close to the edge as with a raw edge. A small margin should be left to accommodate the piping.

The topline is one of those areas where adding extra reinforcement is essential on a wholecut, especially with this leather. That said, the bottom of the lacing is even more critical. This strip of tape absorbs part of the tearing forces, helping prevent the leather from ripping.

Sew the topline. Always start the stitch from the top, regardless of whether it is the right or left shoe, beginning just under the dogtail. Leaving half a stitch of clearance under the dogtail ensures it will turn cleanly and neatly to the other side.

Trimming the backseam is very similar to the derby shoe upper discussed in the previous chapter. The difference here is the use of alligator leather. When selecting the placement of the upper on the hide, alligator leather is not particularly weak or vulnerable, at least when the cutting area can be chosen freely. In this case, there is little control over the back section of the hide. It is likely taken from a softer, weaker side area; therefore, extra attention must be paid to reinforcement, even leaving slightly more excess for the reverse seam.

Add backline tape over the reverse-seam excess. Leaving a small amount of extra at the bottom and turning it to the grain side is highly recommended.

Remove the backing layer from the double-sided tape and pull it against the medial side until the top edges line up perfectly. Check this from the other side as well.

By default, the weakest part of a shoe –is the back area around the top of the counter. This is also a highly vulnerable zone due to delasting, wear, and repeated entry and exit (ideally with the use of a shoehorn). Extra reinforcement is therefore required here. Turn the upper to the correct position and add a piece of reinforcement tape. Placement is critical; otherwise, the tape will not be tight enough and will fail to serve its purpose. Make sure the dogtail is covered from the inside and keep a safe distance from the piping edge. Leaving this narrow strip unreinforced is actually safer for delasting than making the area completely rigid and unable to stretch.

Execute the sewing process starting from the top. On the left shoe, begin at the dogtail; on the right shoe, start with a "stitch in the ditch," directly against the dogtail. As an optional and often preferred step, stitching on both sides of the backseam can be added. This further reinforces the backline and is highly recommended in this case.

Hammer the backseam. This flattens the reverse seam further and makes the stitches look neater. With this step, the upper is ready for binding. This is the last opportunity to lightly flame the topline or dye the edge. In that case, an alcohol-based marker is a good choice. Work very carefully, as alcohol-based dyes create deep stains that are very difficult to remove.

Add cement to the topline. Here is a useful tip for this step: even if a high-quality rubber cement (upper glue) is available, or double-sided tape is preferred–which is also a valid option–the most reliable way to achieve a neat piped edge is to use the strongest adhesive in the workshop: sole glue. The reason is simple: if the layers delaminate during sewing, achieving a clean result becomes very difficult.

Start the binding so the strip slightly extends under the end of the lacing slit. This may feel risky, as the placement might not appear perfect at first, but there is a simple way to manage it. Do not press it down firmly right away. Sole glue bonds with very light contact. This allows the upper to be turned over and checked from the grain side. If the position looks correct, press it down firmly to create a secure starting point for the process.

Use the dominant hand to press the strip and the topline together, and the other hand to guide the strip. The goal is to keep the strip roughly in position and prevent the layers from touching before the correct moment; otherwise, they must be separated, and sole glue might not be well suited for repeated repositioning. The strip can be guided outward, which greatly reduces the risk of unwanted adhesion.

At the corners, the process becomes slightly more challenging. Do not try to press the strip perfectly flat. Even with relief notches, the strip will still tend to curve away. An effective method is to let the edges touch only where the piping is intended to sit. Press a short section before and after this point together with your fingers so the correct amount of piping is available for the corner. From there, simply adjust the piping and proceed.

The piping should appear as a very thin edge, with the raw edge either barely visible or not visible at all. (This guideline is not absolute, and alternative designs are possible.)

Make sure there is no gap between the piping and the topline and that the bond is firm; otherwise, the sewing machine may separate the layers further, resulting in a poor finish.

Add double-sided tape for the lining lap seam between the vamp and the quarter. This is the cleanest method, although liquid glue also works well in this case.

Mark the overlap placement on the vamp lining with a scratch awl. A few points are sufficient.

Remove the backing layer and place the quarter onto the vamp. Make sure there is enough allowance between the two pieces.

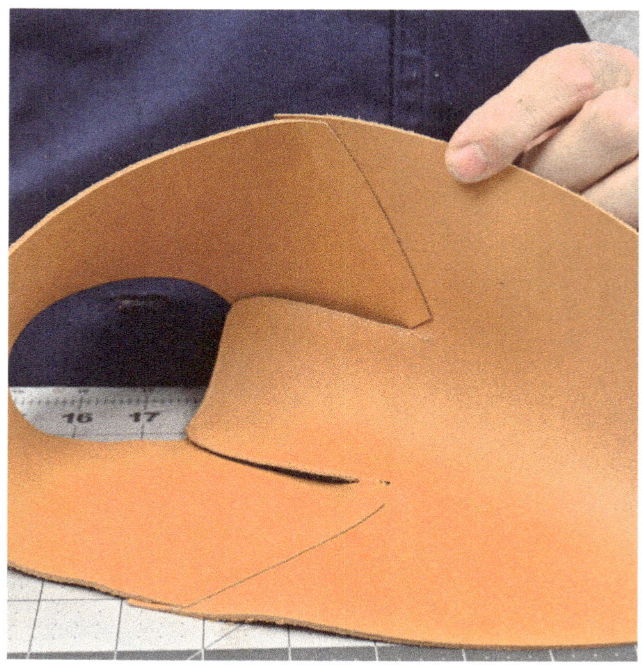

The lining is assembled and ready for sewing. The main advantage of this method is that it provides very good access to the seams, making it particularly suitable for wholecuts.

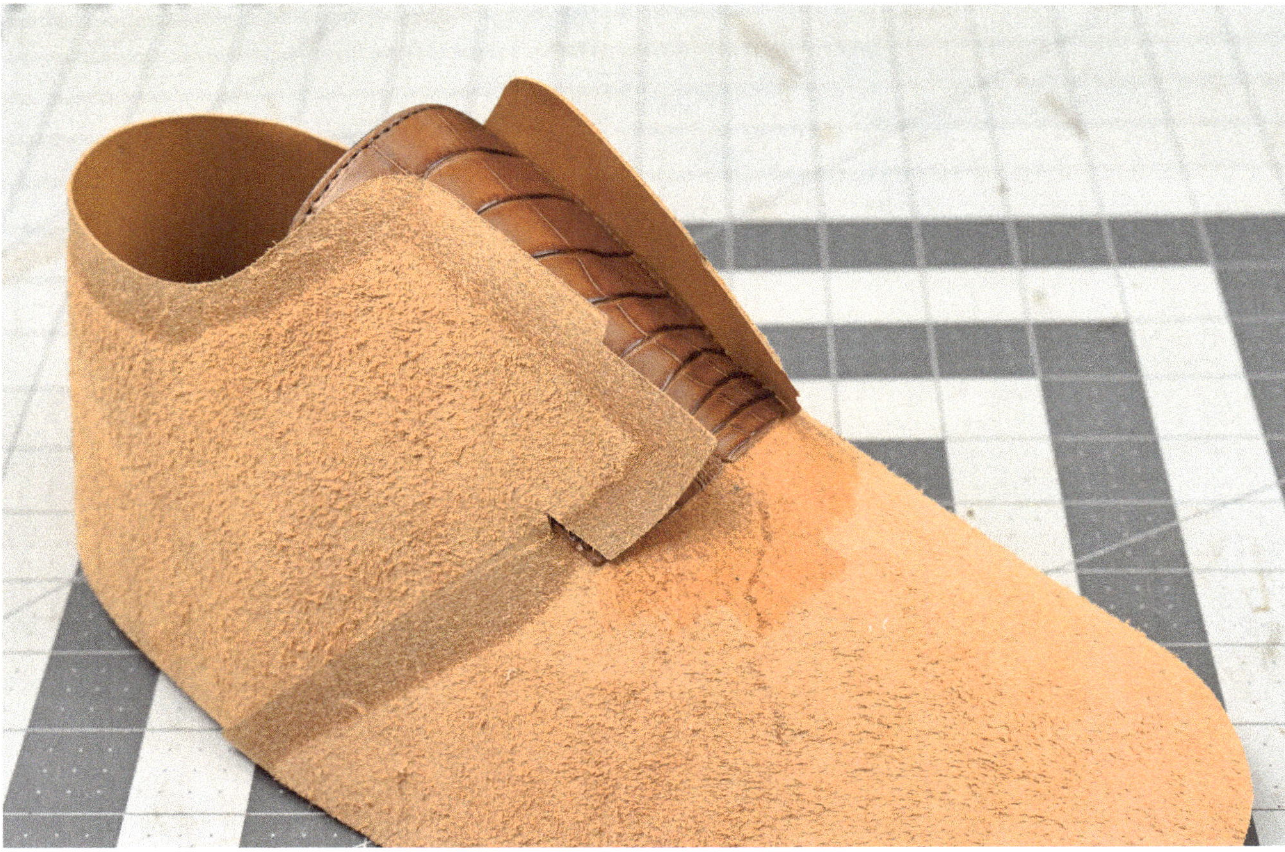

This is how the lining structure looks from the top side. Interestingly, it closely resembles a derby upper, which is no coincidence: this type of lining structure can be used for most classic laced footwear patterns, including derby, blucher, Norwegian, oxford, balmoral, and sneakers.

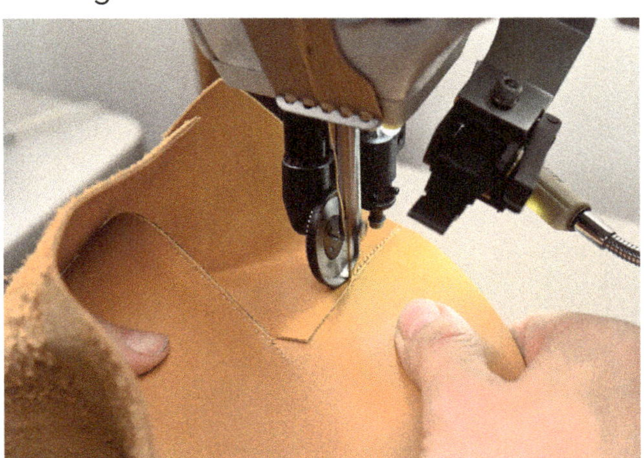

Sew the lining in the usual way, starting at the lasted edge. At the corner, the seam can either be stitched down or left to be held by the straight seam.

Sew very close to the edge; otherwise, constant foot movement will cause the edge to curl back and wear prematurely.

The reverse seam at the back can be sewn and finished either at this point or before the complete lining assembly. In practice, it is easier to do it at this stage. Make sure there is a generous amount of backstitching at both ends.

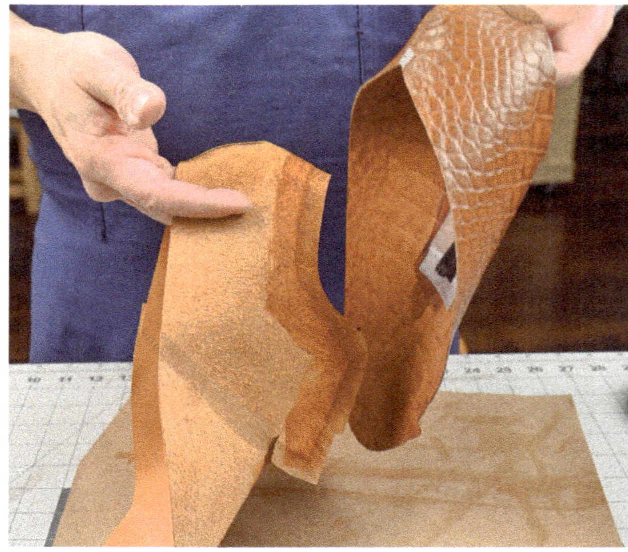

For attaching the lining to the upper, upper cement is probably the best choice. Start from the center of the counter and make sure the pieces line up very accurately. Also ensure that the correct amount of allowance is left throughout, as even a small inaccuracy can lead to significant misplacement at the front.

It can be useful to mark the center point with a silver pen or a small point mark on the grain side of the lining leather. Accuracy is critical here, as even a minor misplacement can affect the allowance toward the vamp lining.

When reaching the lacing, regardless of which side is being worked on, excess allowance indicates a problem with the pattern or the placement. If the 5 mm excess is consistent, the placement is good.

The upper has reached its final structure, although only a temporary bond is present. Check everything again: make sure the tongue is centered and positioned neither too high nor too low. Also ensure there are no gaps between the layers–any small gap can be pushed forward by the roller foot and quickly turn into a major issue.

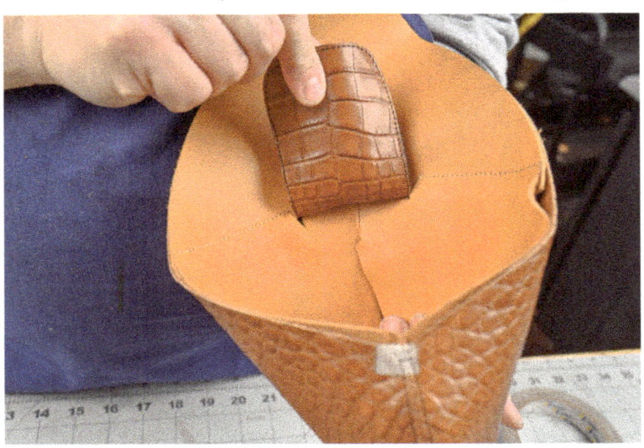

Pulling the tongue away from the lacing stitch is key to success with this assembly order. Check that this is possible. The machine must be able to reach the beginning of the slit.

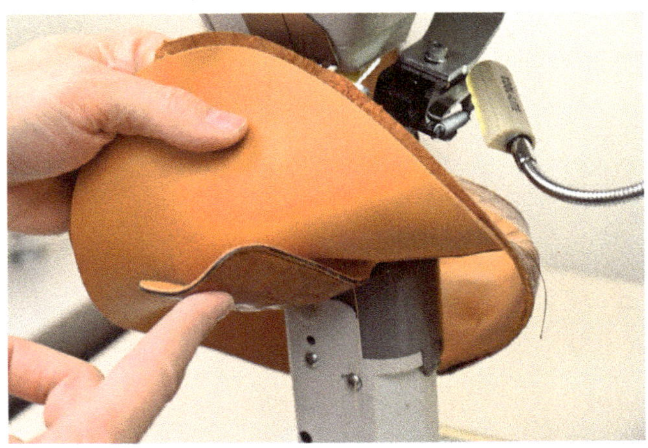

When starting the stitch, the tongue should not pass under the machine at any point. Make sure it is kept pulled away without damaging the lining.

Sew the topline all the way around. Stay close to the edge; otherwise, trimming the lining becomes difficult and the exposed upper leather edge may delaminate over time. The machine should reach the very bottom of the lacing and return to the same point without interruption. This is not easy, even with a perfect pattern, but it is entirely possible with careful handling. This demanding step is often the reason some makers choose alternative methods.

If the piping is narrow and the stitching runs close to the edge, the trimming process should not be very difficult. Use the corner of the workbench if possible, as it provides very good access. Some makers prefer to use curved scissors for this task.

Punch the holes for the lacing, again using the corner of the workbench. The tongue must be pulled away, just as during sewing. The main challenge is usually the bottom one or two holes.

The holes are done. The shoe can be left as is or fitted with hidden eyelets. Traditionally, oxfords and wholecuts use five pairs of eyelets, but creative concepts may dictate otherwise.

One of the final steps is to sew the reinforcement stitch at the bottom of the slit. Ideally, this should cover both the beginning and the end of the topstitch. Use the method discussed in the Balmoral chapter.

Under the tongue, there is no bond between the lining and the upper structure. This is typical for this type of construction. Unlike oxford-style patterns, where an additional stitch line can be added to secure the tongue, this is not an option here, as the goal is to keep the upper as clean as possible. An extra stitch line across the vamp would not be visually appropriate.

Instead, sole glue is used, and in this case the bond should be as strong as possible. Follow the standard process: apply a thin layer of glue, allow the proper drying time, then press the layers together firmly. Glue placement is critical. Applying it too close to the toe can create problems during lasting, as separating the layers for the toe stiffener may cause damage with this type of adhesive.

The upper is complete, and after a quick lacing it is ready for lasting. At this stage, one additional step is recommended: applying a light layer of rubber cement around the topline. This bond is easy to separate later when space is needed for stiffeners.

The topline may also require light cleaning with crepe rubber and/or a rubbing agent to remove loose fibers and achieve a neat finish.

This step is typically reserved for high-end shoes, but an alligator shoe certainly falls into that category.

11 THE BALMORAL BOOT

The balmoral boot – named after Balmoral Castle in Scotland – was first commissioned by Prince Albert in 19th-century Britain. It soon became the preferred style of the aristocracy, for both men and women. It is essentially a higher-top version of the oxford shoe, making it an elegant and sleek alternative to other, more rugged boot designs, allowing the wearer to move from outdoor to indoor settings without sacrificing sophistication.

Like the oxford shoe, the balmoral boot features a closed lacing, with the vamp overlapping the quarters, although there are some stylistic differences between the two. When assembling the boot, it is possible to follow the same order as for an oxford shoe; however, an alternative method will be demonstrated, one that can also be applied to shoe assembly. It is common for the quarters to feature a different, often contrasting material (suede leather or even textiles) from the vamp. This calls for a different edge treatment - rather than elegant piping or folding, a form of binding is preferred to conceal the raw edges of the material. This provides an opportunity to introduce Italian binding, which is a wider version of French binding.

Adjusting the Geometric Pattern

Boots - even short ones - require a different master pattern. Simply adding a shaft to a shoe pattern is a common mistake and will usually result in an uncomfortable, if not unwearable, boot.

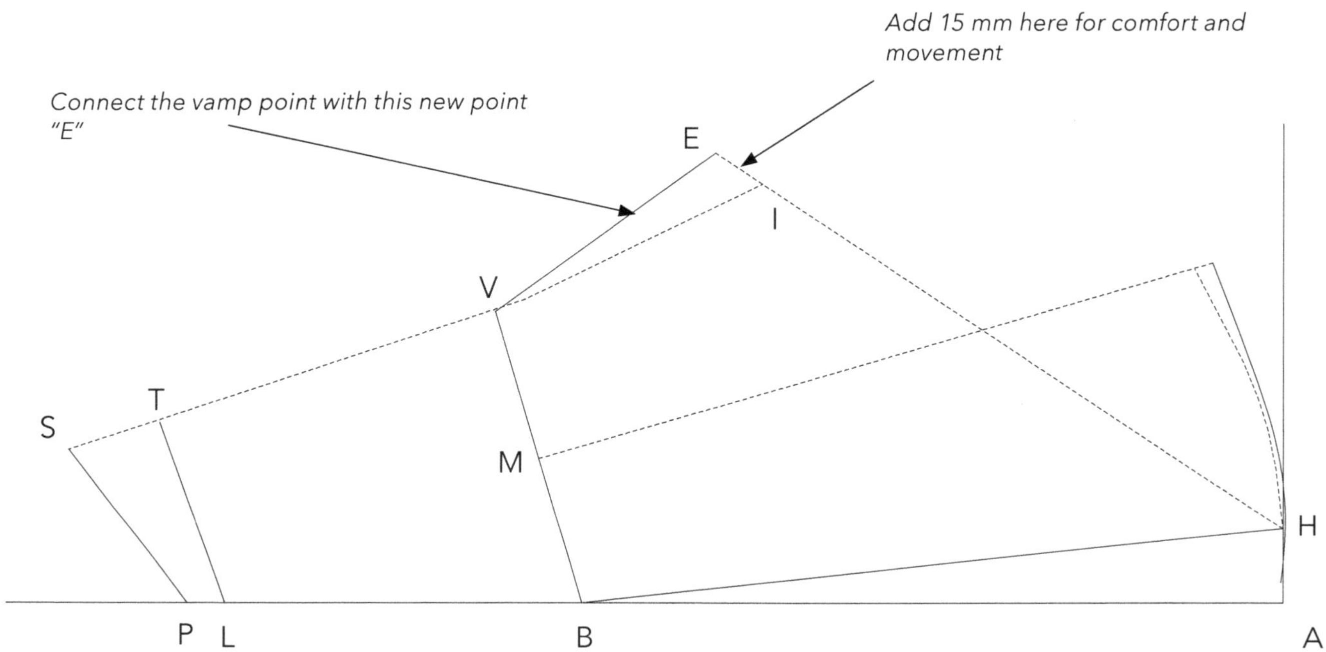

The foot has an active range of motion around the ankle, as this joint moves the foot in multiple directions. This must be accounted for, just like the movement of the tibialis anterior tendon. This tendon lifts and stabilizes the foot during dorsiflexion (the movement in which the foot is pulled upward toward the shin), helping control foot position and preventing the toes from dragging during walking. A low-cut shoe does not need to address this, as it does not cover the tendon and allows it to move freely.

Boots, however, are worn during the same activities – walking, rushing, driving – and these actions rely on the tendon moving freely, especially when pressing pedals and similar actions.

Different boot types with different openings – for example, flexible openings (Chelsea) or lacing systems (such as this one or a derby-style opening) – behave differently, depending also on the material and the intended use of the boot. This is a standard starting point and can, of course, be adjusted as design intent or product testing requires.

The next step is to draw three lines perpendicular to the baseline. The first one starts from "E". The next is drawn from the midpoint of the extended short heel line (the two red lines are the same length). The third line comes from the extended counter line – we add 5 mm here. For the record, this applies to boot models with openings. A slip-on boot pattern would look very different.

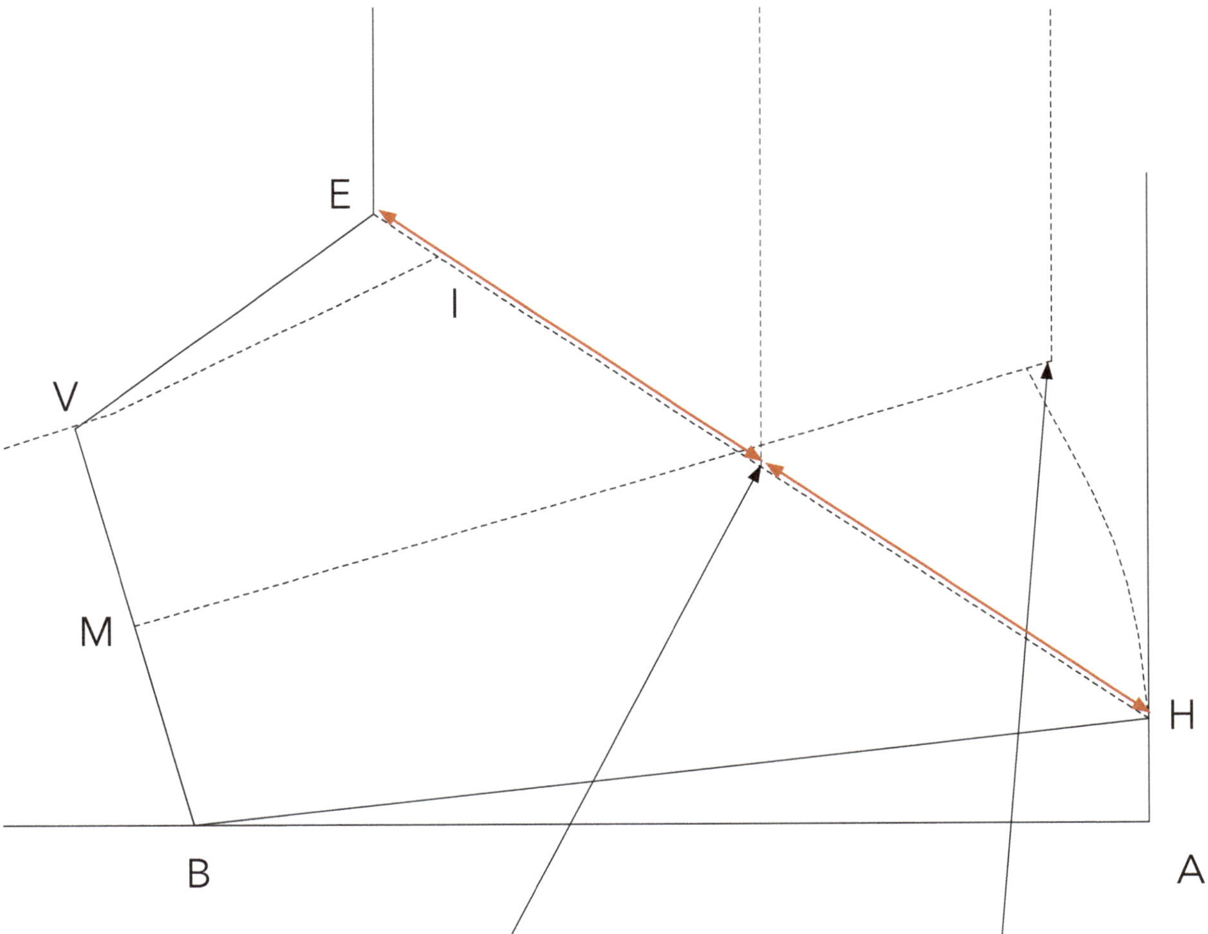

This is the midpoint of the extended short heel line

extend the counter point with 5 mm towards the back

Topline of an Ankle Boot

First and most importantly, there is no universally correct or incorrect height for a short boot. The classic height is based on a geometric principle, making it a traditional solution, but there are also many examples of higher designs, especially in Victorian-era footwear. In the standard method, the height is set at one-third of the SLL, measured upward from the midpoint of the short heel line.

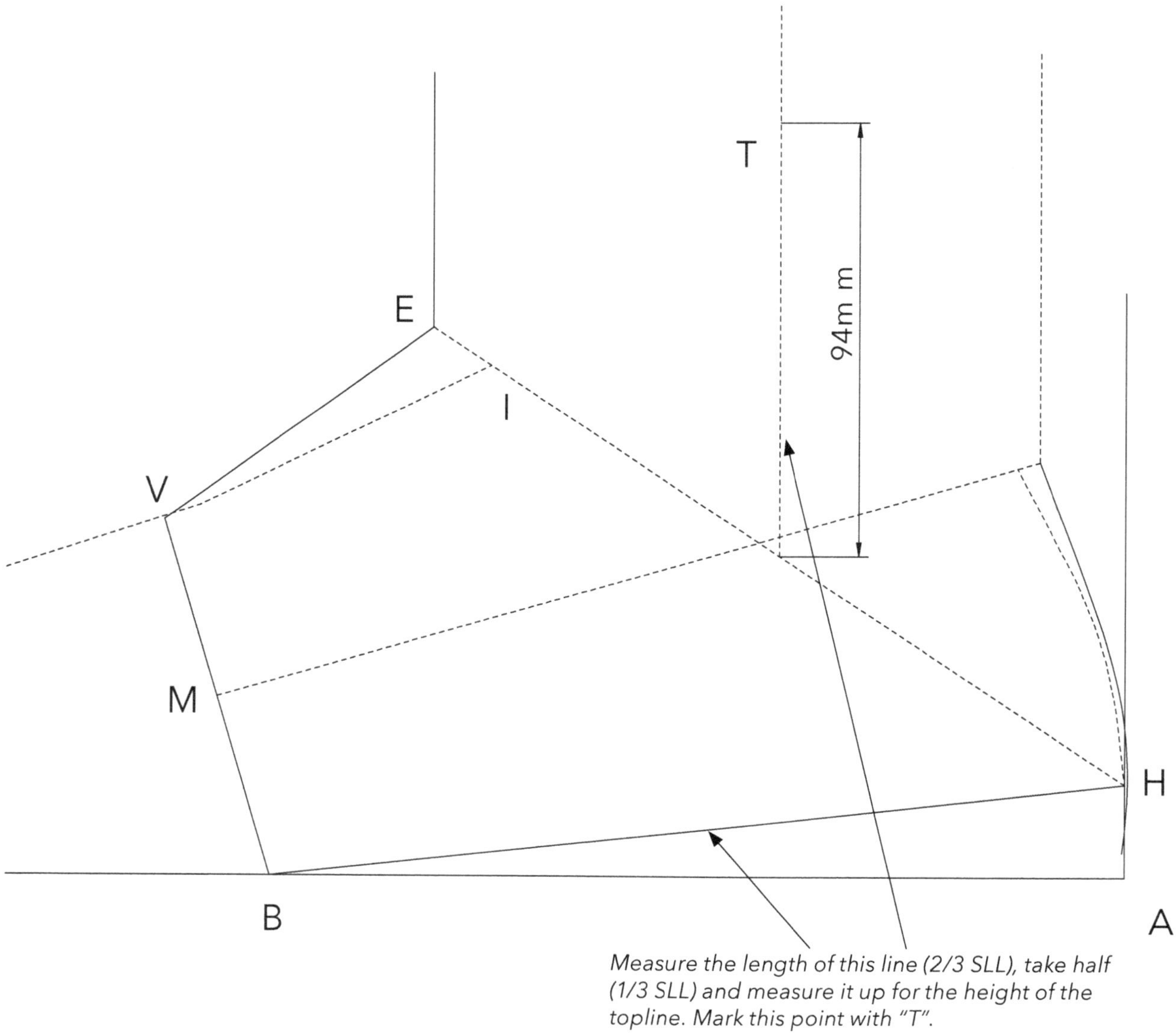

The Topline Angle

The topline is set at an angle of approximately 80-85°, measured from point "T." This angle is only indicative and is used to ensure that the topline meets the backline perpendicularly.

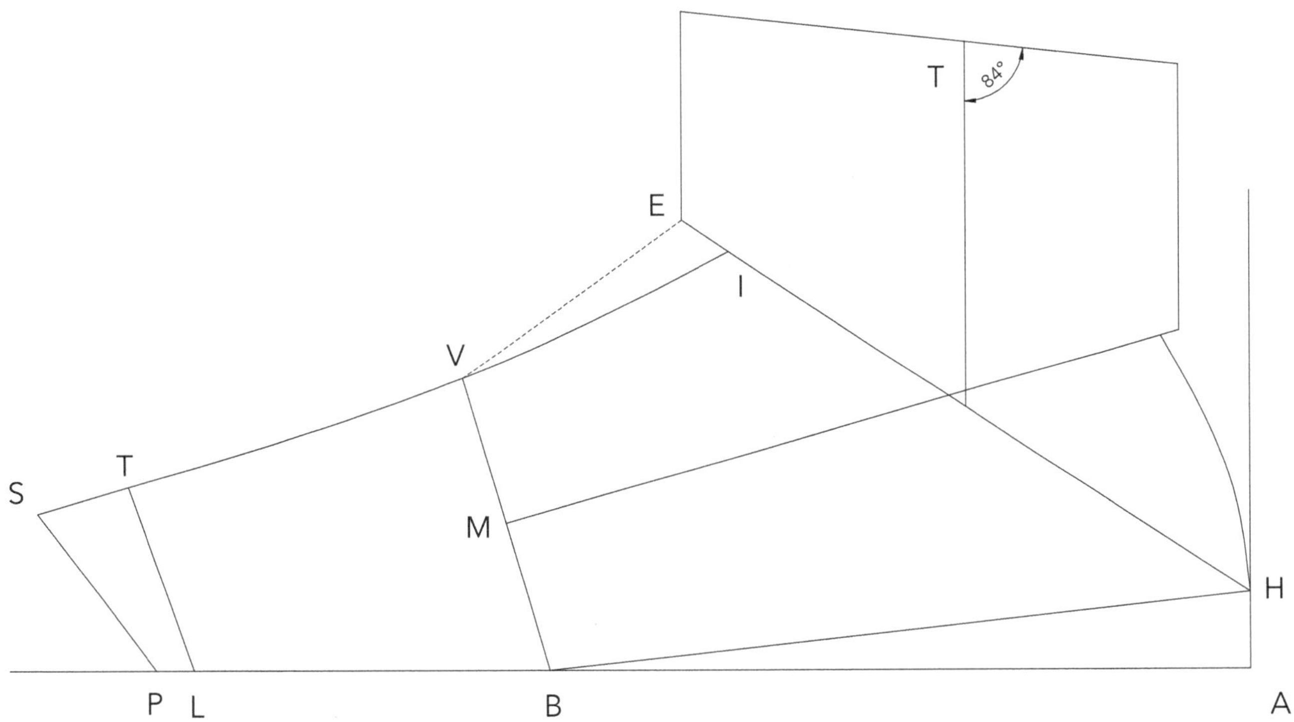

For reference, this is a very classic design line. A true square angle would also function correctly, as would any angle that properly secures the foot. However, since this chapter focuses on a classic balmoral, this specific angle was selected to reflect traditional proportions and line placement as accurately as possible.

Topline Girth

At this stage, the topline length – or the girth of the upper portion – is fine-tuned. This can be set by direct measurement, for example:

$$\frac{\textbf{measured ankle girth} - \textbf{desired opening}}{\textbf{2}}$$

If the desired lacing opening for a balmoral is zero, simply divide the total girth by two. For example, if the girth is 224 mm, the value taken from the vertical line will be 112 mm.

For other boot types, this value is usually proportional to the surrounding measurements. By following the geometric method, the resulting topline length will fall within a range that works well for most wearers.

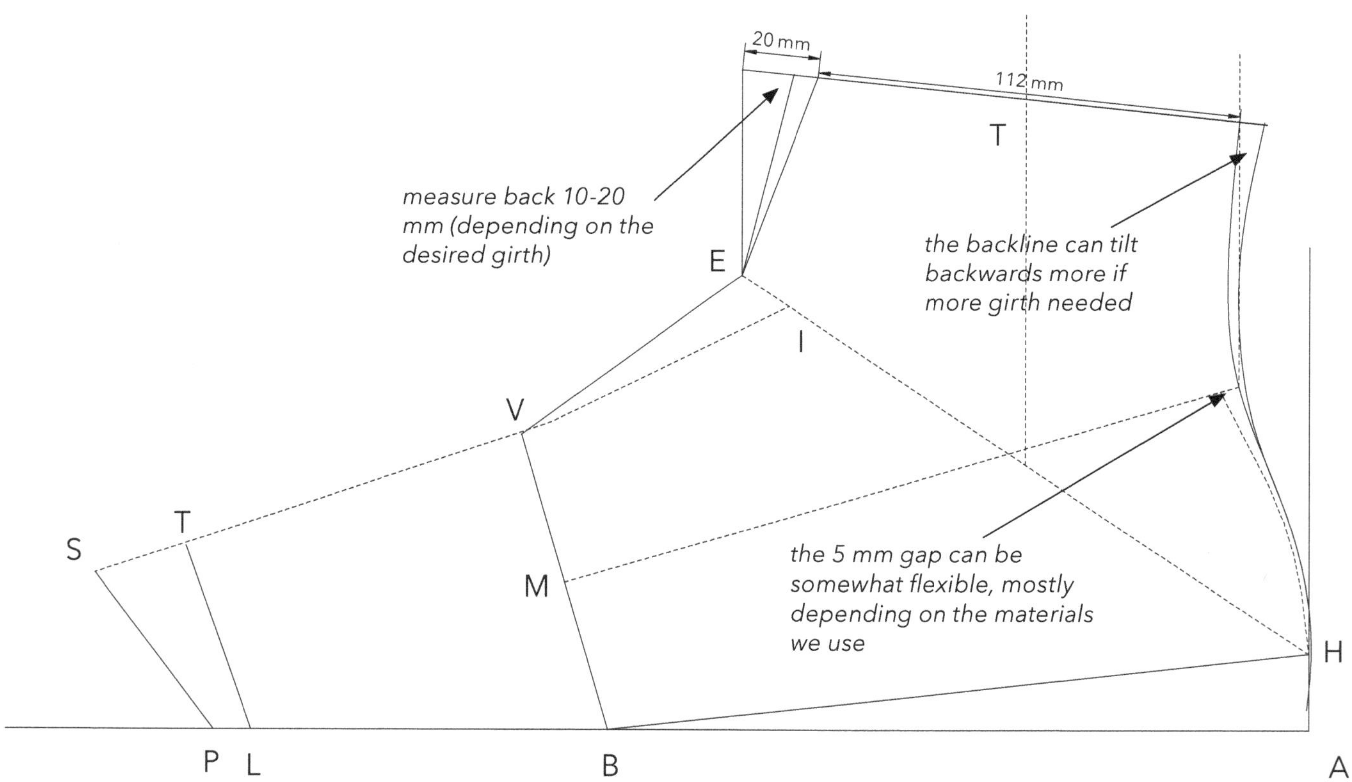

Design and Details

The geometric shape is complete, it's time to add a last copy and begin placing the first design lines. The topline is straightforward: it is essentially a smoothed version of the straight reference line. Keep in mind the planned edge treatment – any corner radius must correspond to the finishing method.

The topline should meet the backline at a perpendicular angle. If it does not, the center of the topline can be raised slightly to correct the angle and achieve the proper intersection.

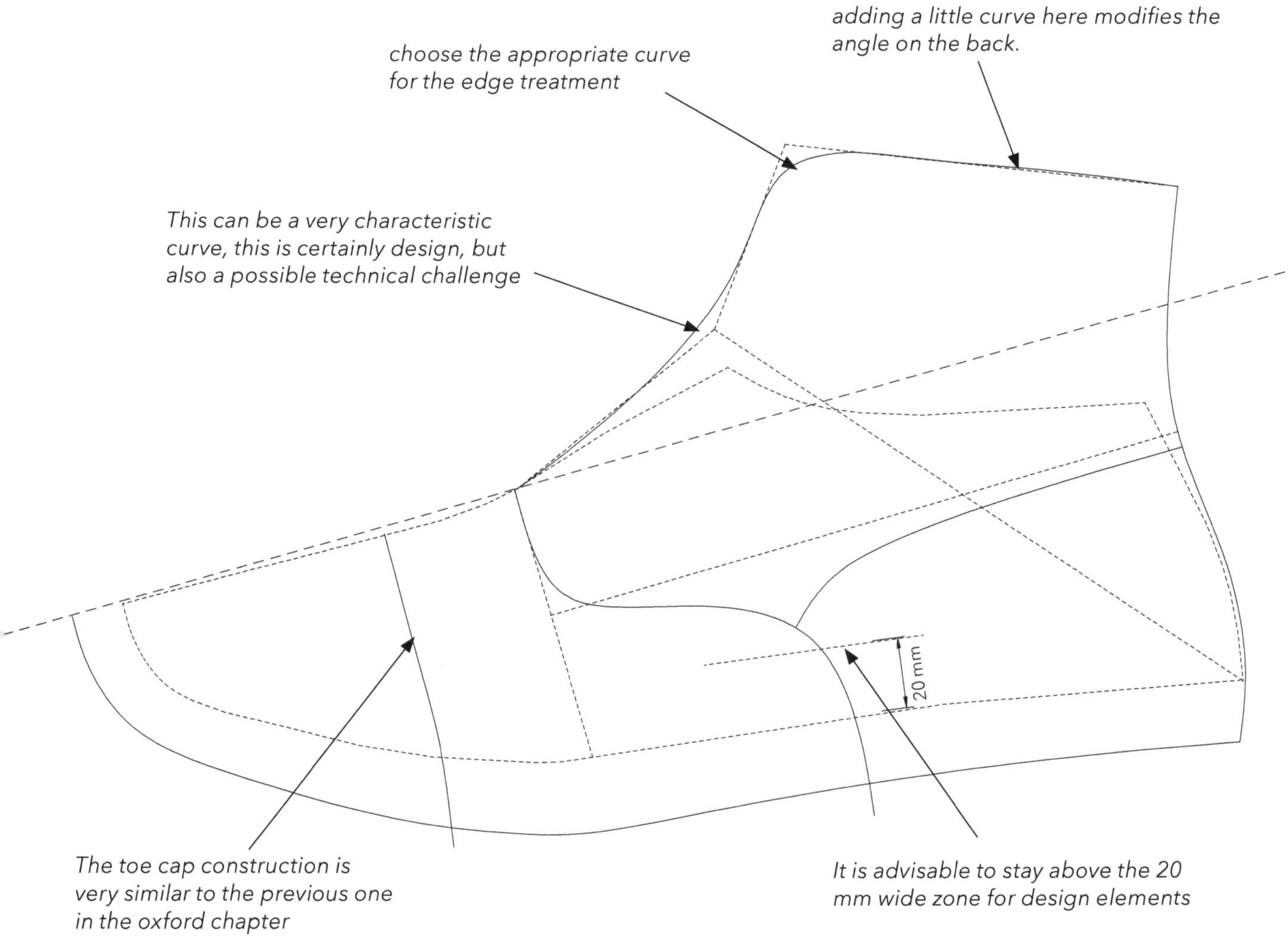

The Tongue

The balmoral tongue is generally straightforward to design; the only consistent challenge is determining the correct length, which is true for most boots built on a similar structure.

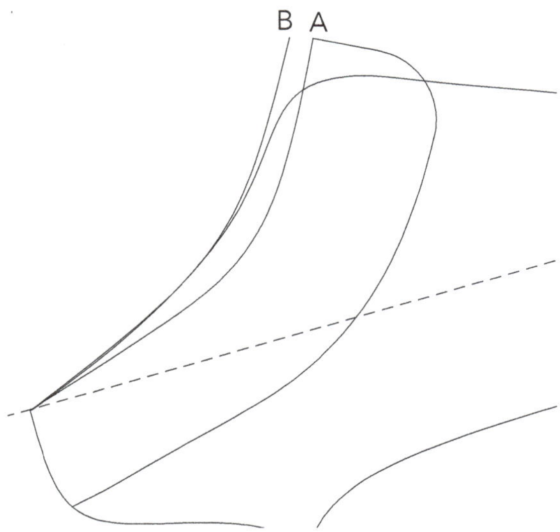

Line B represents the theoretical tongue line if it were to follow the exact front-cone profile of the last. However, this is not possible for two reasons. First, the combined thickness of the upper, lining, and laces creates noticeable bulk, preventing the tongue from sitting precisely on that profile. Second, shaping introduces additional deviation: a tongue must be molded directly on the boot to achieve an ideal contour, and even then it is unlikely to align perfectly with the last's cone line.

Line A is the realistic path the tongue will follow. This line is slightly longer than B, and that extra length must be accounted for in accurate pattern design.

Pivoting the Tongue

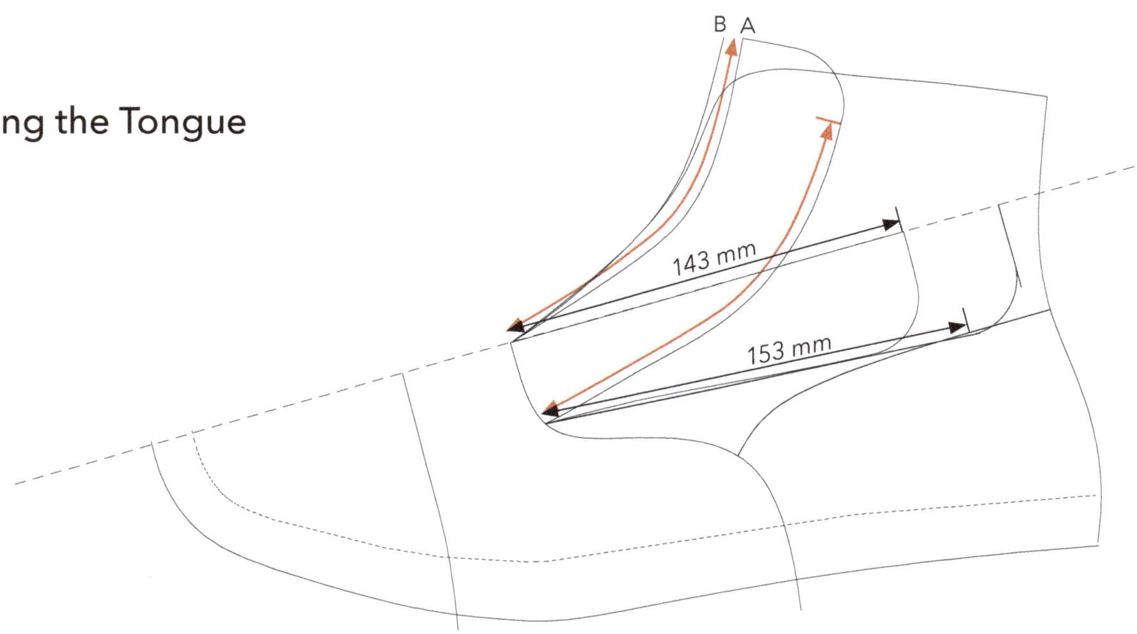

This topic was addressed in the pivoting chapter and appears here directly on the pattern. When flattened on the master pattern, the front edge of the tongue measures approximately 143 mm, while the straight section of the outer edge measures 153 mm. The two pivoted tongue examples demonstrate the two methods: the shorter preserves the 143 mm centerline, while the longer preserves the 153 mm outer edge. In practice, neither line remains fully unchanged, as both require stretching or compression of the leather.

This can be made more visual by placing reference points at every 20 mm along the centerline, dividing the line into approximately seven sections. Using these same distances, the curved line is then flattened onto a straight folding axis. The connecting lines indicate that the same number of points is used, and therefore the measured distances remain identical.

This exercise illustrates pivoting a piece while keeping one line at a constant length, as discussed earlier. The result, however, is incorrect: the tongue will almost certainly be too short. It would only fit properly if the outer line could be stretched without compressing the centerline, which does not occur in practice. Additionally, crimping such a narrow piece would be extremely difficult without damage.

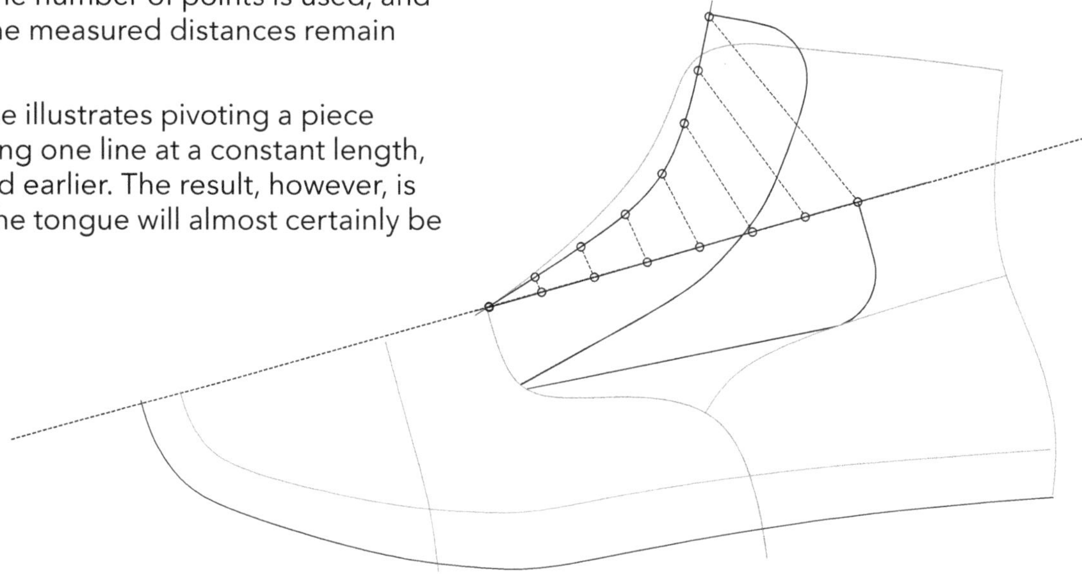

Using the mid-line as the pivot line will slightly elongate the centerline and shorten the outer line – essentially what happens in practice. In reality, the final tongue length depends on multiple factors, so fine-tuning a pattern like this almost always comes down to making a prototype.

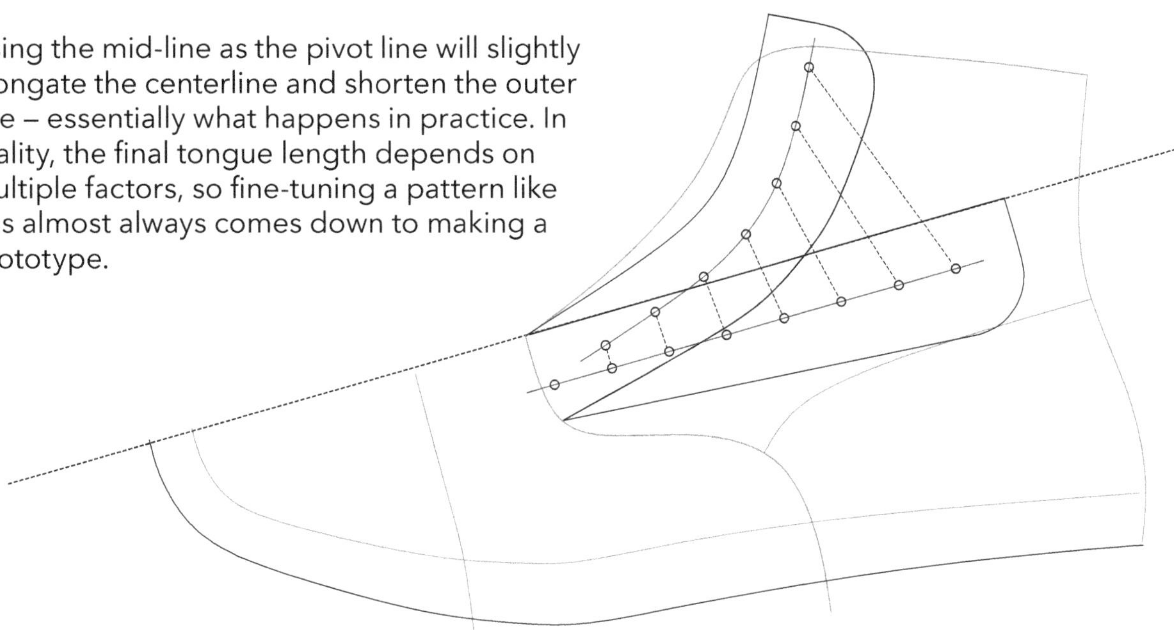

Spacing the Lace Holes

Lace holes must be evenly spaced. In lower-cut shoes, the general preference is for the spacing between the holes to be roughly equal to the distance from the edge. In boots, however, this rule shifts for practical reasons. Even spacing would result in a high number of eyelets, as seen on many vintage boots. More eyelets require a much longer lace (typically 7–10 cm per pair) and significantly more time to lace the boot properly.

Once a boot reaches six to seven pairs of eyelets, there is also a real chance that the laces must be pulled completely out of the top few holes to allow the foot to slip in, which is highly inconvenient. While fast-lacing systems exist, they can look out of place on a classic style such as a balmoral. The simplest and most effective solution is to increase the spacing between the eyelets so that fewer are required. This approach is very common on boots designed with functional eyelets.

The Parallel Shifting Error

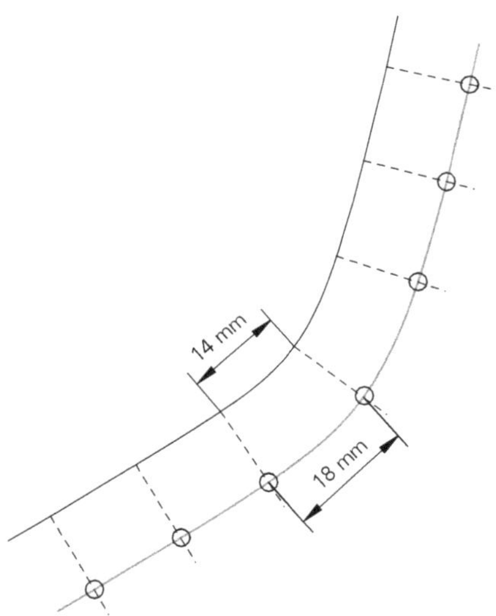

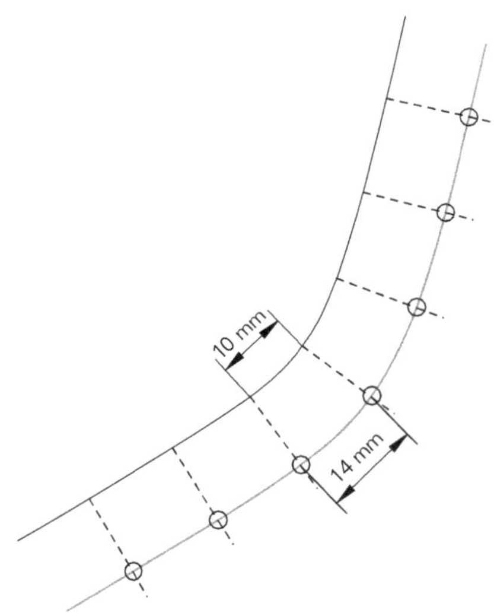

This example shows what happens when the eyelets are spaced by measuring along the edge and then shifted parallel to the intended lace-hole line. In the curved section, the spacing inevitably increases, creating a mismatch. This method works reliably only on straight lines.

The proper way to achieve accurate spacing is to establish the actual lace-hole line first and then measure the distances directly on that line. The projection of these points onto the edge may shift – and this will be visible once the laces sit along that edge – but the spacing itself will be correct. This issue also suggests avoiding overly aggressive curves in the lacing line.

The Starting and Ending Point

Lacing is also part of the visual appearance and must be considered carefully. Avoid placing laces over open areas or too close to the vamp. Even if the lacing appears to start unusually far from the topline, the reasoning is simple: the laces should not cross a gap.

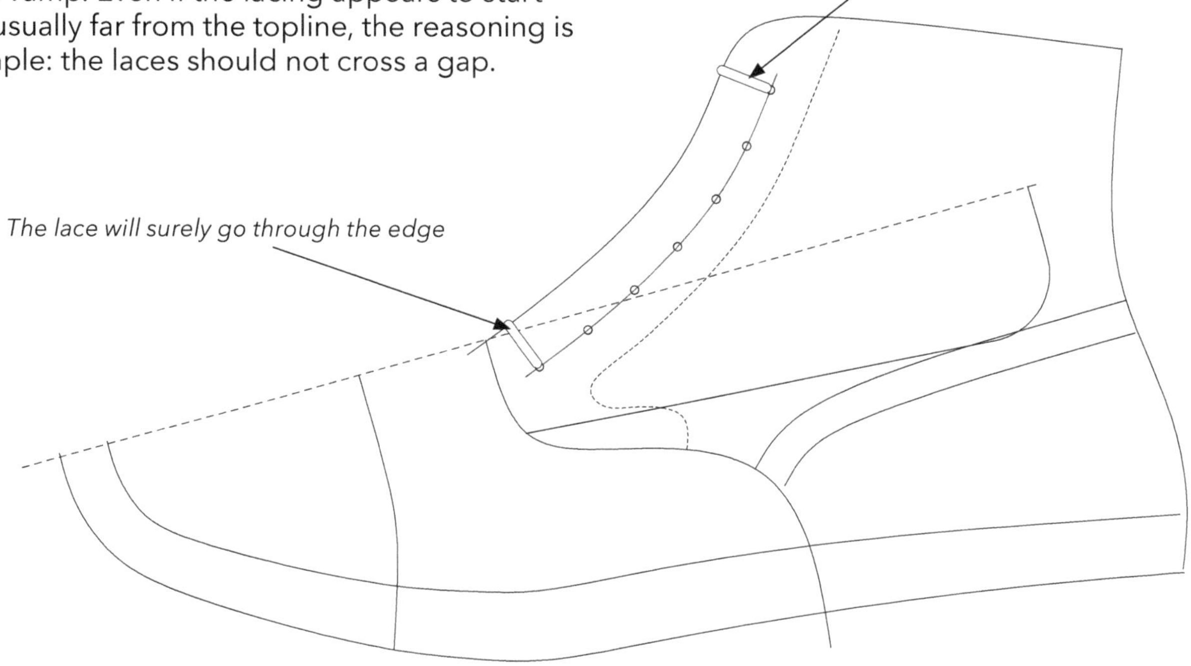

Perception Error With The Lace Holes

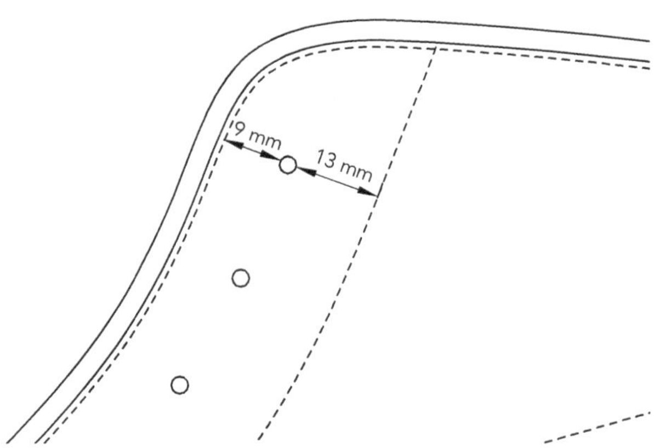

Even when the lace holes sit exactly centered between the edge and the stitch line, they are often perceived as off-center.

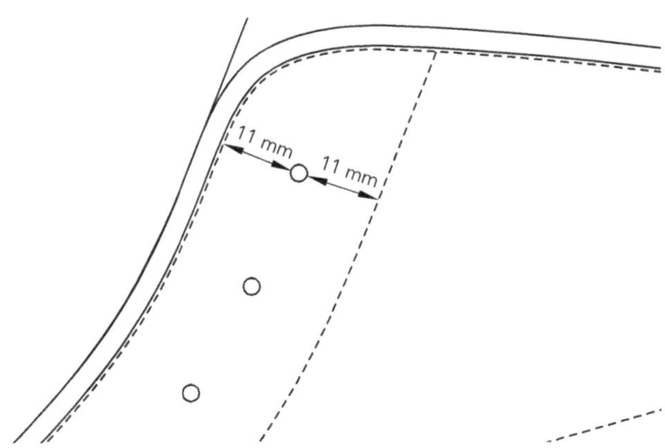

By taking the edge elements into account–the binding and the stitch line – the lace holes can be positioned at the visual center, even though this places them slightly off the technical center.

Stitchline Variations

A balmoral boot is a classic style well suited to showcasing refined stitch design. These are just two traditional examples, but a skilled designer or pattern maker can easily develop additional variations.

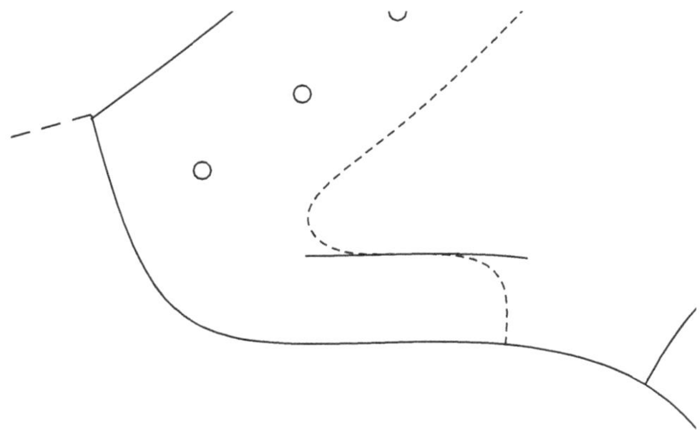

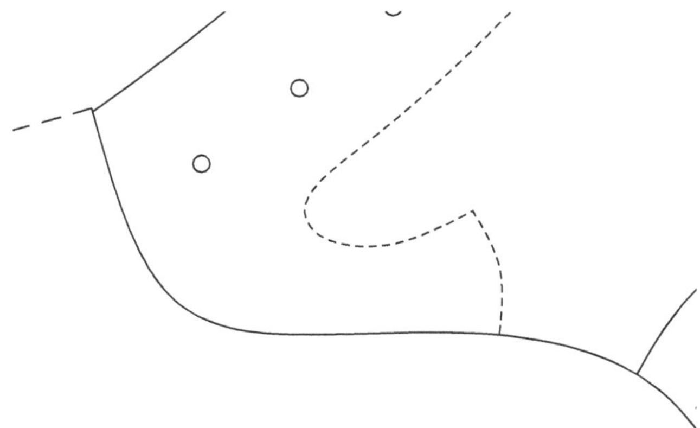

Reducing the Vampline Tension

Since the angle between the lacing line and the vamp folding axis is too steep, unwanted tension will develop in this area. If brogueing were placed along this edge, the issue would become even more serious, as the tension could overstretch the margin and distort the brogueing holes.

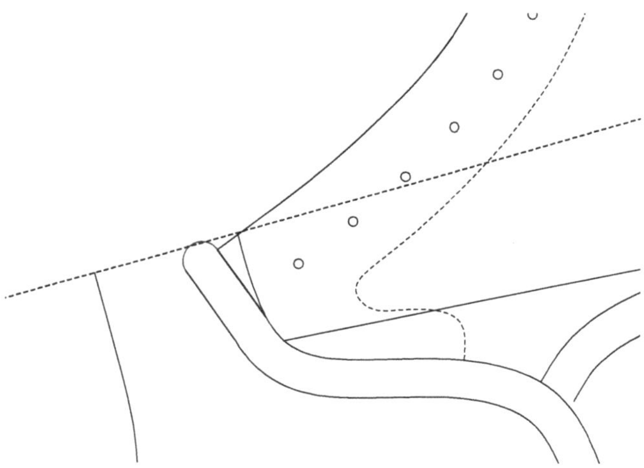

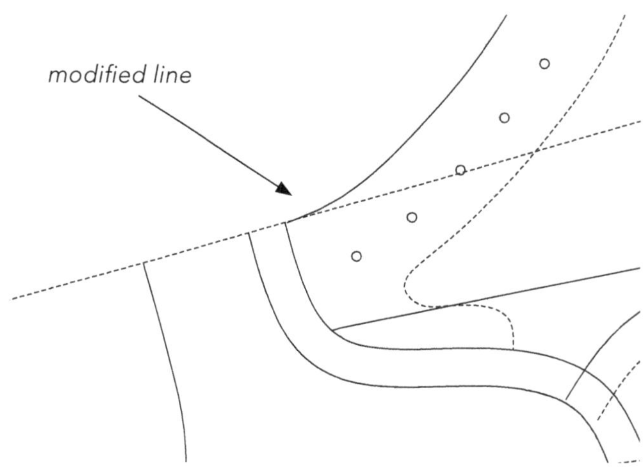

Solution 1: use the same method discussed in the oxford chapter. By extending the quarters, we add a small amount of extra material that compensates for the three-dimensional shape of the foot. This reduces tension and minimizes distortion.

Solution 2: modifying the quarter edge can also reduce tension at the vamp point, since the line already meets the area at the proper angle.

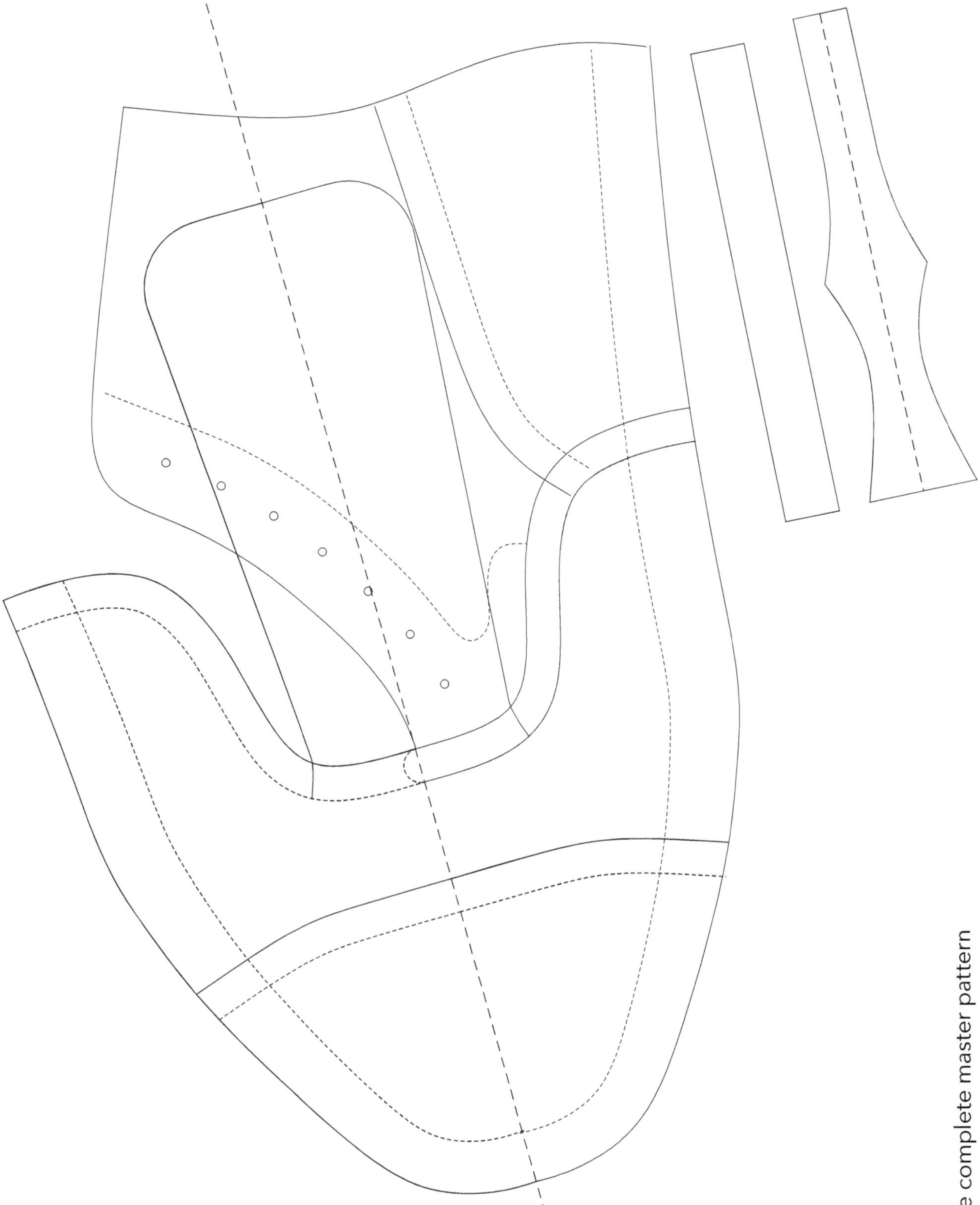

the complete master pattern

Upper pieces

The pattern pieces can now be worked out. The quarter requires overlap allowances along its bottom edges, since both the vamp and the counter overlap this section. The topline treatment is a design decision; because Italian binding is used, no additional allowance is needed here. The backline is joined with a zigzag seam, so there is also no need to add extra allowance along this edge.

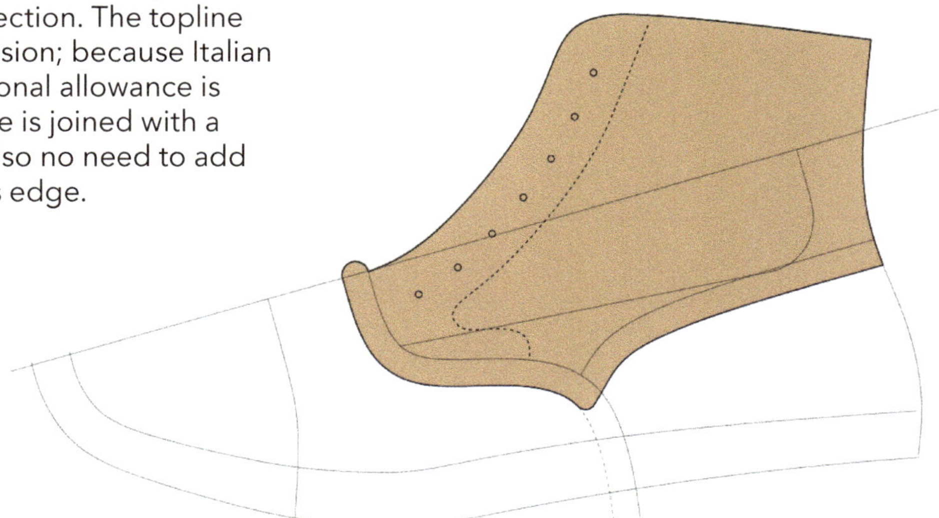

The vamp pieces are very straightforward. The vamp overlaps the quarters. An allowance is added only when an edge treatment other than a raw cut is used (in this example, a raw cut is applied). The front edge includes an overlap allowance so the toe cap can overlap it.

Counter Variations

Two different variations for the counter are discussed. In the first example, the counter ends at the backline and is assembled with a zigzag seam, then covered with a backstrap. This is the version used in the sample assembly.

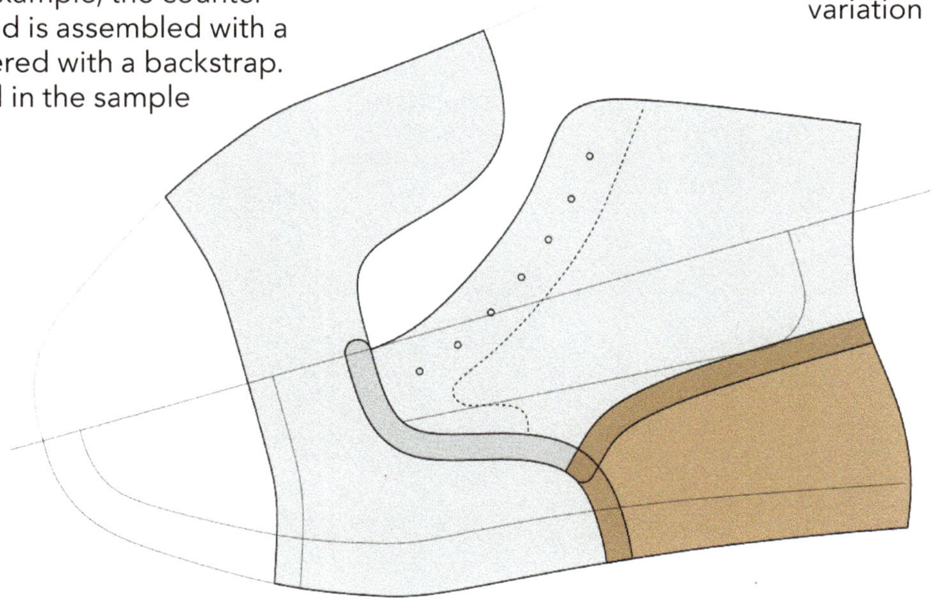

variation 1.

The second variation features a one-piece counter with a small reverse seam at the bottom. Note that the axis is slightly shifted from the backline—by the thickness of the quarter leather. This subtle adjustment keeps the axis visually aligned after lasting, ensuring that the reverse seam sits cleanly without drifting toward the backline. The reverse seam curve can also be adjusted, as it can follow the last curvature better than a backstrap.

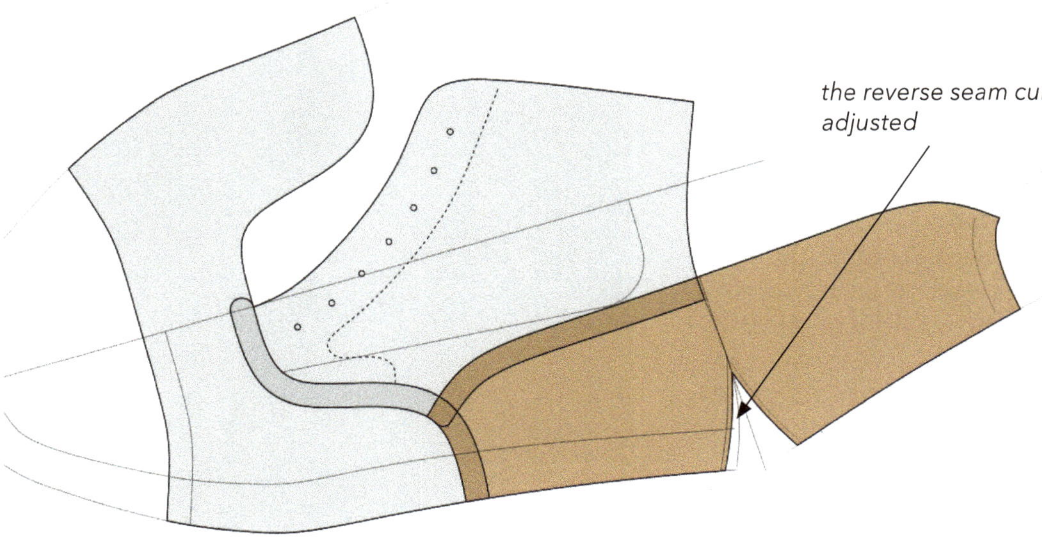

variation 2.

Tongue and the Backstrap

The tongue is ready with the adjusted length. With full closure, the actual length after shaping may depend on other factors, so only a test using the actual or a very similar material can confirm whether this length is appropriate.

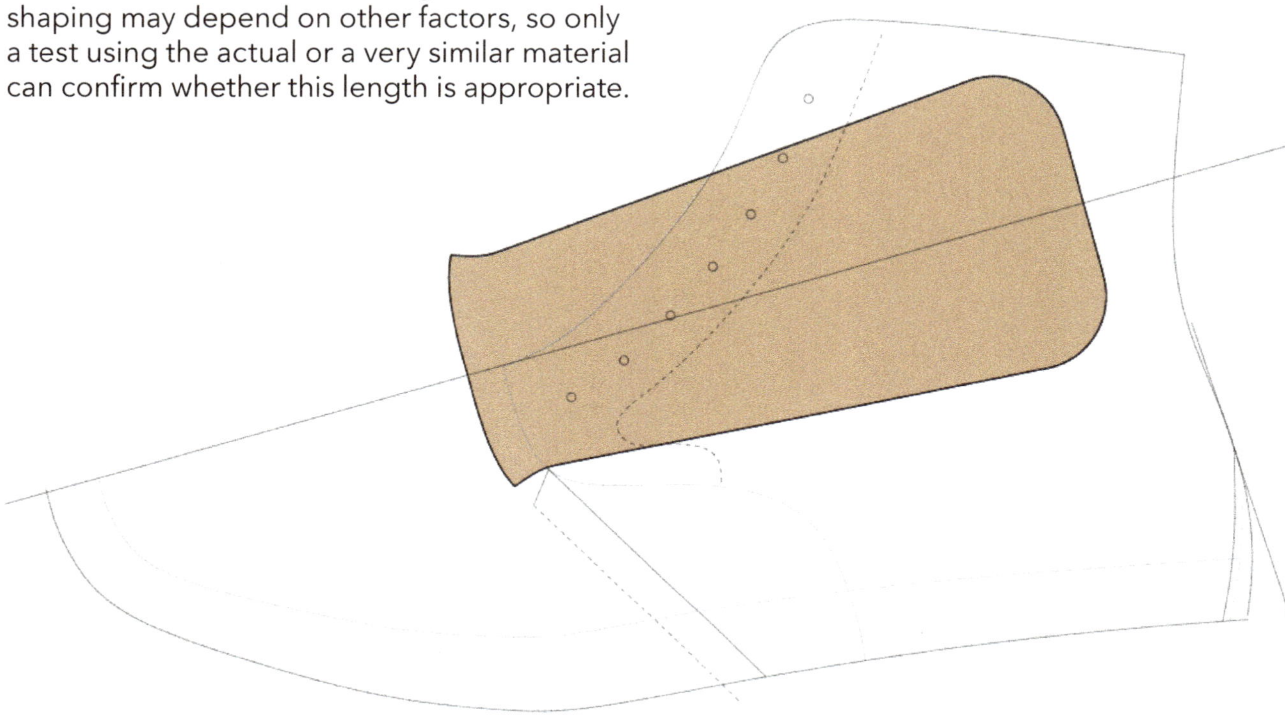

The backstrap can be quite simple, typically 15-20 mm wide, as long as it fully covers the zigzag seam. A straight strap is the most common and technically forgiving option, but it is not the only one. A more classic, decorative shape can also be used; however, this introduces an additional technical challenge. Unlike a straight strap, an ornamental strap must align precisely in its designated position. The corners must land exactly at the edge of the counter, making this variation better suited to an advanced upper maker.

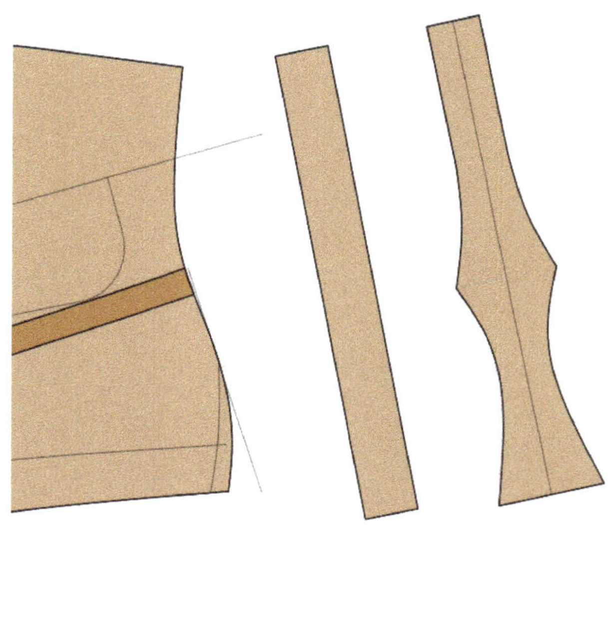

Lining Pieces

The vamp lining includes both the vamp and the tongue, as usual. In this construction, the same axis is used for the tongue relative to the upper structure. If the tongue were cut as a separate piece, the axis could be adjusted independently. Surprisingly, that solution often provides greater accuracy, as the tongue is shaped and its position can be fine-tuned more precisely.

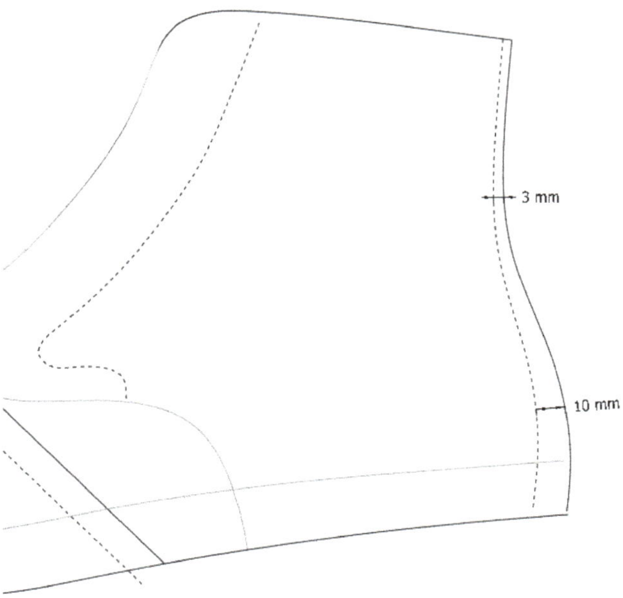

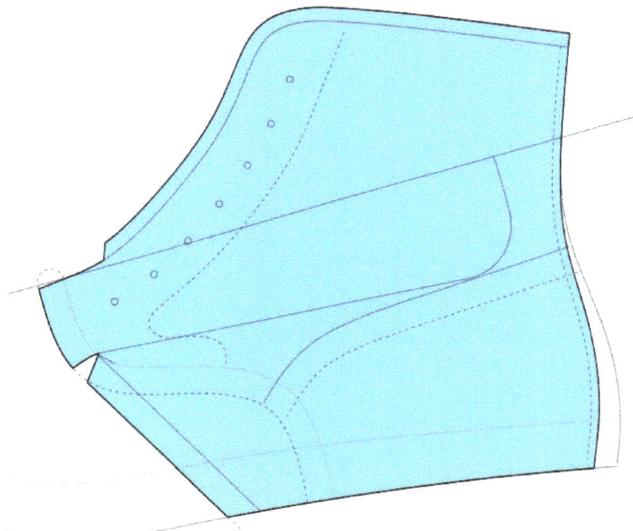

Two different variations for the quarter lining are explored. This is the simpler option, using a reverse seam at the back. The compensation for the orange-peel effect is the same as used on lower-cut shoes: 3 mm at the top and 10 mm at the bottom.

While this solution is very common due to its simplicity, it has one major shortcoming: the bulk of the reverse seam sits directly underneath the backstrap and above the counter stiffener. These layers add up noticeably, both visually and to the touch, which argues strongly against using this method.

The line created here serves as a guideline for the second, more refined solution.

The Two-Axes Quarter Lining Solution

This solution, although slightly more challenging, produces a very clean and refined result: a seamless back lining combined with an additional heel counter piece. The counter piece can be turned suede-side in or simply cut as another grain-side piece. The suede-side-in version can sometimes pull down the sock during wear, so a wear test should determine whether this option is suitable. For completeness, the grain-side-in version is fully functional and avoids this issue.

Two axes are added to follow the backline shape. These axes can reproduce the curve quite accurately across most of the height, except for the bottom few centimeters, where a reverse seam can be used as usual to manage the transition.

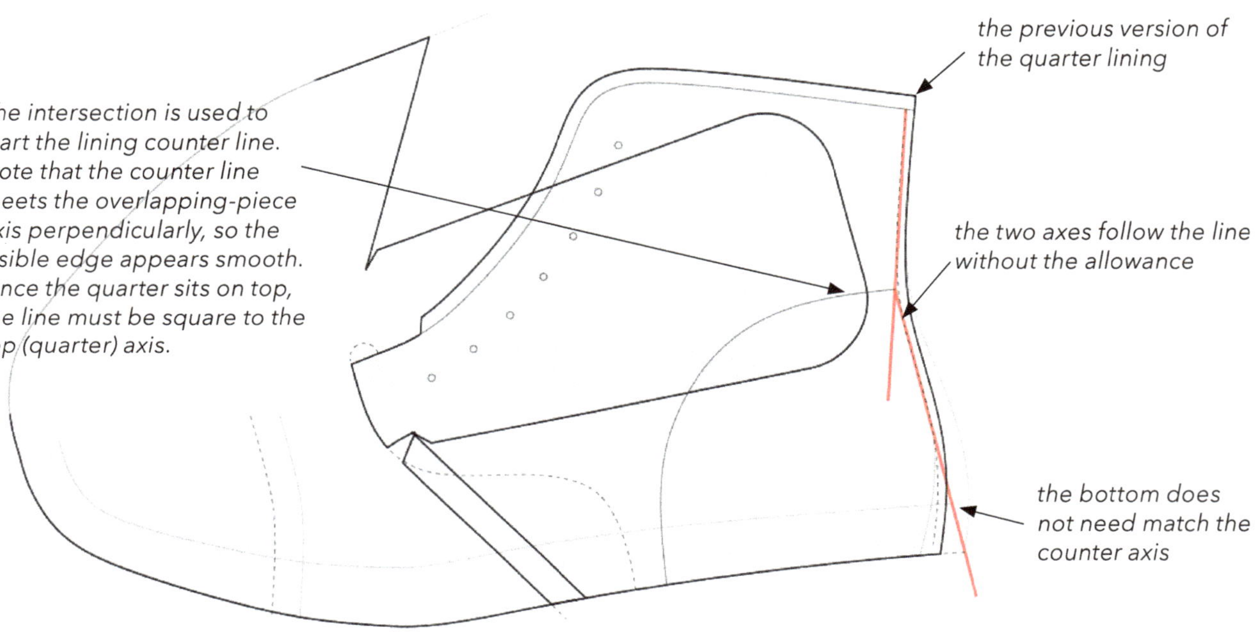

This is how the two pieces look when opened along their own axes. In this flat position, they may appear mismatched, but in three dimensions they will form the shape of the backline accurately once the seam is completed.

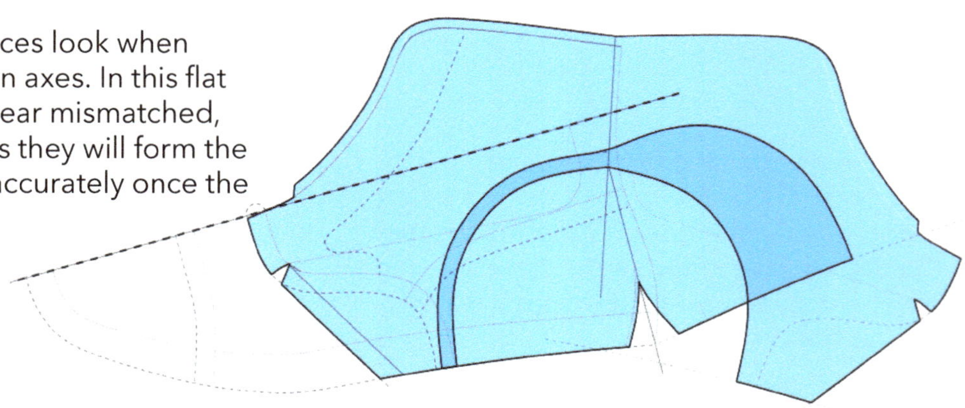

The Complete Pattern

Here is the complete pattern laid out, with the upper and lining pieces separated for clarity.

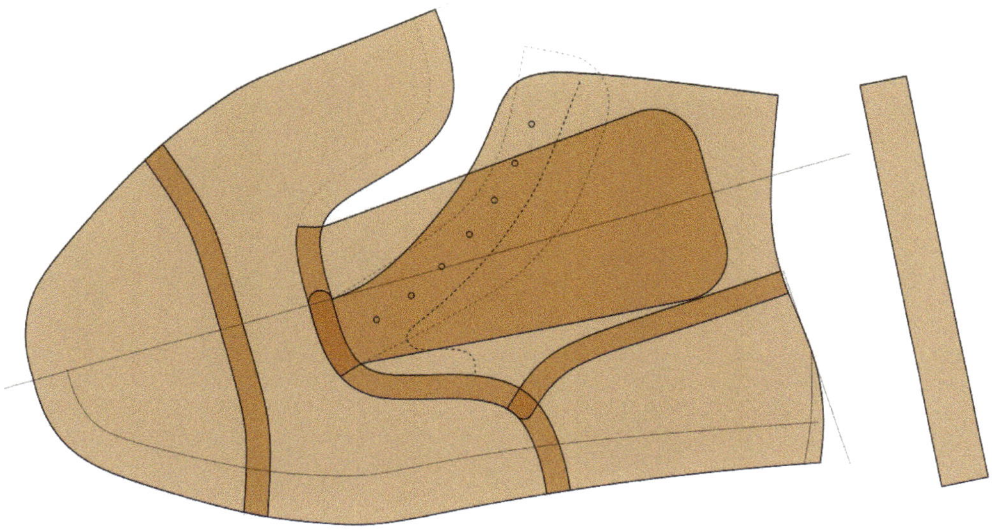

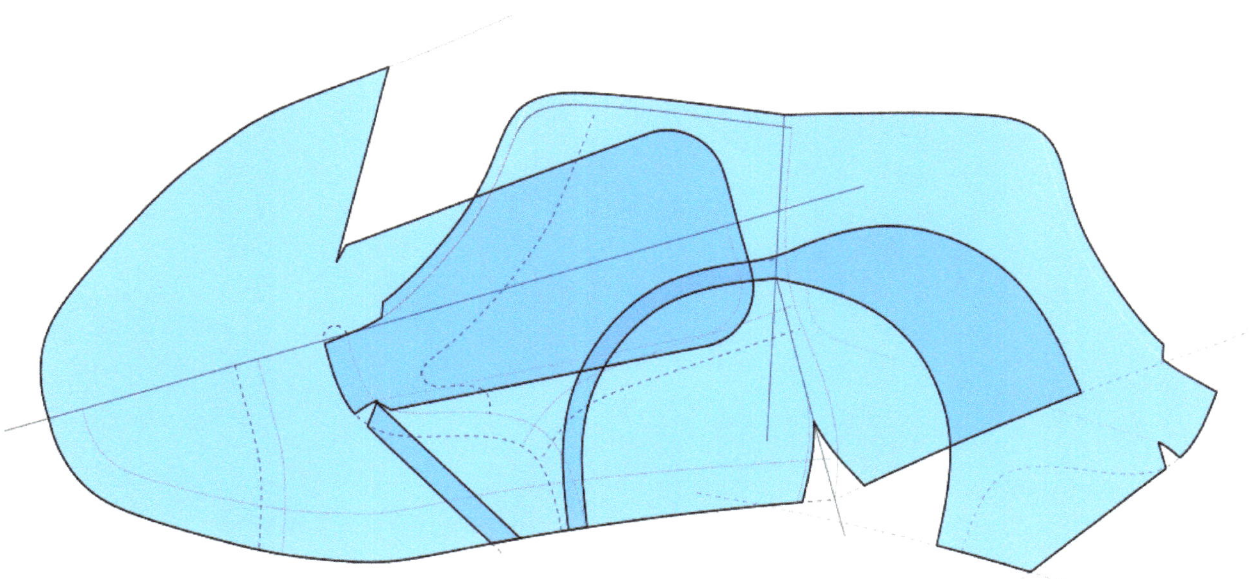

1	NO SKIVING
2	SHORT SKIVING HALF THICKNESS
3	MEDIUM LONG SKIVING HALF THICKNESS
4	LONG SKIVING HALF THICKNESS
5	LONG SKIVING NO EDGE THICKNESS
6	SPLITTING

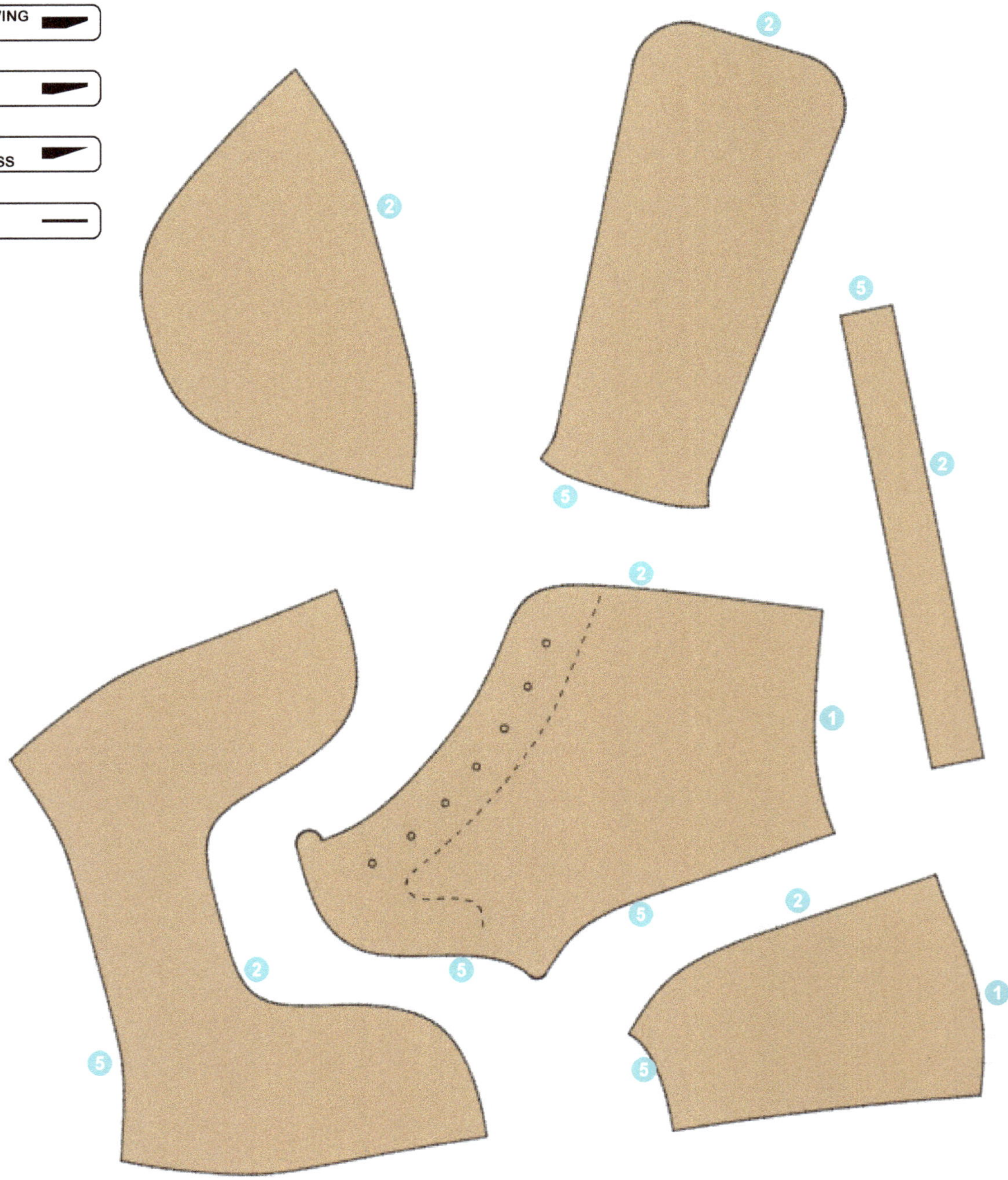

Marcell Mrsan

3
3
5
3
4
2
5
3

Assembly order

1. Clicking, skiving, marking if necessary
2. Sand and dye the edges
3. Cement the counter piece to the quarter
4. Cement the tongue to the vamp lining
5. Cement the toe cap to the vamp
6. Cement the lining quarter to the lining counter
7. Sew the upper counter to the quarter
8. Sew the toe cap to the vamp
9. Sew the tongue topline, then trim the lining allowance around the tongue
10. Sew the lining counter to the lining quarter
11. Shape the tongue
12. Zigzag the quarter panels together
13. Cement the backstrap
14. Sew the backstrap
15. Sew the decorative seam on the quarter
16. Sew the binding strip to the quarters
17. Turn the binding strip
18. Tack the quarter pieces together
19. Cement the vamp upper in place
20. Sew the vamp to the quarters
21. Cement the quarter lining to the upper
22. Sew the topline
23. Trim the lining allowance
24. Add the eyelets
25. Sew the reinforcement seam
26. Cement the vamp lining in place along the lap seams
27. Sew the lap seams
28. Cement the vamp layers together
29. Sew the tongue in

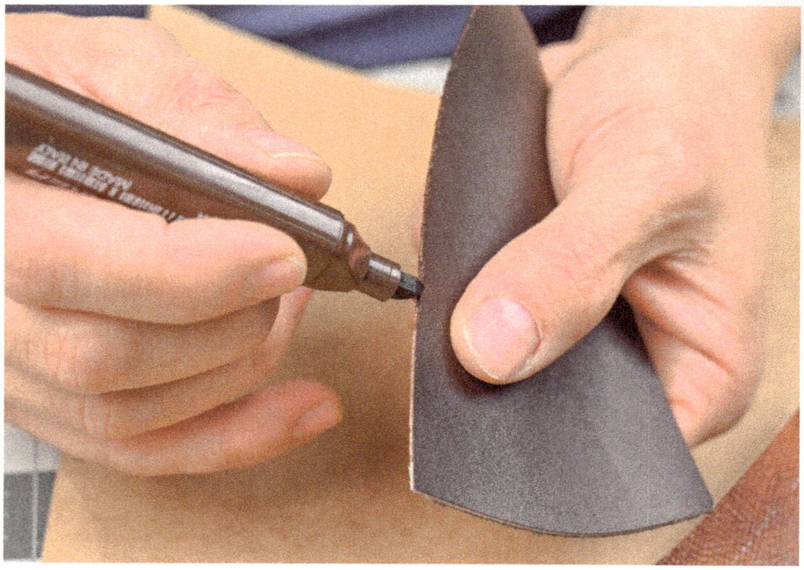

After skiving, a few additional steps can help elevate the upper and the final appearance of the boot. Chrome-tanned leather has a white inner layer because the leather is almost waterproof; therefore, the dye cannot penetrate fully during the tanning process. To eliminate this thin white line, either an edge treatment–such as folding or binding–can be used, or an alcohol-based marker can be applied to dye the edges. A dauber with alcohol dye is also an option, but the risk of drips and irreversible damage makes the marker the safer choice. In addition, lightly burning the edges with the flame of a lighter to create a smooth finish adds a professional touch.

Some leathers are difficult to bond with any cement. Lightly sanding the surface helps create a strong bond between the layers, reducing the chance that they will shift under the sewing machine. It is recommended to use a piece of fine sandpaper and avoid sanding too close to the marked edges. Very light sanding is sufficient, so there is no need to spend time overdoing this step.

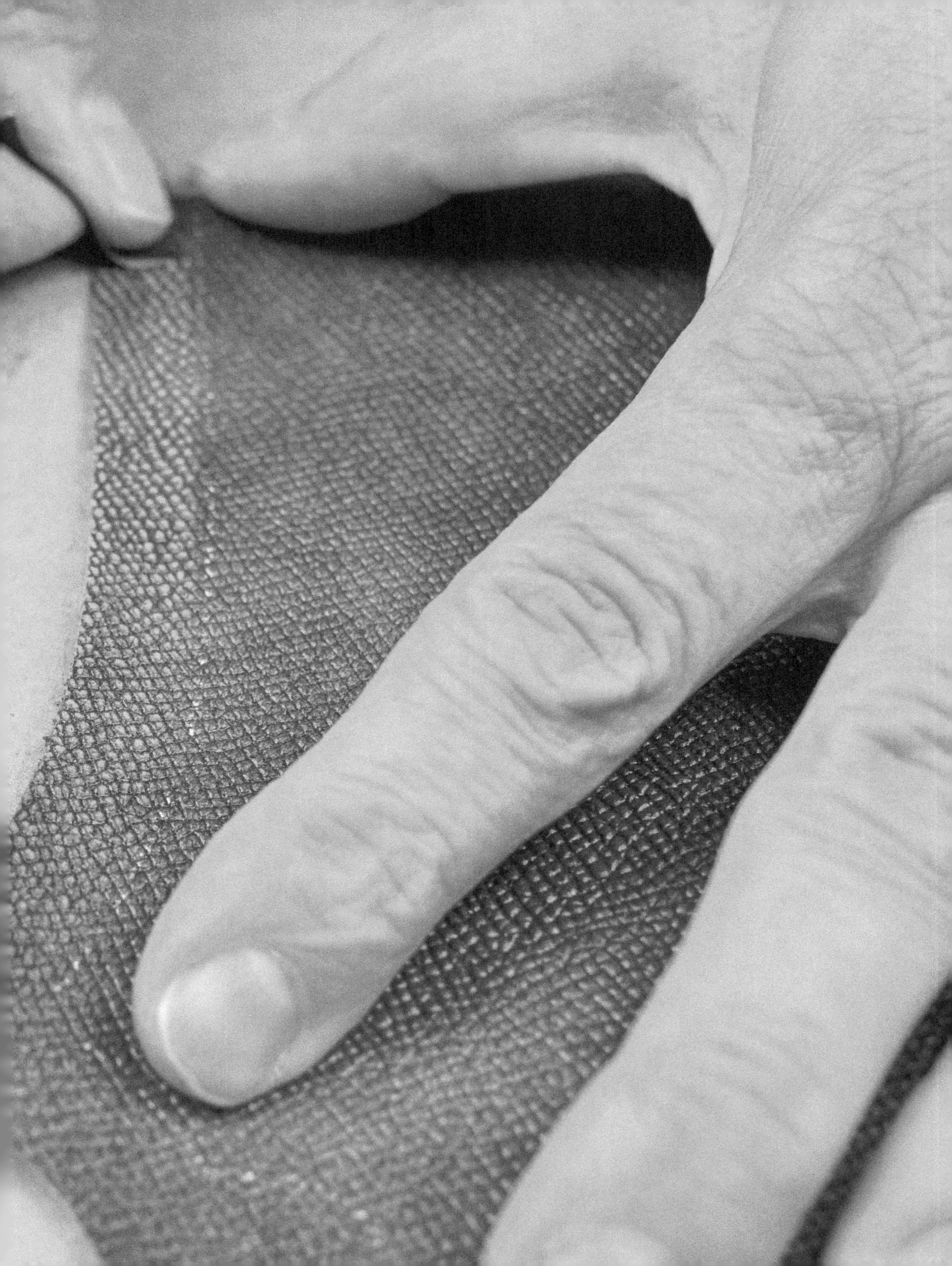

Attach the pieces needed for the first sewing step, applying a light layer of glue to both overlapping surfaces. In this example–following a classic balmoral construction with a counter piece - the counter overlaps the quarter, so it must be assembled first. In general, the goal is to minimize unnecessary movement between the worktable and the sewing machine to maintain an efficient workflow.

The next pieces to attach are the tongue and the vamp lining. This piece requires some additional shaping, which is addressed later in this chapter. The vamp lining includes a 5 mm trimming allowance, but it is intentionally left unmarked. There is no need for a guide here, as the tongue is easy to center visually, and markings on the flesh side are difficult to remove cleanly.

Next come the vamp and the toe cap. This seam is highly visible, so the marked placement line must be followed precisely. Ideally, the edge of the toe cap should lie completely flat. Because two skived layers meet here, this is more difficult than it sounds, and any small dent can complicate the sewing. Lightly hammering the overlap before sewing is recommended to create a firm temporary bond, leading to cleaner and easier sewing.

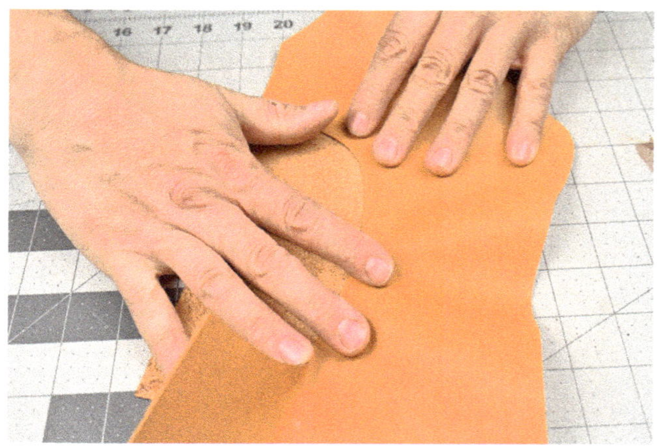

The quarter lining and the counter lining form a difficult attachment, as they use two axes and cannot be joined completely flat. If the pattern is accurate, the attachment can begin at one corner and progress toward the opposite side. Starting from the center is also effective, as it allows flexibility for minor pattern imperfections or slight misalignments.

Sewing begins with the quarter panel. As usual, the roller foot runs on the upper piece, which is the counter in this case. For a classic, elegant style, a small stitch placed fairly close to the edge is the best choice. Both ends of this seam are later covered by another piece, so the backstitches should remain within that hidden area, about 10–12 mm from the cut edge.

After the first stitch line is completed, both thread ends can be trimmed at the same time. This leaves a small excess, which will be hidden and does not affect the construction, while saving time in the process.

The next stitch line is sewn close to the first. Ideally, the first line should sit centered between the second line and the edge. From this point on, this stitching style should be used consistently.

Before executing this stitch, it is important to decide whether additional reinforcement is needed beneath the layers. This is unnecessary if the leather is thick and firm enough. Otherwise, adding a piece of reinforcement tape is highly beneficial—not only for increased strength but also for a cleaner appearance.

The next step is to sew the tongue together with the vamp lining. Run the stitch close to the edge – otherwise the layers can separate.

At the end of the seam, pull the threads toward the grain (top) side. Although this may seem unusual, this side will be hidden.

Carefully trim off the allowance with a sharp clicking knife, holding it at an angle to avoid cutting into the upper leather.

Hammer the seam all the way around. Even though this is not a highly exposed seam, this step contributes to durability.

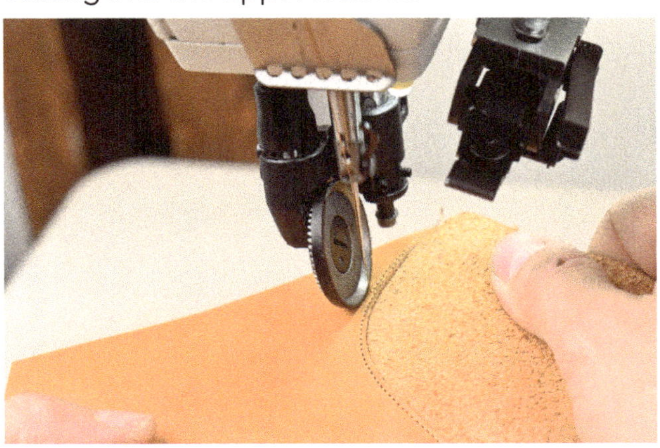

Sew the counter with a double line. This seam has the highest risk of wear, so a second line is essential.

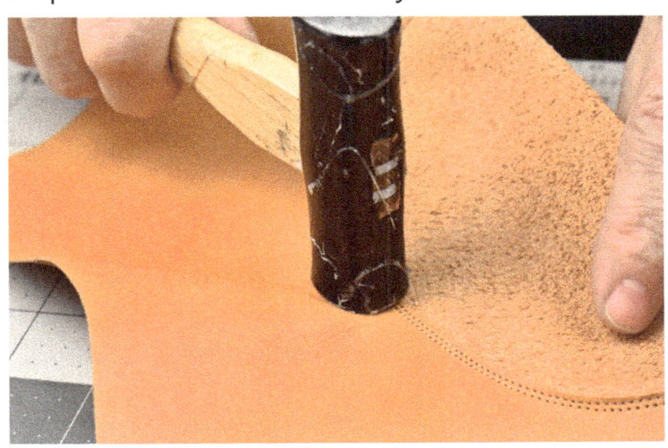

Hammer the thread into the lining so the threads are better protected against friction from wear.

Shaping the tongue is very useful for comfort and, fortunately, an uncomplicated part of the process. The key is to moisten the piece and use a wide (8–10 cm or 4 in) crimping shape. Very little effort is required to shape a tongue. Fix the wet piece in place with one or two leather straps and allow it to dry.

Fix the top edges of the upper quarters together for the zigzag seam.

Start the zigzag seam at the top and make sure to feed the pieces evenly.

Holding the pieces this way can help keep the edges close and feed them evenly into the machine.

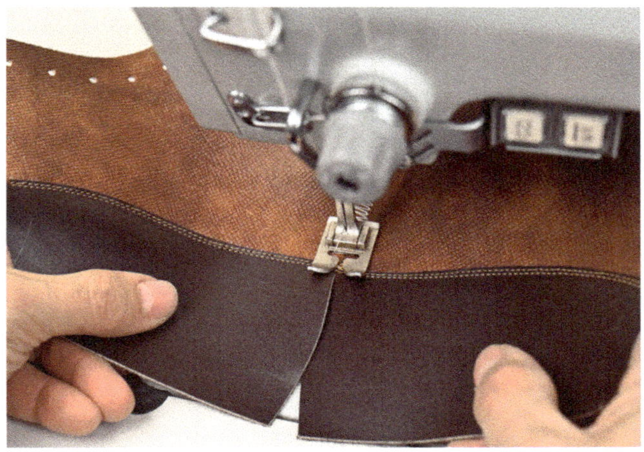

The backseam curves make this process a bit challenging. Focus on the next few stitches.

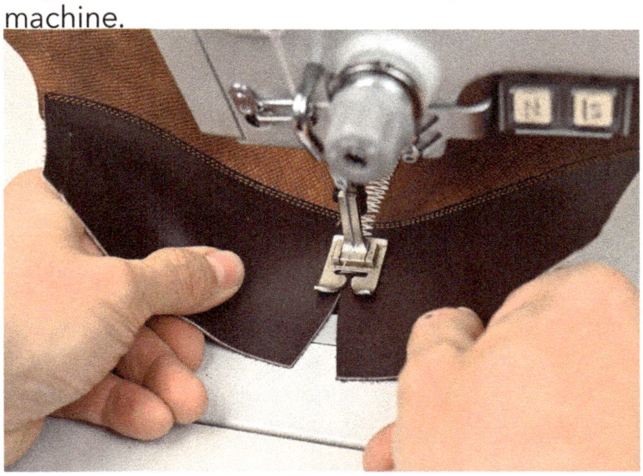

When reaching the bottom curve, cup the pieces upward as shown to achieve a tight seam.

The finished backseam, before turning it to the correct shape.

The backstrap is a relatively narrow piece (2 cm or 3/4 in), yet its placement can be challenging because the surface is highly curved. First, apply double-sided tape to avoid the mess associated with gluing. This eliminates unnecessary cleanup during this demanding step.

Start the attachment at the top. Flatten the strap as much as possible and press it downward. Then focus on the heel curve and apply a slight stretch. These two steps help secure the strap firmly to the quarter panels.

Place the completed quarters on a shaped wooden block (or the last) and lightly hammer the strap. This strengthens the adhesion and prevents the strap from shifting during sewing. Whenever pieces must be bent while held together only by cement or glue, securing them is essential. In this case, even the small flat plate of a post-bed machine can cause the layers to shift or bend, so stabilizing this seam before sewing is recommended.

During sewing, the way the piece is held must be adjusted. The top section is relatively flat, so it can be stitched without difficulty. However, beginners may find it helpful to mark the stitch line in advance. The geometry of this piece can cause the roller foot to drift inward, so an additional guideline helps prevent issues before they arise.

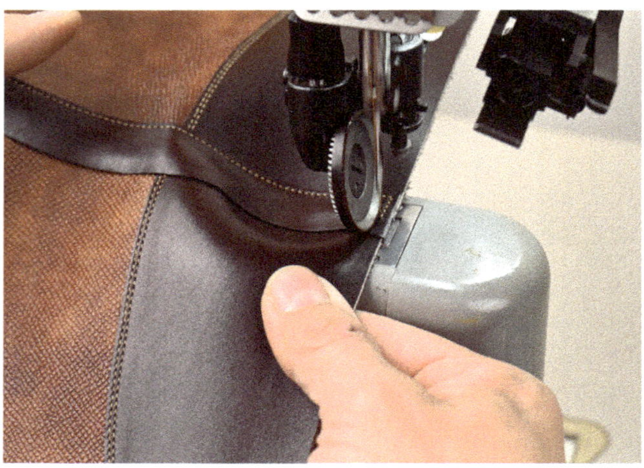

When reaching the counter, turn the piece as pictured so there is better access to the seam. This will also force the strap to shift, so thorough preparation with double-sided tape will pay off.

Turn at the bottom and sew the other side. This U-shaped seam leaves the top untouched, as it will receive a different edge treatment at a later stage, so no additional stitch line is needed.

Add a lace reinforcement tape to the topline and secure it firmly to the leather. This helps prevent future stretching and, more importantly, creates a cleaner, more rounded topline shape by adding structure to the leather.

After the reinforcements are in place, add the decorative seam. This seam can be done before or after assembling the panels. The choice depends on the maker's preference and on whether the design calls for stitching through the backstrap.

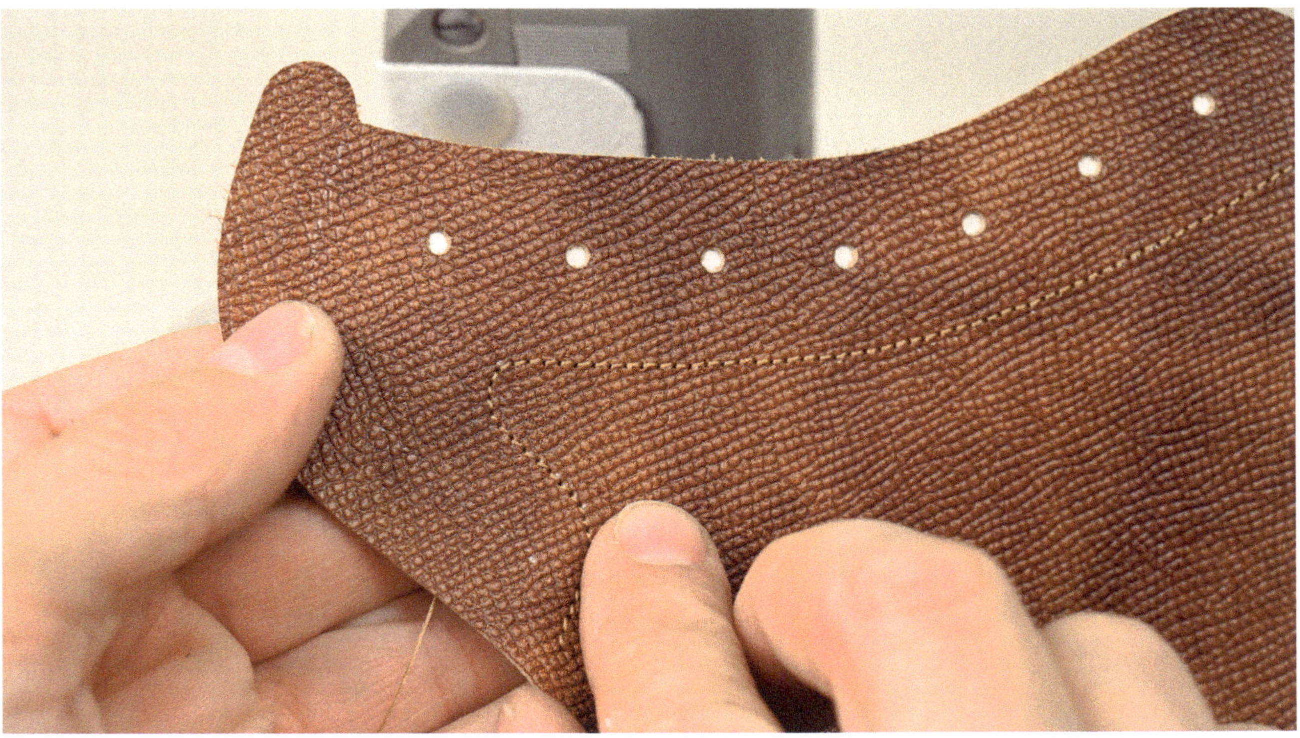

This is a very traditional decorative seam, and there are a few points worth keeping in mind. First, it can be done with the lining attached; however, in this case, the assembly order dictated otherwise. Second, the appearance of the seam is always improved with a layer of material underneath. Finally, tight turns introduce the risk of a clearly visible error known as turn distortion.

The Turn-distortion

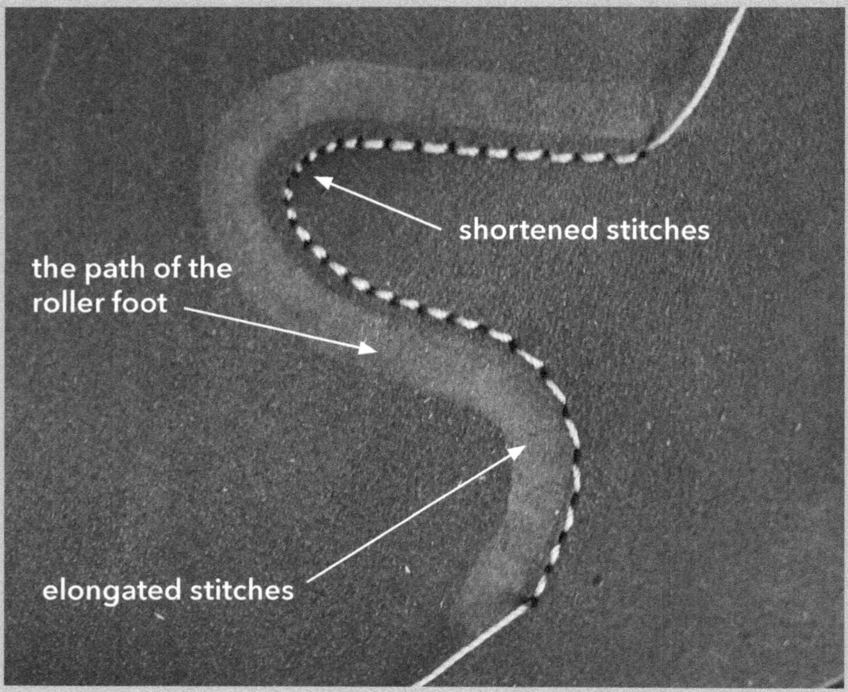

This picture illustrates turn distortion. The cause is that when making a tight turn on a roller-foot machine, the center of rotation becomes the right edge of the roller foot, causing the stitch length to change depending on the direction of the turn. Stitch length is determined by the feed wheel, and along the inside of the curve this results in shorter stitches (as seen at the top of the picture), while the outside produces longer stitches (at the bottom of the picture). In extreme cases, the shortened stitches can perforate or even cut the leather.

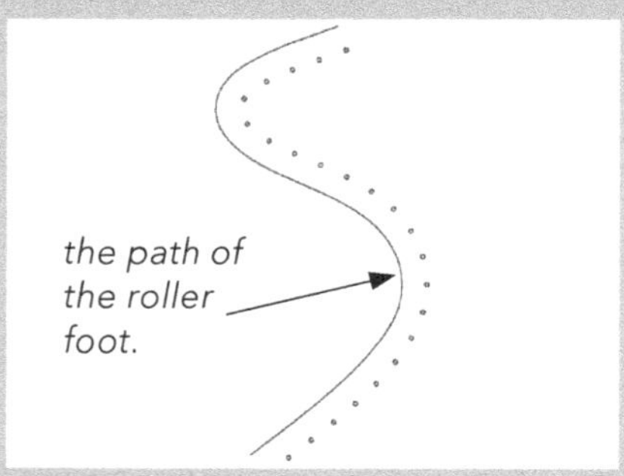

Ideally, a seam should have even stitches, regardless of how tight the curve is. Since every stitch is straight – a simple thread between two holes – this should be entirely achievable. In fact, programmable stitching automats, where the seam is essentially a series of preprogrammed hole positions, can execute this perfectly.

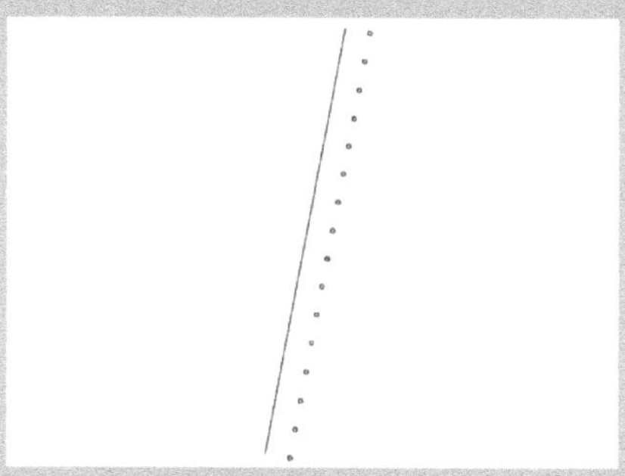

The issue lies in the way material is fed on roller-foot machines. As long as the feed mechanism moves the material along a path that is close to parallel and straight relative to the seam, no problem occurs. The key point is that the material-feed mechanism is not located at the needle but offset to the left, and this is where stitch length is actually measured.

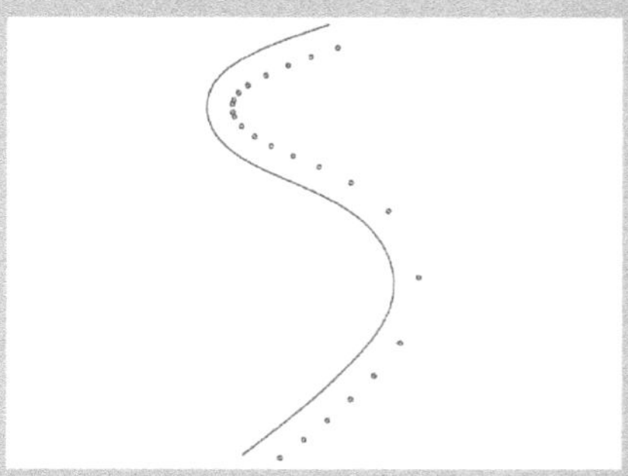

When the turn goes counterclockwise, the stitches become shorter, and the tighter the curve, the more the stitch length is reduced. When the turn goes clockwise, the stitches become longer. These issues only need correction on tight curves–longer, gradual turns will not show visible distortion. When the stitch length becomes noticeably altered, it is time to apply a few corrective techniques.

Correcting the Turn-distortion

This error can be corrected with a simple sequence. The most important rule throughout the process is that either the needle or the roller foot must always be fully down to maintain consistent stitch length.

1. Sew normally until the curve begins.

2. When the stitch line starts to turn, stop with the needle fully down.

3. Manually turn the flywheel toward yourself to loosen the top thread—otherwise, high tension can pull the bobbin thread to the surface.

4. Press the knee pad to lift the roller foot and, with the needle halfway down, rotate the material toward the next hole. Remember that the stitches themselves are not curved, so aim for the next straight stitch hole. Tight curves may require two, three, or even four stitches. Make sure the material is not under-turned.

5. Release the knee pad and make one stitch, either by hand-cranking or using the pedal.

6. Repeat from step 2 as many times as needed.

7. Once the next straight (or nearly straight) section is reached, sewing can continue normally.

A word of advice: practice on scrap material first and inspect the underside to ensure the tension is correct. A poorly adjusted machine may pull the top thread to the underside—or vice versa—which must be corrected before sewing the actual piece.

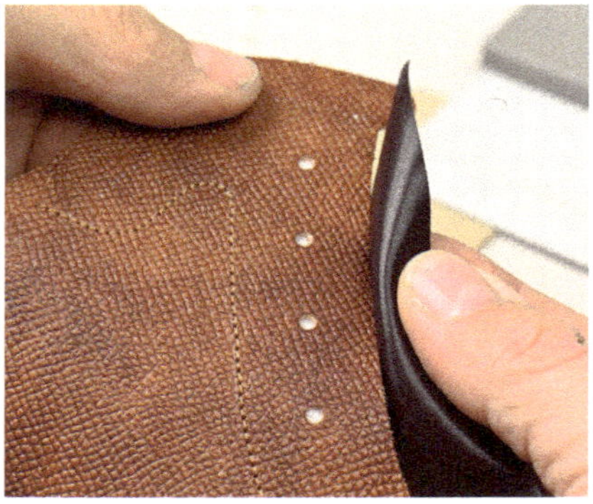

Turn the binding strip and the upper grain-side to grain-side, as shown. When folding the binding strip, both grain sides will be exposed, as intended. It is surprisingly common to position the binding strip incorrectly, so pay close attention to this step.

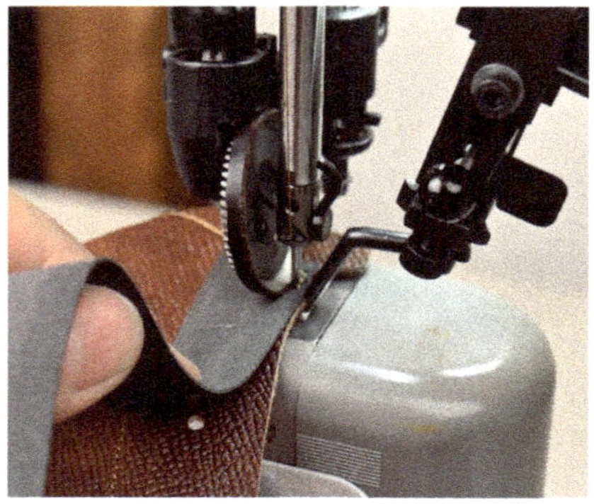

The beginning of the seam should have a solid amount of double stitching, but make sure to hit the original holes rather than creating new ones. New holes weaken the material, and the seam may tear later during the turning process.

An edge guide can be helpful in maintaining a constant distance from the edge. When placing the strip, leaving a very small sliver visible from the upper makes it easier to spot if the upper leather starts to slide away.

At tight turns, binding can become tricky without the special binding attachments used in factories. When turning clockwise (as pictured), the strip tends to curl upward, making sewing more difficult. When this happens, use a pointed (but not sharp) awl or a similar tool to push the material down. Hold the tool two to three stitches away from the actual stitch hole to avoid interfering with the needle. Sew this section very slowly, hand-cranking the machine and making one stitch at a time if necessary.

The first step of the binding is complete. Make sure to include a sufficient number of backstitches at the end of the seam. Of the many different binding treatments, this is the simplest version. The stitch line must be at an even distance from the edge. If this is not the case, there is still an opportunity to correct it with careful trimming.

The binding strip will not sit flat around the corner, which is to be expected. However, this will not have a negative effect on the final appearance.

Apply glue to the top side of the strip over a 6-8 mm wide area. This step is optional, but it results in a more secure and successful binding.

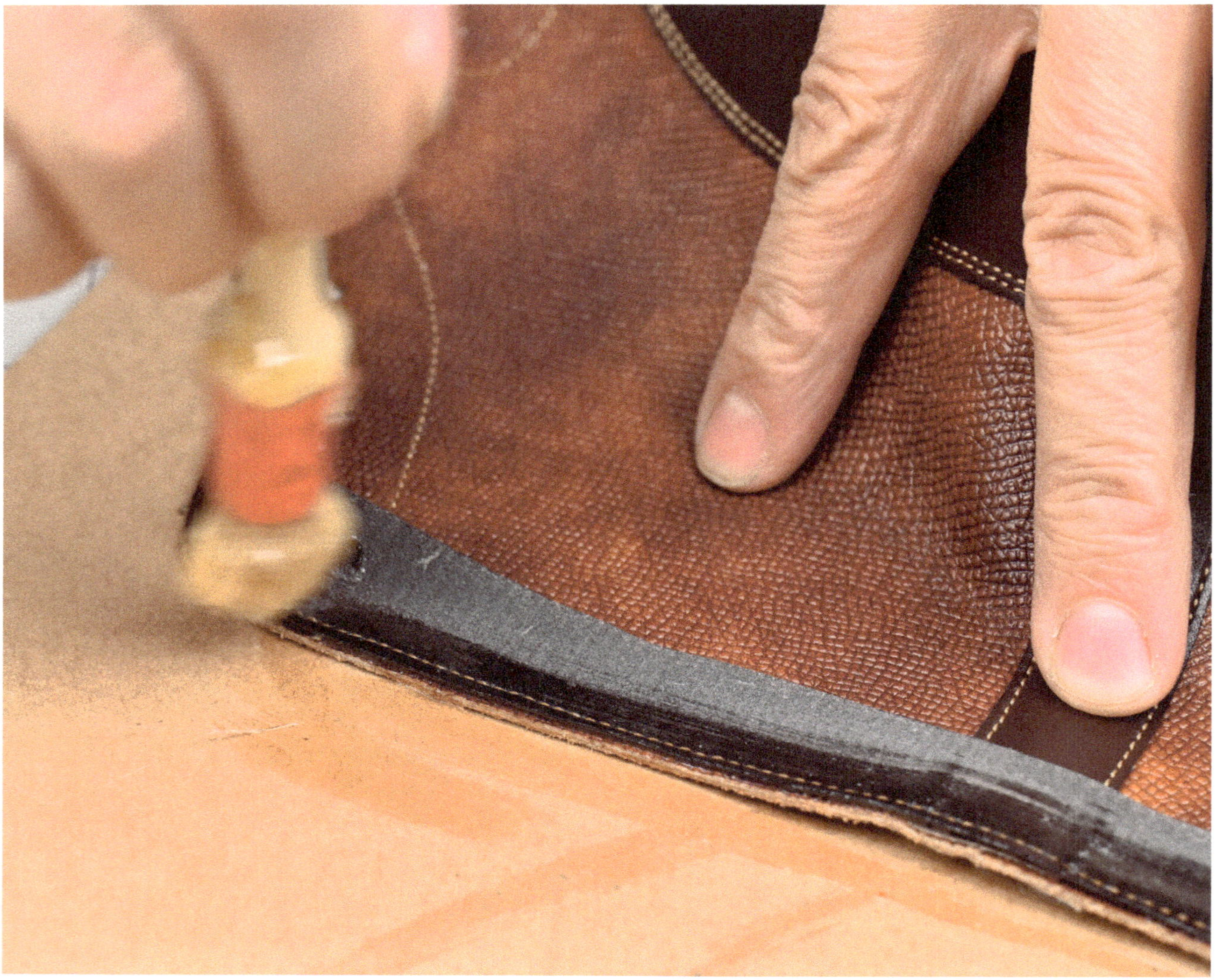

Once the glue has properly dried, begin folding the binding strip onto itself. Make sure to apply even tension throughout.

Leave the corner for last in the process. Gently hammer the folded section onto the topline. Adjust as needed if the tension is uneven.

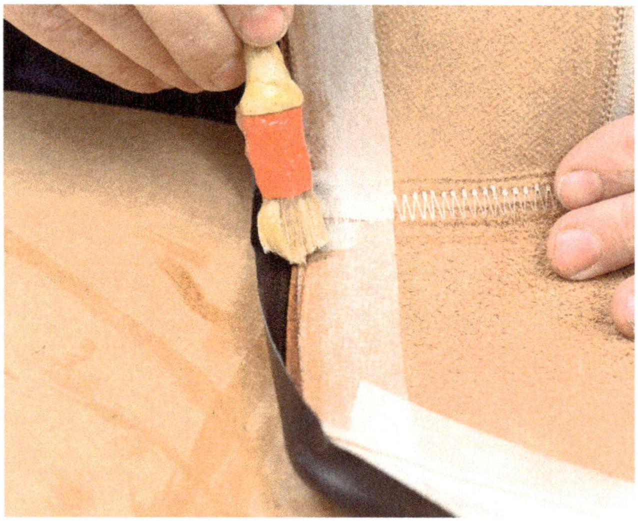

Add glue to the flesh side all the way up to the topline. Make sure to apply glue to both surfaces. The binding strip may curl slightly, which actually makes cleanup easier, as there is little chance of glue contaminating the other side. The glue layer must be thin and even, without lumps or thicker spots.

Folding the corner is similar to any regular folded edge. The only difference is that, since the topline is sewn at a distance from the edge, the allowance is slightly longer, which causes the leather to pile up in small pleats. However, handling this step with precision results in a clean-looking binding from the outside.

On concave curves, the leather may come up too short. Making small incisions helps the strip lie flat. These cuts should not extend too far, or they may become visible later. When using a stretchier leather for the strip, this step is unnecessary, as the material will stretch sufficiently on its own.

Hammering the strip from the grain side is essential. This is when the Italian binding takes its final shape – flat and narrow, rather than the rounded profile of the French version. This step must be carried out immediately after the first folding process, while the cement is still slightly active beneath the layer.

Before moving on to sewing the vamp topline, the two sides of the quarter panel must be fixed so they are not pushed apart. This is a very common mistake, but it is easily preventable.

Add glue to the overlap allowance on the vamp and the quarters. At this stage, the pieces become fully three-dimensional, so the bond must be firm to prevent the sewing machine from pushing them apart and altering the overlap allowance, which would affect symmetry. The center point on the quarters is obvious, but it is worth marking it on the vamp as well. Do not be discouraged if a slight misalignment remains despite the marking—the lasting allowance can absorb a small amount of excess, although it is always best to work as precisely as possible.

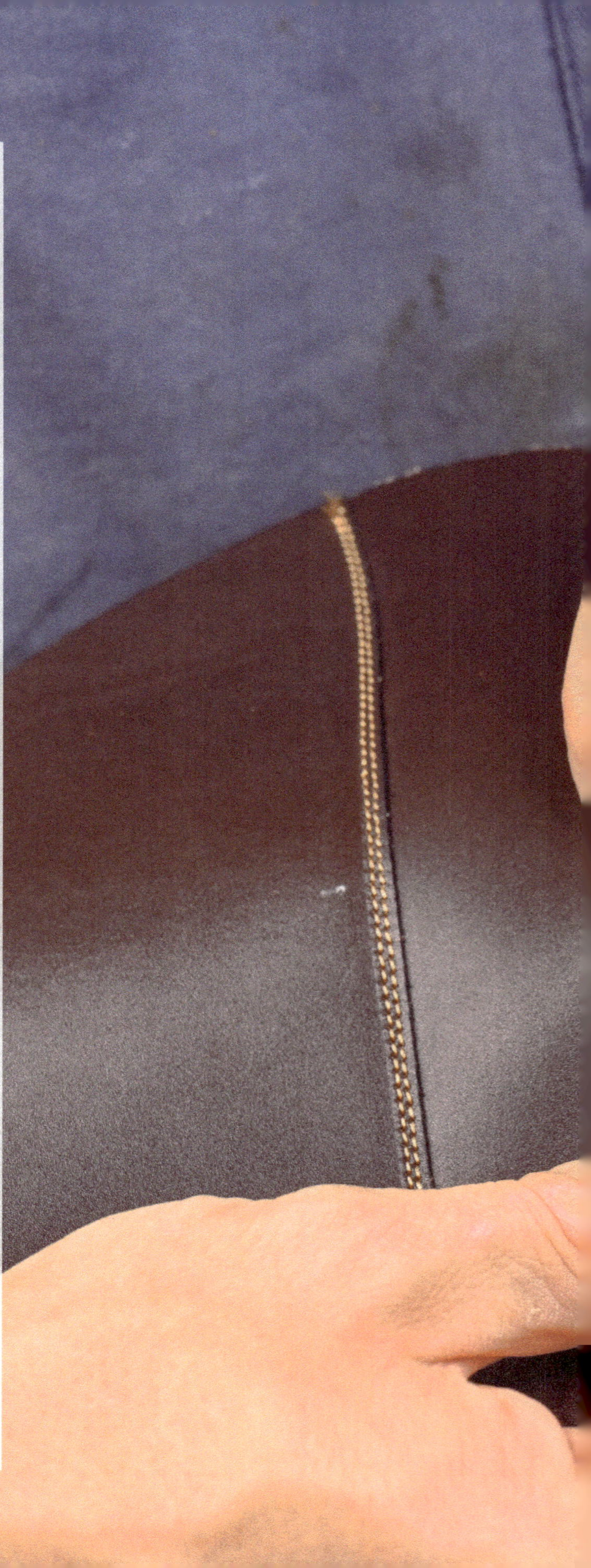

Start adhering the pieces from the center. It is possible to begin from the edge, but even the thickness of the leather can contribute to misalignment; therefore, it is more practical to distribute this evenly to both sides rather than concentrating it on one side only. This approach also provides better control over the overlap. Based on the pattern, the overlap is more generous in this area.

Sew the pieces together by running the wheel on the vamp. This seam is one of the more challenging ones. Maintaining an even distance close to the edge becomes especially difficult toward the center, where the thickness changes drastically. Slowing the machine down and hand-cranking the flywheel is a sound approach. The changing thickness can misguide the machine or even break the needle.

Add a sufficient amount of backstitching at both ends. After completing the first line, a second line can be added, which is very much part of this style.

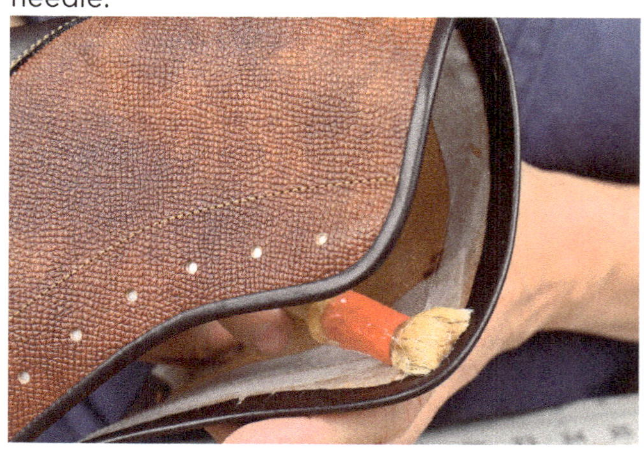

Add cement to the topline from the inside. Do not apply glue all the way to the edge. Repeat this process on the lining quarters and keep the trimming allowance clean.

Place the lining from the open bottom side of the upper. Match the center marks. Make sure the two pieces do not touch prematurely, as contact cement will bond immediately and create a mess.

Start the bonding process from the center. In addition to accurate placement, it is important not to leave any pockets. If the topline is not firmly attached all around, the sewing process will become problematic.

This is one of those "moments of truth," when it becomes clear whether the lining pattern is correct. Ideally the trimming allowance should be even all the way around.

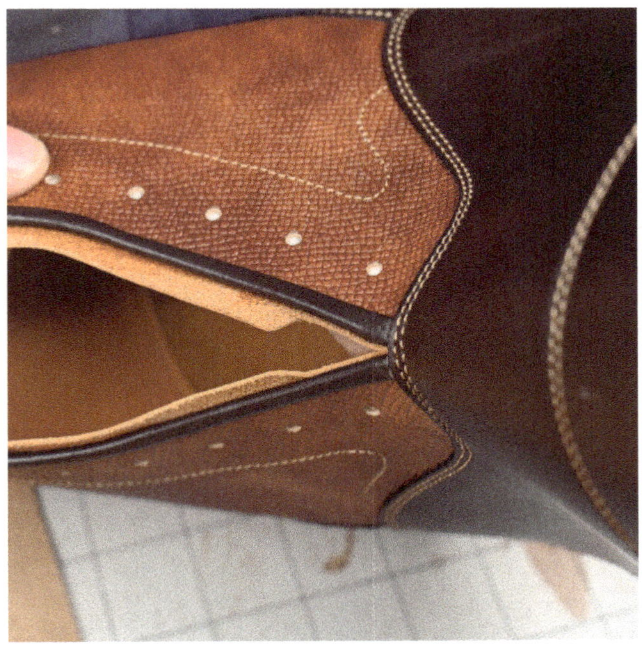

The front of the lining has zero trimming allowance, yet it still cannot sit exactly on the edge. Some excess is still required to sew both layers securely, without the lining slipping out.

Pro Tip!

This topstitch is where the post-bed machine really comes in handy, as the entire upper can be laid on top easily. A useful technique is to use a thread pig (as demonstrated earlier) for a few stitches, allowing the threads to be held securely. There are several thread ends to manage, and without a thread pig, operating the backstitch lever would be quite challenging.

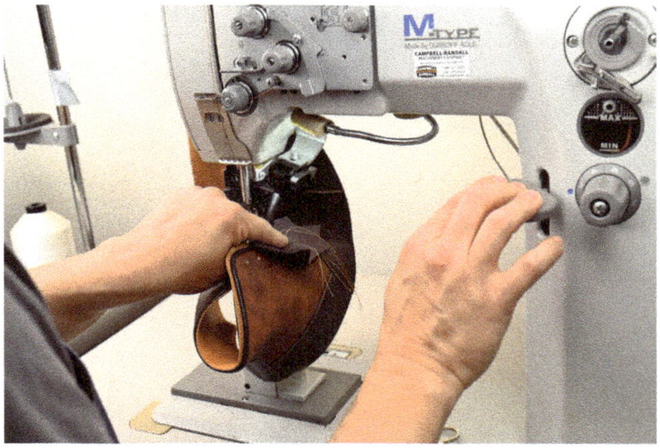

Start with a few backstitches to secure the seam. Hold the thread pig and the upper with the left hand.

After the backstitches are completed, continue sewing as usual.

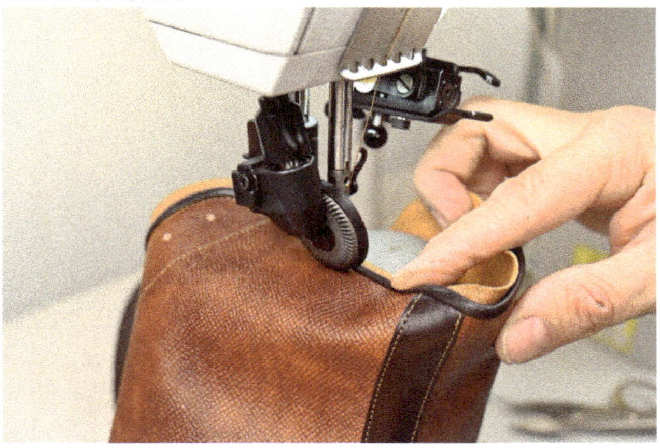

When reaching the topline, bend the top of the upper toward the post. The cement should hold the pieces together without shifting.

The top of the quarter is usually the easy part. The backseam can pose a challenge, as the thickness changes.

Manually adjust the machine and guide the upper.

After the topline, the descending section is nothing new. Focus only on the next few stitches.

Trimming the lining of a balmoral boot is one of the more difficult trimming processes. One way to do this is with a pair of small, curved scissors. Regardless of the curve orientation, small cuts will produce a smooth result. Carefully pushing the scissors is also an effective method, but only if the tool is sharp enough.

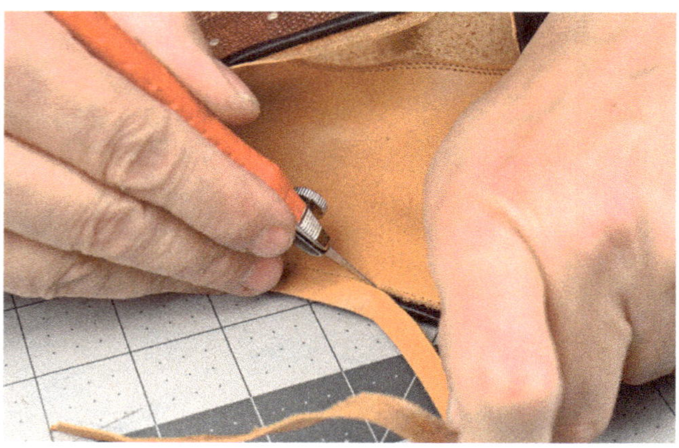

Alternatively, a sharpened clicking knife can also be used, although it requires more skill. The key is to pull the trimming allowance continuously away from the cut. Control is crucial to avoid damaging the upper.

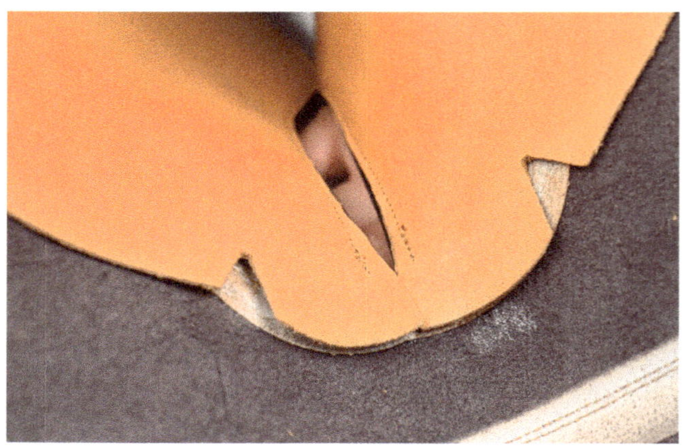

The vamp does not require special trimming, even though the edge might suggest otherwise. If the two sides meet properly without a gap, the result will be cleaner from the grain side.

A preferred method for setting the eyelet is from the inside. This creates a visible frame that is slightly larger than the one on the outside. Since the lining leather is softer, the result is generally more durable.

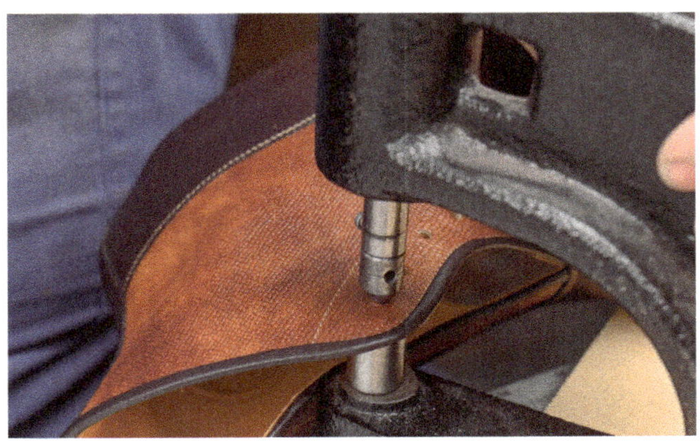

Although hand tools are available for this task, a small eyelet-setting machine provides a more controlled and consistent result, especially with this type of eyelet.

Eyelets

Surface finishes span a broad spectrum, including matte and mirror-polished metallic finishes, lacquered coatings, and modern treatments such as PVD. These finishes influence appearance, wear resistance, and durability.

Eyelet setting can be performed using hand tools, simple mechanical devices, or high-volume industrial eyelet setters. Advanced machines are capable of punching the lace hole and setting the eyelet in a single operation, completing the process in a fraction of a second while maintaining consistent quality.

Eyelets play a critical role in footwear construction. They reinforce lace holes, distribute the tensile forces generated during lacing, and significantly extend the service life of the footwear. Eyelets are available in a wide range of shapes, sizes, materials, and surface finishes, and their selection has both functional and aesthetic implications.

For classic footwear, brass eyelets are generally preferred due to their corrosion resistance, ease of forming, and long-term reliability, although their cost is noticeably higher than that of many alternatives. Steel eyelets provide high mechanical strength but depend on effective surface protection to prevent corrosion. Various metal alloys are also used as more economical options, particularly in high-volume production. Material quality is critical, as poorly finished or cracked edges can abrade the laces and lead to premature failure.

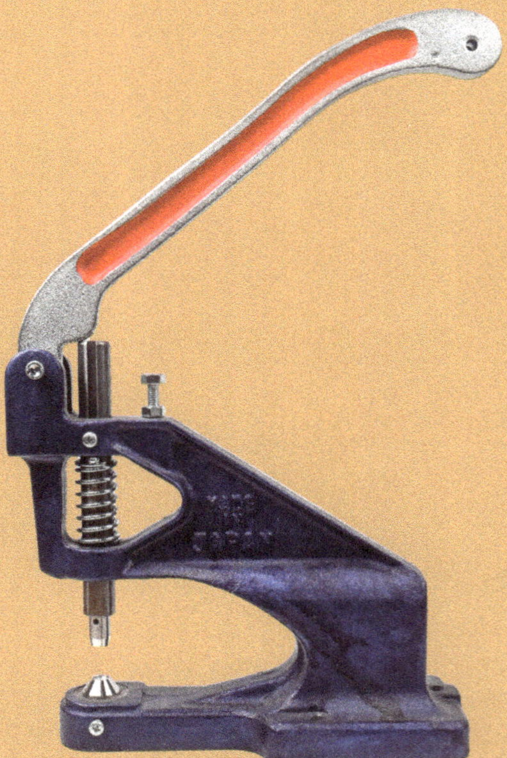

The eyelet setting is complete, and the upper is ready for the next stage. It is important to set the eyelets after the lining is attached but before the tongue, as access is better at this point.

Another detail to address before attaching the tongue and the vamp lining is the reinforcement stitch at the bottom of the quarters.

A small hole-punching tool is preferred for this task, but a small awl with a rounded profile can also be used. The holes should grip the thread firmly.

Create a bundle by looping the thread seven to eight times, depending on the thread thickness and the desired appearance.

The aim of this detail is as much decorative as it is functional. Each knot should have even tension and no twisting.

Loop the thread, pulling the needle through the middle where there is enough space. Form the knot, then push it to one side.

When all the knots are finished, pull the needle to the inside and secure the thread on the lining side. Use a piece of tape if necessary.

Tie a knot with the thread ends and lightly melt them to secure it.

The reinforcement stitch is complete.

While the rest of the upper was being worked on, the tongue has taken shape. Even a few hours is sufficient for the tongue to develop a three-dimensional form. It is important to do this with all boots so the shaping process is controlled and does not occur entirely on the wearer's foot, which could cause inconvenience or discomfort.

Add a thin layer of glue to the overlap allowance on the quarter and the vamp. Double-sided tape also works well for this task.

Check the placement at the tip of the toe and bring the allowances together. If all placements up to this point have been precise, this should also be accurate.

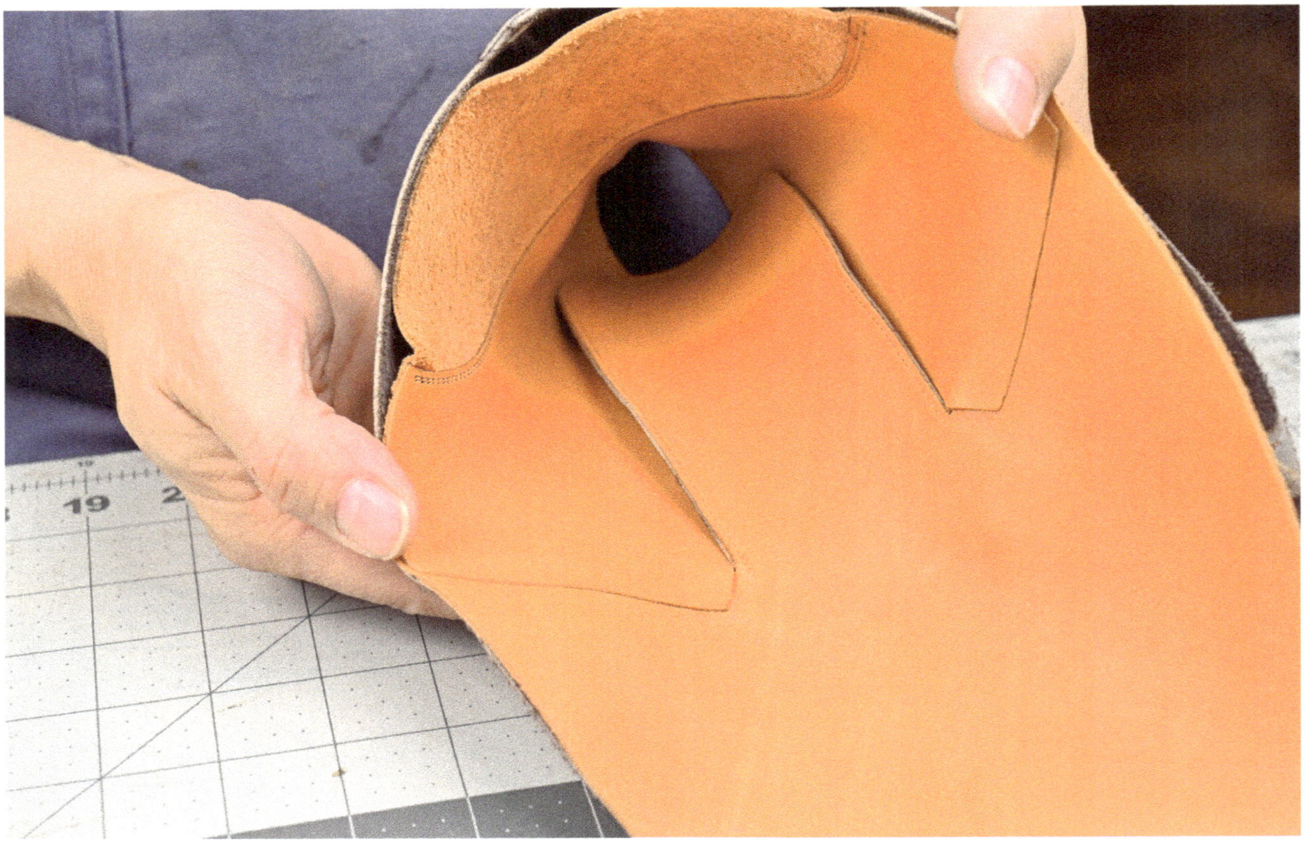

If an inaccuracy appears at this stage, it is usually caused by the scissor cut. Surprisingly, even on a small curve such as the top of the instep, a noticeable orange-peel-related effect can occur. Re-cutting the scissor cut as needed will usually resolve the issue. This typically means a slightly longer (2–3 mm) and slightly wider cut.

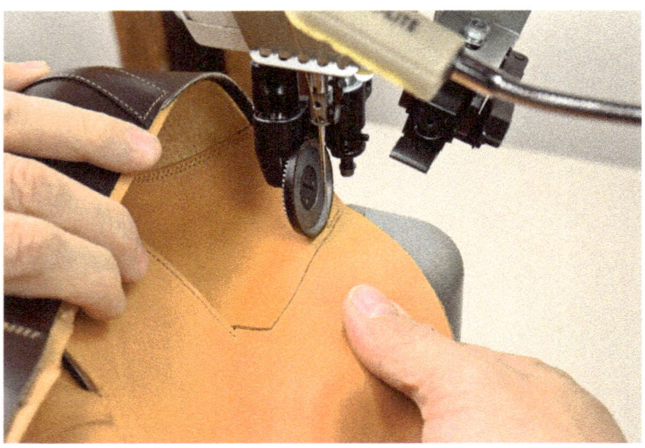

The lining stitch holds the quarter and the vamp lining together. Sew close to the edge and maintain a consistent distance (approximately 1 mm). Some makers opt for a double stitch.

Without cutting the thread, jump to the other side. This way, the thread is held firmly by the previous seam. Finish the seam with a sufficient number of backstitches.

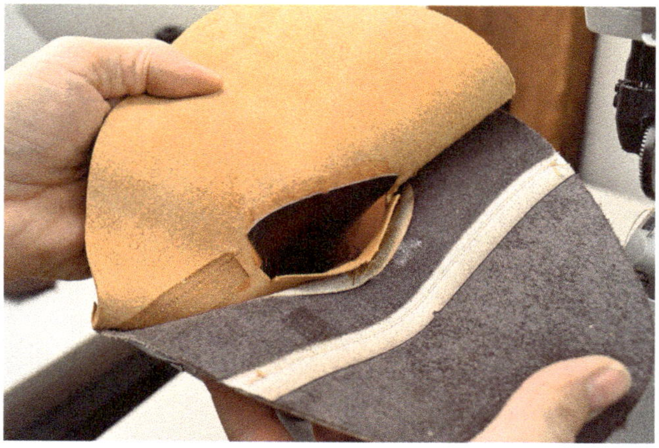

Note that the vamp and the upper around the bottom of the tongue are still separate. This is perfectly normal, and even if left this way, it would not cause much trouble. However, thorough work requires closing this gap.

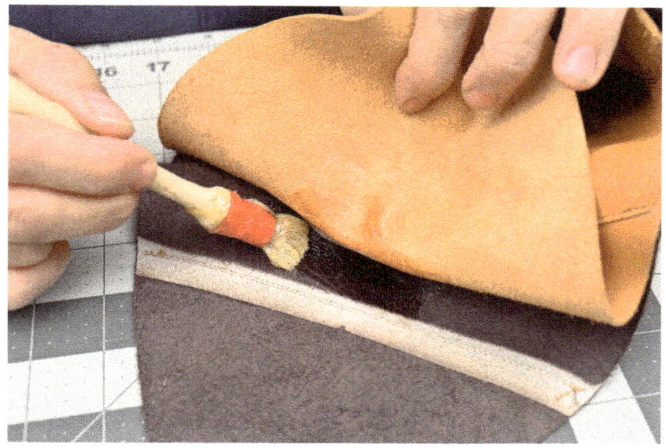

An easy and fast way to do this is by gluing the pieces together. After the proper drying time, press them together to achieve a strong bond. This alone is sufficient to make the boot functional.

A good way to hold the layers together is to apply glue up to the topline. This helps the boot take and retain its shape, and also improves durability by securing the two layers.

Only the counter area needs to be glued down, as the construction process requires the layers to be separated once more and then glued again.

In addition to the gluing, sewing the tongue (and all layers) on the quarter, close to the vamp is another way of closing this opening. Only tongue section needs to be sewn, so mark its position accurately. This is an almost invisible stitch, going right next to the edge of the vamp, under the reinforcement stitch.

With this final stitch, the upper is finished. Placing it on a shoe tree or last, as shown, is a good way to ensure that the topline develops a clean, curved form and that the tongue retains its shape until the upper is ready for lasting.

Acknowledgements
Thank you to my mother, who is the start of everything
Thank you to my wife, who is literally my better half

www.ingramcontent.com/pod-product-compliance
Ingram Content Group UK Ltd.
Pitfield, Milton Keynes, MK11 3LW, UK
UKHW062000041026
15591UKWH00016B/1106

9 798986 322711